Saint Peter's University Library
Withdrawn

W. Larcher

Physiological Plant Ecology

Corrected Printing of the Second Edition

Translated by
M. A. Biederman-Thorson

With 193 Figures

Springer-Verlag
Berlin Heidelberg New York Tokyo 1983

Professor Dr. WALTER LARCHER
Institut für Allgemeine Botanik der Universität Innsbruck
A-6020 Innsbruck, Sternwartestr. 15

MARGUERITE A. BIEDERMAN-THORSON, Ph.D.
The Old Marlborough Arms, Combe, Oxford, England

Translated and revised from the German edition "Walter Larcher, Ökologie
der Pflanzen", first published 1973 by Eugen Ulmer, Stuttgart. © 1973 by
Eugen Ulmer

ISBN 3-540-09795-3 2. Aufl.
Springer-Verlag Berlin Heidelberg New York
ISBN 0-387-09795-3 2nd ed.
Springer-Verlag New York Heidelberg Berlin

ISBN 3-540-07336-1 1. Aufl. Springer-Verlag Berlin Heidelberg New York
ISBN 0-387-07336-1 1st ed. Springer-Verlag New York Heidelberg Berlin

Larcher, Walter, 1929 — Physiological plant ecology. Translation of Ökologie
der Pflanzen. Bibliography: p. Includes index. 1. Botany — Ecology. 2. Plant
physiology. I. Title. QK901.L3513 1980 581.5 79-26396.
This work is subject to copyright. All rights are reserved, whether the whole or
part of the material is concerned, specifically those of translation, reprinting,
re-use of illustrations, broadcasting, reproduction by photocopying machine
or similar means, and storage in data banks. Under § 54 of the German
Copyright Law, where copies are made for other than private use, a fee is pay-
able to the publisher, the amount of the fee to be determined by agreement with
the publisher.
© by Springer-Verlag Berlin · Heidelberg 1975 and 1980.
Printed in Germany.
The use of registered names, trademarks, etc. in this publication does not
imply, even in the absence of a specific statement, that such names are exempt
from the relevant protective laws and regulations and therefore free for general
use.
Typesetting, printing, and bookbinding: v. Starck'sche Druckereigesellschaft,
Wiesbaden. 2131/3130-543210

QK
901
.L 3513
1980

Preface to the Second Edition

Since the first edition of this book appeared the field of plant physiology and ecology has advanced so far as to call for a complete revision of the material. To the extent that they could be accommodated in an introductory text, the requests and critical comments of users and reviewers of the original edition have been taken into account. The chapters have been rearranged; a survey of the effects of the physical environmental factors radiation and heat, and the plants' responses to them, is given in Chapter 2, whereas the chemical factors are treated in the chapters on metabolism and the turnover of matter. The ecophysiology of tropical plants and of plants growing in arid regions has been more strongly emphasized. The environmental influences affecting growth and development are mentioned occasionally, but are not discussed in such detail as those affecting metabolism, for as a rule the former are treated extensively in textbooks of plant physiology. References to ecosystems have been omitted to permit a sharper focus upon physiological relationships. I hope that the book will continue to be useful in this new form. It is not intended as — nor could it be — a comprehensive textbook of plant ecology; it is one of many possible ways of presenting what is known in the field.

Many readers of the first edition have asked for expansion of the reference list. The data and interpretations on which the text and many tables are based were drawn from thousands of original publications, so that it is impossible to document each item of information with the appropriate citation. To provide the reader with better access to the literature and a greater opportunity to advance from the general to the concrete, most of the tables and figures are accompanied by additional references to reviews as well as papers on special topics. The list has thus grown to comprise almost 800 references. Nevertheless, many publications of proved value could not be included; I beg the understanding of the authors.

While revising the text I have received the generous cooperation of many colleagues in providing advice and documentation. Dr. M. A. Biederman-Thorson again conscientiously translated the new sections, and Mr. R. Gapp prepared most of the new illustrations. Dr. K. F. Springer and his coworkers responded to my wishes with helpful understanding. To all of them I extend my sincere thanks.

Innsbruck, April 1980   W. LARCHER

Preface to the First Edition

Ecology is the science of the relationship between living organisms and their environment. It is concerned with the web of interactions involved in the circulation of matter and the flow of energy that makes possible life on earth, and with the adaptations of organism to the conditions under which they survice. Given the multitude of diverse organisms, the plant ecologist focuses upon the plants, investigating the influence of environmental factors on the character of the vegetation and the behavior of the individual plant species.

Plant ecophysiology, a discipline within plant ecology, is concerned fundamentally with the physiology of plants as it is modified by fluctuating external influences. The aim of this book is to convey the conceptual framework upon which this discipline is based, to offer insights into the basic mechanisms and interactions within the system "plant and environment", and to present examples of current problems in this rapidly developing area. Among the topics discussed are the vital processes of plants, their metabolism and energy transformations as they are affected by environmental factors, and the ability of these organisms to adapt to such factors. It is assumed that the reader has a background in the fundamentals of plant physiology; the physiological bases of the phenomena of interest will be mentioned only to the extent necessary for an understanding of the ecological relationships.

Ecology is very much a modern field, but by no means a recent innovation. I have tried to portray this rich historical background in the choice of illustrations and tabular material; the results presented reflect the broadness of vision, the struggles and the successes of the pioneering experimental ecologists in the first half of this century, as well as the advances in knowledge made most recently. Moreover, the student of ecology must bear in mind the particular characteristics of different localities; I have tried to include a broad selection of examples illustrating the ecophysiological behavior of plants in the greatest possible variety of habitats.

My first thanks are due to Dr. K. F. Springer; his publication of this English edition has made the textbook accessible to a wider circle of readers. I am grateful to the publisher of the original German edition, Roland Ulmer, for his cooperation. In particular, I thank Dr. Marguerite Biederman-Thorson for her thoughtful and sympathetic translation into English of the German text.

Above all, however, I should like to express my thanks to the pioneers of experimental ecology — Arthur Pisek, Otto Stocker, Heinrich Walter, and the late Bruno Huber. They inspired my enthusiasm for this difficult, but so attractive, field, and allowed me to benefit from their experience.

Innsbruck, September 1975   W. LARCHER

Contents

X

Abbreviations, Symbols and Conversion Factors

A Area

Acc   Acceptor molecule

ADP   Adenosine diphosphate

ATP   Adenosine triphosphate

B Plant biomass (also called phytomass, the mass of a stand of plants)

ΔB Change in biomass (positive for a growing stand)

bar   Unit of pressure (1 bar = 10^5 Pascal)

°C   Degree Celsius; relative measure of temperature

C Concentration

C_a Concentration of CO_2 and H_2O in the air outside a leaf

C_i Concentration of CO_2 and H_2O in the intercellular system of a leaf

cal   Calorie, a unit of energy (1 cal = 4.1868 joule = $4.1868 \cdot 10^7$ erg)

CAM   Crassulacean acid metabolism

Chl   Chlorophyll

D Molecular diffusion coefficient ($m^2 \cdot s^{-1}$)

d   Day as a unit of time

d Diameter

DL_{50} Drought lethality (degree of dryness causing 50% injury)

DM   Dry matter

dm^2 Unit of area; for leaves, it refers to one (projected) surface

$dm_2{}^2$ Unit of leaf area referring to the entire surface (upper and lower)

dyn   Measure of force (1 dyn = 10^{-5} N)

E Amount of water transpired

E   Einstein; amount of light quanta (1 E = 1 mol photons)

E_p Evaporative power of the air; potential evaporation

erg   Unit of energy or work (1 erg = 1 dyn · cm)

F Photosynthesis

F_g Rate of gross photosynthesis (true photosynthesis)

F_n Rate of net photosynthesis (apparent photosynthesis)

g Gram; unit of mass

G Grazing (loss of dry matter to consumers)

GAP   Glyceraldehyde-3-phosphate

h   Hour

ha   Hectare (1 ha = 10^4 m²)

h Planck's constant ($6.625 \cdot 10^{-34}$ J · s)

I Irradiance; the radiation flux at a given level within a stand of plants or body of water

I_0 Maximum radiation flux; that incident upon a stand of plants or body of water

I_a Long-wavelength radiation from the atmosphere

I_{abs} — Absorbed radiation

I_K — Compensation light intensity (at which $F = R$)

$\bar{I}_l$ — Long-wavelength radiation balance

$\bar{I}_s$ — Short-wavelength radiation balance

I_S — Light intensity at which photosynthesis is saturated

IAA — Auxin, indole acetic acid

IR — Infrared radiation (> 750 nm)

J — Joule; unit of energy ($1\ J = 1\ N \cdot m$)

J — Flux, mass flow

K — Kelvin; unit of temperature

k — Coefficient, conversion factor

k_F — Photosynthetic efficiency coefficient

k_M — Recycling factor for mineral nutrients in a stand of plants

k_{PP} — Productivity coefficient

k_T — Reaction rate of biochemical processes at a given temperature

kcal — Kilocalorie ($1\ kcal = 10^3\ cal$)

kg — Kilogram; unit of mass

kJ — Kilojoule ($1\ kJ = 10^3\ joule$)

kLx — Kilolux ($1\ kLx = 10^3\ lux$)

kW — Kilowatt ($1\ kW = 10^3\ watt$)

l — Liter; unit of volume

L — Loss of organic dry matter as detritus

L_E — Water loss via evapotranspiration

L_I — Water loss via interception

L_O — Water loss via runoff and percolation

l — Wavelength (radiation)

λ — Latent heat of vaporization of water

LAI — Leaf-area index

LAR — Leaf area ratio

lx — Lux; photometric unit of illuminance

m — Meter; unit of length

M — Molar; measure for concentration

M_{abs} — Quantity of minerals absorbed

M_B — Mineral content of a stand of plants

M_G — Loss of minerals via grazing

M_i — Quantity of minerals incorporated

M_L — Loss of minerals as detritus

M_r — Minerals lost in inorganic form ("recretion")

mg — Milligram ($1\ mg = 10^{-3}\ g$)

min — Minute

ml — Milliliter ($1\ ml = 10^{-3}\ l = 1\ cm^3$)

mm — Millimeter; measure of length ($1\ mm = 10^{-3}\ m$) and measure of precipitation (1 mm precipitation = 1 liter water $\cdot$ m^{-2} of ground)

mol — Mole; unit of quantity

μm — Micrometer ($1\ \mu m = 10^{-6}\ m$)

n — Number of particles

N — Newton; unit of force ($1\ N = 1\ kg \cdot m \cdot s^{-2}$)

NAD^+ — Nicotinamide-adenine-dinucleotide, reduced form: $NADH + H^+$ (simplified notation $NADH_2$); reduction system

$NADP^+$ — Nicotinamide-adenine-dinucleotide-phosphate, reduced form: $NADPH + H^+$ ($NADPH_2$); reduction system

NAR	Net assimilation rate (= unit leaf rate)	Q_H	Energy conversion associated with convection
nm	Nanometer (1 nm = 10^{-9} m)	Q_I	Energy conversion associated with radiation from the sun and reradiation
OAA	Oxalacetate		
ω	Water use efficiency	Q_M	Energy conversion associated with metabolism
P	Turgor pressure		
P	Production of vegetation	Q_P	Energy conversion in plant communities
P_g	Gross productivity		
P_i	Inorganic phosphate	Q_{Soil}	Energy conversion in the soil
P_n	Net productivity		
π	Osmotic pressure	Q_{10}	Temperature coefficient of biochemical and physiological process
Pa	Pascal; unit of pressure (1 Pa = 1 N · m^{-2} = 10^{-5} bar)		
		r	Transport or diffusion resistance
PEP	Phosphoenol pyruvate		
PGA	3-phosphoglyceric acid	r_s	Stomatal diffusion resistance
pH	Negative logarithm of the hydrogen ion concentration		
		R	Gas constant ($R = 8.3$ J · K^{-1} · mol^{-1})
PhAR	Photosynthetically active radiation (400–700 nm)		
		R	Respiration
PP	Primary production (yield of a stand)	R_d	Dark respiration
		R_l	Respiration in the light
ppm	Parts per million	*RH*	Relative humidity
PPR	Primary production rate (yield of a stand per unit time)	RuBP	Ribulose-1,5-bisphosphate
		RuP	Ribulose 5-phosphate
		RWC	Relative water content
PR	Production rate (yield of a plant per unit time)	s	Second; unit of time
		σ	Surface tension of water
Pr	Precipitation (total falling on a stand of plants)	*SLA*	Specific leaf area
		t	Time (point in time or duration)
Pr_n	Precipitation reaching the ground beneath a plant canopy		
		t	Ton (Metric; 1 t = 10^3 kg)
π^*	Potential osmotic pressure		
Φ	Quantum yield [mol O$_2$ · Einstein^{-1}]	*T*	Temperature (all temperature data in °C)
		τ	Matric pressure or potential
Ψ	Water potential		
PWP	Permanent wilting percentage	TCA	Tricarboxylic acids
		TL$_{50}$	Temperature-stress lethality (the temperature at which 50% of plants are killed by heat or cold)
Py	Pyruvate		
Q	Energy flow		
Q_E	Energy conversion associated with evaporation and condensation		
		torr	Unit of pressure (1 torr = 1.33 · 10^{-3} bar $\doteq$ a 1-mm column of Hg)

Tr	Transpiration	W_{FC}	Water content of soil at field capacity
UV	Ultraviolet radiation (< 400 nm)	W_{PWP}	Water content of soil at permanent wilting percentage
W	Watt; unit of power (1 W = 1 J · s⁻¹)		

Tr   Transpiration

UV   Ultraviolet radiation (< 400 nm)

W   Watt; unit of power (1 W = 1 J · s⁻¹)

W Weight

W_{abs} Quantity of water absorbed

W_{act} Actual water content (when sample is taken)

W_{av} Available water

W_d Dry weight

W_{FC} Water content of soil at field capacity

W_{PWP} Water content of soil at permanent wilting percentage

W_f Fresh weight

W_s Water content in saturated state

WSD   Water saturation deficit

yr   Year

z Relative height or depth

$\doteq$ Approximately equal to

$>$ Larger than

$<$ Smaller than

Equivalents

Energy (work)

$1 \text{ J} = 1 \text{ N} \cdot \text{m} = 1 \text{ kg} \cdot \text{m}^2 \cdot \text{s}^{-2} = 1 \text{ W} \cdot \text{s} = 0.239 \text{ cal} = 10^7 \text{ erg}$

$1 \text{ W} \cdot \text{h} = 3.6 \text{ kW} \cdot \text{s} = 3.6 \text{ kJ} = 0.86 \text{ kcal}$

$1 \text{ MJ} = 0.278 \text{ kWh}$

$1 \text{ cal} = 4.1868 \text{ J}$

$1 \text{ cal (thermochemical)} = 4.184 \text{ J}$

$1 \text{ kcal} = 1.163 \text{ W} \cdot \text{h}$

Energy Consumption in the Evaporation of Water

Heat of vaporization at $0° C = 2.50 \text{ kJ} \cdot \text{g}^{-1} \text{ H}_2\text{O}$ (597 cal · g⁻¹ H₂O)

at $10° C = 2.48 \text{ kJ} \cdot \text{g}^{-1}$ (592 cal · g⁻¹)

at $20° C = 2.45 \text{ kJ} \cdot \text{g}^{-1}$ (586 cal · g⁻¹)

at $30° C = 2.43 \text{ kJ} \cdot \text{g}^{-1}$ (580 cal · g⁻¹)

Radiation

$1 \text{ W} \cdot \text{m}^{-2} = 1 \text{ J} \cdot \text{m}^{-2} \cdot \text{s}^{-1} = 1.43 \cdot 10^{-3} \text{ cal} \cdot \text{cm}^{-2} \cdot \text{min}^{-1}$

$1 \text{ cal} \cdot \text{cm}^{-2} \cdot \text{min}^{-1} = 6.98 \cdot 10^2 \text{ W} \cdot \text{m}^{-2} = 6.98 \cdot 10^{-5} \text{ erg} \cdot \text{cm}^{-2} \cdot \text{s}^{-1}$

$1 \text{ erg} \cdot \text{cm}^{-2} \cdot \text{s}^{-1} = 1.43 \cdot 10^{-6} \text{ cal} \cdot \text{cm}^{-2} \text{ min}^{-1} = 10^{-3} \text{ W} \cdot \text{m}^{-2}$

$1 \text{ klx} \doteq 4\text{--}10 \text{ W} \cdot \text{m}^{-2}$ (depending on light source)

$1 \text{ W} \cdot \text{m}^{-2} (\text{PhAR}) \doteq 3\text{--}5 \text{ μmol photons} \cdot \text{m}^{-2} = 30\text{--}50 \text{ nmol photons} \cdot \text{cm}^{-2} \cdot \text{s}^{-1}$

$1 \text{ mol photons} = 1.7 \cdot 10^5 \text{ J}$ (at $\lambda = 700$ nm) to $3 \cdot 10^5 \text{ J}$ (at $\lambda = 400$ nm)

$1 \text{ fc (foot candle, obsolete)} = 10.76 \text{ lux}$

$1 \text{ ly (langley, obsolete)} = 1 \text{ cal} \cdot \text{cm}^{-2}$

Pressure

$1 \text{ MPa} = 10^6 \text{ Pa} = 10 \text{ bar}$
$1 \text{ bar} = 10^5 \text{ N} \cdot \text{m}^{-2} = 10^5 \text{ Pa} = 100 \text{ J} \cdot \text{kg}^{-1} = 10^6 \text{ erg} \cdot \text{cm}^{-3}$
$1 \text{ bar} = 750 \text{ torr} = 0.9869 \text{ atm}$
$1 \text{ torr} = 1.33 \cdot 10^{-3} \text{ bar} \doteq 1\text{-mm column of mercury}$
$1 \text{ atm} = 1.0132 \text{ bar} = 760 \text{ torr}$

Phytomass

$1 \text{ g DM} \cdot \text{m}^{-2} = 10^{-2} \text{ t} \cdot \text{ha}^{-1}$
$1 \text{ g org. DM} \doteq 0.45 \text{ g C} \doteq 1.5 \text{ g CO}_2$
$1 \text{ g C} \doteq 2.2 \text{ g org. DM} \doteq 3.4 \text{ g CO}_2$
$1 \text{ g CO}_2 \doteq 0.65 \text{ g org. DM} \doteq 0.30 \text{ g C}$

Gas Exchange

$1 \text{ μmol CO}_2 \cdot \text{m}^{-2} \cdot \text{s}^{-1} = 0.044 \text{ mg CO}_2 \cdot \text{m}^{-2} \cdot \text{s}^{-1}$
$1 \text{ mg CO}_2 \cdot \text{m}^{-2} \cdot \text{s}^{-1} = 22{,}7 \text{ μmol CO}_2 \cdot \text{m}^{-2} \cdot \text{s}^{-1}$
$1 \text{ mg CO}_2 \cdot \text{dm}^{-2} \cdot \text{h}^{-1} = 0.028 \text{ mg CO}_2 \cdot \text{m}^{-2} \cdot \text{s}^{-1} = 0.63 \text{ μmol CO}_2 \cdot \text{m}^{-2} \cdot \text{s}^{-1}$
$1 \text{ g CO}_2 \text{ turnover} \doteq 0.73 \text{ g O}_2 \text{ turnover} (RQ : CO_2/O_2 = 1)$
$1 \text{ g O}_2 \text{ turnover} \doteq 1.38 \text{ g CO}_2 \text{ turnover}$
$D_{CO_2} = 0.64 \ D_{H_2O}$
$D_{H_2O} = 1.56 \ D_{CO_2}$

Further aids to conversion can be found in the manuals of methods by Šesták et al. (1971), Slavík (1974), O'Connor and Woodford (1975), Rose (1979), Incoll et al. (1977), Savage (1979), Bell and Rose (1981), as well as in volumes of physiological and biological tables.

1 The Environment of Plants

Plants have colonized nearly all regions of the earth, including the oceans and inland waters; on land they can be found even in such inhospitable places as deserts and fields of ice. Far back in geological time, when the first land plants were evolving, they encountered a world of water, air and stone. That is, their environment consisted of the hydrosphere, atmosphere and lithosphere. Later, as the cover of vegetation gradually closed, and with the assistance of microorganisms and animals, there developed the most important substrate of plants: the soil—the pedosphere.

1.1 The Hydrosphere

The hydrosphere comprises the *oceans* of the world, which cover an impressive 71% of the earth's surface, as well as the *inland waters* and the *groundwater*. Great differences exist in the chemical compositions of these bodies of water (Fig. 1.1). Sea water, rich in Na^+, Mg^{2+}, Cl^- and SO_4^{-2} and with an average salt content of 35 g $\cdot$ l^{-1}, differs fundamentally from fresh water, which usually contains more Ca^{2+} and HCO_3^-; but there are local differences as well, depending on the nature of the inflowing waters and the degree of mixing. Moreover, *currents* have an effect upon temperature gradients. Where there are no currents, the strong absorption of radiation in the upper levels of the water leads to a characteristic layering with respect to temperature and density; this has a marked influence upon nutrition, productivity and distribution of aquatic organisms.

1.2 The Atmosphere

The *air enveloping the earth* provides plants with carbon dioxide and oxygen. It also mediates the balance of water through the processes of rain, condensation and "evapotranspiration". Continual movement of the air ensures that its composition remains fairly constant—79% nitrogen (by volume), 21% oxygen and 0.03% carbon dioxide, water vapor and noble gases (Fig. 1.1). In addition the air contains gaseous, liquid and solid impurities; these are primarily sulfur dioxide, unstable nitrogen compounds, halogen compounds, dust, and soot.

The part of the atmosphere with which plants come into contact is the *troposphere*, the weather zone of the earth's envelope of air. The nature of this zone varies over short distances and is characterized in several ways: (1) by the *weather* (short-term events such

1

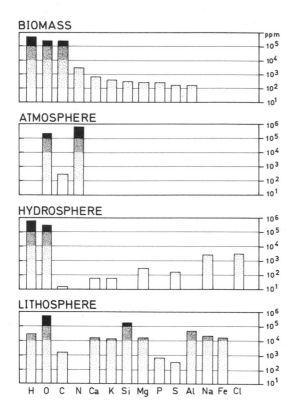

BIOMASS

ATMOSPHERE

HYDROSPHERE

LITHOSPHERE

H O C N Ca K Si Mg P S Al Na Fe Cl

Fig. 1.1. Composition of the biomass, atmosphere, hydrosphere and lithosphere, in terms of the relative numbers of atoms (atoms per million atoms, not the proportion by weight) of the various chemical elements. The composition of living organisms is clearly distinct from that of the three components of their environment; they select from the available elements, according to their needs. The scale of the ordinate is logarithmic. For example, in the biomass H, O, C and N are present in the greatest proportions: $4.98 \cdot 10^5$ atoms per million (i.e., about 50% of all atoms) are hydrogen atoms; oxygen and carbon atoms each comprise $2.49 \cdot 10^5$ atoms per million (about 25%), and $2.7 \cdot 10^3$ (about 0.3%) are nitrogen atoms. After Deevey (1970)

as showers, thunderstorms, and gusts of wind), (2) by meteorological events of intermediate duration such as periods of rain or frost and (3) by the *climate* (the average state and ordinary long-term fluctuations in meteorological factors at a given place). Depending on the terrain and on the density, height and type of vegetation, individual climatic regions of different sizes are formed. Within the large-scale "*macroclimate*" measured by the network of meteorological stations, one may distinguish "*microclimates*" that prevail in specific places such as certain slopes or narrow valleys, the *bioclimate* in (for example) stands of vegetation, and an "interface" climate—in the layer of air near the ground and the surface of leaves. Thus the parts of plants above ground are exposed to variability, in space and time, with respect to radiation, temperature, humidity, precipitation, and air motion; any of these can from time to time represent a threat to the organism.

1.3 The Lithosphere and the Soil

The **earth's crust** is the inexhaustible reservoir of the variety of chemical elements of which organisms are composed (Fig. 1.1). The lithosphere exchanges matter with the hydrosphere, and also affects the composition of the atmosphere through volcanic action and the products of radioactive decay. Primarily, however, it is the basic material for the formation of the soil.

Soil is more than just superficially loosened lithosphere. It is the product of the *transformation and mingling of mineral and organic substances*. Soils are produced with the assistance of organisms and under the influence of environmental conditions. They are subject to continual change: soils grow, mature, and can age and perish. Physical and chemical weathering continuously frees mineral substances from the rocky substrate, and there is an unceasing decay of plant remains and dead organisms. These decay products, together with the excrement of soil animals, gradually turn into humus, which forms complexes with the mineral products of weathering. In natural soils a profile is established of more or less horizontal layers (*"horizons"*): between the strata of litter and humus, and the stratum where weathering of the parent rock occurs, there are transitional zones with varying proportions of humus. The types and thicknesses of the horizons in such a profile are characteristic of a given *type of soil* and reflect the influence of climate, plant cover, soil organisms, underlying rock and the activity of man. *Pedology* is the science of soil formation and composition, and of the classification of soil types. Knowledge of the fundamentals of this field is an absolute prerequisite to understanding plant ecology.

The solid particles in the ground stick together to form aggregates, leaving small open spaces. Together these form a *system of pores* penetrating the entire soil, filled partly with air and partly with water. Thus the soil is a multiple-phase system in which solid, liquid, and gas phases are intermingled. It has an enormous capacity for uptake and storage, and is particularly suited for the *buffering* of physical and chemical influences. Below the top centimeters of soil the prevailing climate is *more stable* than that of the atmosphere; radiation is essentially unable to penetrate, there are no sharp gradients of temperature, and the processes of exchange are slow, occurring by diffusion. Therefore the soil is the most suitable habitat for many organisms. The roots—in many respects the most vulnerable organ system of the higher plants—are entirely adapted to life in the soil. A landscape without soil is a life-repelling "lunar" landscape. Only a few remarkable plants such as aerial algae, lichens and mosses can actually thrive on bare stone or sand.

1.4 The Ecosphere

Atmosphere, hydrosphere and lithosphere existed before there was life on earth, but it is of course only through the appearance of living organisms that they became significant as an "environment". *Environment*, as defined by A. F. Thienemann, is "the totality of external conditions affecting a living organism or a community (biocenosis) of organisms in its habitat (its biotope)". In this strict sense, only living beings have an environment. It comprises not only the influences exerted by the abiotic surroundings, but also those due to the other organisms present.

The part of the earth which supports life is called the *ecosphere*. Within the biogeosphere (the terrestrial part of the ecosphere), the *biosphere* (in the terminology of J. Lamarck, all the organisms on earth) is restricted to a narrow interface region between

3

atmosphere and lithosphere; trees are at most 70—100 m high, and the part of the ground penetrated by roots and inhabited by animals and microorganisms is only a few meters deep. Within the biohydrosphere (oceans and inland waters), the upper, euphotic layer of water is the most densely populated. Only specialized animals and heterotrophic microorganisms can live permanently at depths below 100 m.

Among living organisms, the plants are of prime significance. They are capable of capturing and storing by photosynthesis the energy from outer space—that of sunlight—and in terms of mass they far exceed all other organisms; about 99% of the total mass of living beings (the biomass) on earth is accounted for by the plants (the phytomass). Because of this enormous mass, the plant cover is a stabilizing factor in the cycling of matter and has a crucial effect upon the climate.

2 Radiation and Temperature: Energy, Information, Stress

2.1 Radiation

All life on earth is supported by the stream of *energy* radiated by the sun and flowing into the biosphere. Even the relatively small amount of radiant energy bound in the form of *latent chemical energy* by the photosynthesis of plants suffices to maintain the biomass and the vital processes of all members of the food chain. By far the larger fraction of radiation absorbed is transformed immediately into *heat*; part of this fraction is used in the *evaporation* of water and the rest produces an increase in the temperature on the earth's surface. Radiation is thus the source of energy underlying the distribution of heat, water and organic substances. It creates the prerequisites for an environment adequate to sustain life.

2.1.1 Radiation Within the Atmosphere

2.1.1.1 Attenuation of Radiation by the Atmosphere

At the outer limits of the earth's atmosphere the intensity of radiation is $1.39 \, kW \cdot m^{-2}$ (the *solar constant*). Of this, however, only an average of 47% reaches the earth's surface. More than half is lost; it is immediately cast back into space, as a result of *refraction and diffraction* in the high atmosphere and, in particular, by *reflection* from clouds, or it is *scattered and absorbed* due to clouds and particles suspended in the air. Of the radiation that does reach the ground or plant cover, about half has passed directly through the atmosphere while the remainder is diffused by the air and clouds. At sea level a horizontal surface at intermediate latitudes at noon receives *total* (*direct plus diffuse*) radiation amounting to as much as $900 \, W \cdot m^{-2}$ ($1.3 \, cal \cdot cm^{-2} \cdot min^{-1}$). Depending on the latitude of a site, its altitude above sea level, the nature of the terrain and the frequency of clouds, there are large regional and local differences in the supply of radiation (Fig. 2.1). Thus, the *high-pressure regions* in the tropics, where clouds are few, receive a greater than average quantity of solar radiation; rather than 47%, an average of 70% of the incident radiation penetrates the clear envelope of air above the dry regions. Moreover, at *greater altitudes*, owing to the shorter optical path of the rays and the lesser degree of air turbidity, more radiation reaches the ground than in lower-lying places.

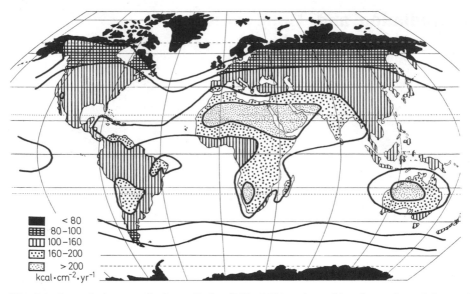

	< 80
	80–100
	100–160
	160–200
	> 200
	kcal·cm⁻²·yr⁻¹

Fig. 2.1. Annual average of solar radiation (300—2200 nm) reaching the surface of the earth (Geiger, 1965). Radiation maps based on satellite measurements are given by Raschke et al. (1973)

2.1.1.2 The Distribution of Radiation in the Plant Cover

In stands of plants, photosynthesis occurs within a stacked arrangement of leaves which partially overlap and shadow one another. The incident light is absorbed progressively in its passage through these many layers, so that most of it is utilized (Fig. 2.2). To facilitate description of the levels of illumination in or below a stand of plants, J. Wiesner (1907) introduced the notion of "relative irradiance" (*relativer Lichtgenuß*), expressed as the average percentage of the external light. In deciduous forests and open stands of conifers of the temperate zone, an average of 10—20% of the incident radiation reaches the herbaceous stratum during the growing season; when the trees are bare this figure increases to 50—70%. In dense coniferous forests, and in tropical forests with their abundance of species, the relative irradiance at the ground falls to a few percent, or even less than 1%. As a rule, the limit for the existence of vascular plants lies in this range. Thallophytes can survive with still less light (about 0.5% relative irradiance in general, as low as 0.1% for aerial algae).

The *attenuation of radiation in a stand* of plants depends chiefly on the density of the foliage and the arrangement of the leaves. The foliage density can be expressed quantitatively by the *leaf-area index* (*LAI*; D. J. Watson, 1947). The cumulative *LAI* indicates the total surface area of the leaves above a certain area of ground:

$$LAI = \frac{\text{Total leaf area}}{\text{Ground area}}. \tag{2.1}$$

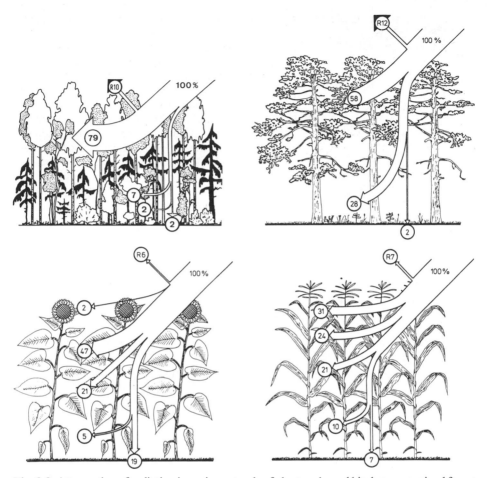

Fig. 2.2. Attenuation of radiation in various stands of plants: a boreal birch-spruce mixed forest (Kairiukštis, 1967), a pine forest (Cernusca, 1977), and fields of sunflowers (Hiroi and Monsi, 1966) and maize (Allen et al., 1964). Of the incident photosynthetically active radiation 6%—12% is reflected (R) at the surface of the stand; most of the radiation is absorbed in the stratum where the foliage is most dense, and the remainder reaches the surface of the ground. Depending on density, arrangement, and inclination of the leaves, quite characteristic differences arise in the distribution of radiation within the stand. Further examples: tropical rain forest (Odum and Pigeon, 1970; Allen et al., 1972), cocoa plantation (Alvim, 1977), Mediterranean sclerophyll stands (Eckardt et al., 1977), dwarf-shrub heaths (Cernusca, 1976), wheat (Baldy, 1973), rice (Udagawa et al., 1974) reeds (Dykyjová and Hradečká, 1976), sweet potatoes (Bonhomme, 1969)

Ordinarily the units used for leaf area and ground area are identical (m²), so that LAI is actually a dimensionless measure of the amount of cover. With a LAI of 4, a given area of ground would be covered by four times that area of leaves—arranged in several layers, of course. On its way through the plant canopy, the radiation must pass these successive layers of leaves. In the process, its intensity decreases almost exponentially

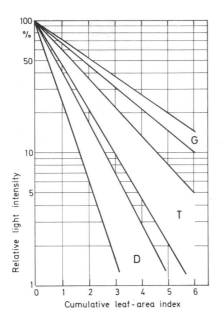

Fig. 2.3. The exponential decrease of light intensity in different stands of plants as a function of leaf-area index. The cumulative *LAI* is derived by summation of the index values for the individual horizontal layers of assimilation surface in the stand. In broad-leaved dicotyledonous communities (*D*), the attenuation of light is considerable even with a low *LAI*, whereas in grass communities (*G*) attenuation occurs more gradually; stands of trees (*T*) represent an intermediate position. After Monsi and Saeki (1953) and Kira et al. (1969)

with increasing amount of cover, in accordance with the *Lambert-Beer Extinction Law*. If the layering of the foliage is taken to be *homogeneous*, the fall-off of radiation can be computed from Monsi and Saeki's (1953) modification of the extinction equation,

$$I = I_0 \cdot e^{-k \cdot LAI} \tag{2.2}$$

where I is the intensity of the radiation at a certain distance from the top of the plant canopy (the irradiance); I_0 is the radiation incident on the top of the canopy; k is the extinction coefficient for this particular plant community; *LAI* is the total leaf area above the level at which I is estimated, per unit ground area (the cumulative *LAI*).

The *extinction coefficient* indicates the degree of attenuation of light within the canopy for a given area index. In grain fields, meadows and clumps of reeds, where the leaves tend to have an upright orientation (more than $^3/_4$ of the leaves are at an angle of more than 45° from the horizontal), the extinction coefficient is less than 0.5, and in the middle of the stand the light intensity is still at least half that of the external light (Fig. 2.3: *graminaceous type*). In contrast, for plant communities with broad, horizontal leaves such as fields of clover, sunflower plantations, or stands of tall perennial herbs, the extinction coefficient is greater than 0.7, and at "half height" $^2/_3$ to $^3/_4$ of the incident light has been absorbed (Fig. 2.3: *dicotyledon type*).

Forests with closely packed crowns and dense foliage swallow up so much radiation that near the trunks and on the ground there is very little. In such *forests* the attenuation of light is similar to, or even more abrupt than, that under dicotyledonous herbs. Woods comprising tree species with sparse crowns (birches, oaks, Scotch pines, eucalyptus) and open stands of trees, on the other hand, attenuate the light as gradually as do grass communities.

The interception of radiation in a stand can also be characterized in terms of the space taken up by the foliage *(leaf area density)*; here the *LAI* is referred not to ground area but to the volume occupied by the stand (m^2 leaf area per m^3 stand volume). This takes into account differences in *arrangement* and *position* of the leaves.

For precise studies—especially the mathematical analysis and simulation of the radiation climate in stands of plants—still other parameters must be specified quantitatively: the angles of inclination of the leaves (frequency distribution), location and extent of gaps in the plant cover, incident radiation and scattered light.

2.1.1.3 The Radiation Climate in Bodies of Water

In water, radiation is more strongly attenuated than in the atmosphere. Long-wavelength heat radiation is absorbed in the upper few millimeters, and infrared radiation in the uppermost centimeters. UV penetrates the top decimeters or meters. Light reaches greater depths, where blue-green twilight predominates in the ocean and yellow-green in lakes. The radiation available in bodies of water depends on the following factors:

a) The intensity and nature of *illumination above the water level.*

b) The amount of *reflection* and *backward scattering* of light at or near the water surface; on the average, when the sun is high, a smooth surface reflects 6% of the incident light and a surface with pronounced waves reflects about 10%. When the sun is low, on the other hand, reflection is considerably increased, so that a large fraction of the light fails to enter the water. The consequence is that, under water, the "day" is shorter than on the land.

c) The *attenuation* as the rays pass through the water; the intensity of radiation decreases exponentially with increasing depth. Radiation is absorbed and scattered by the water itself as well as by dissolved materials, suspended particles of soil and detritus, and plankton. In turbid, flowing waters the light can decrease to 7%—a value comparable to that beneath the crowns of the trees in a spruce forest—at depths of little more than 50 cm. In clear lakes 1% of the incident radiation reaches depths between 5 and 10 m, so that vascular plants can exist at 5 m, and sessile algae as deep as 20—30 m. The layer of water above the limit for existence of autotrophic plants is called the *euphotic zone*. In the open ocean the illuminated euphotic zone is deeper than in lakes; in the Mediterranean near the coast 1% of the radiation penetrates to 60 m, and in the clear water of the oceans the corresponding depth is as great as 140 m.

2.1.2 Absorption of Radiation by Plants

The ecosphere receives *solar radiation* at wavelengths ranging from 290 nm to about 3000 nm. Radiation at shorter wavelengths is absorbed in the upper atmosphere by the ozone and oxygen in the air; the long-wavelength limit is determined by the water-vapor and carbon-dioxide content of the atmosphere. About 40—50% of the solar energy received falls in the spectral region 380—780 nm, which we perceive as *visible light*. This region is bounded on the short-wavelength end by *ultraviolet radiation* (UV-A, 315—380 nm; UV-B, 280—315 nm; UV-C from artificial sources of radiation,

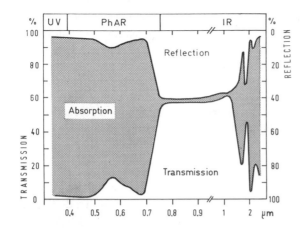

Fig. 2.4. Relative reflection, transmission and absorption of a poplar leaf (*Populus deltoides*) as a function of the wavelength of the incident radiation. After Gates (1965). Further examples are given by Wolley (1971); Gausman and Allen (1973); Alekseev (1975) and Jarvis et al. (1976)

100—280 nm) and at the other end, by *infrared radiation* (780—3000 nm). In addition, plants absorb *heat radiation* (long-wave infrared, at wavelengths from 3000 to 10^5 nm), and they themselves give off such radiation.

Some of the *radiation incident on a plant* is reflected at the surface, some is absorbed so as to be physiologically effective, and the remainder is transmitted. The degree of reflection, absorption and transmission in plant tissues depends on the wavelength of the radiation (Fig. 2.4).

Reflection. In the infrared region, leaves reflect 70% of radiation incident perpendicularly, whereas in the visible range an average of only 6—12% is reflected. Green light is more strongly reflected (10—20%), orange and red light less so (3—10%). Little ultraviolet radiation is reflected; as a rule, in the UV range, leaves reflect no more than 3%. Certain flowers, however, display marked UV reflection that can be detected by insects and serves as an attractive target.

The capacity to reflect light depends upon the nature of the *leaf surface*; for example, a dense covering of hairs can increase reflection of visible light and the near infrared by a factor of two or three.

Absorption. Radiation penetrating a leaf is to a great extent absorbed. Most of the UV is retained by the waxy, cuticular, and suberized outer layers of the epidermis, as well as by phenolic compounds in the cell sap in the outermost layers of cells, so that 2—5% at most, and usually less than 1% of the UV radiation enters the deeper levels of the leaf (see Fig. 2.7). Thus the epidermis is an effective UV filter, protecting the parenchyma in which photosynthesis occurs. Absorption of visible light is determined primarily by the *chloroplast pigments*. Correspondingly, the spectral absorption curves of leaves show maxima wherever the absorption maxima of chlorophylls and carotenoids occur (cf. Figs. 2.4 and 2.6). About 70% of the photosynthetically utilizable radiation entering the mesophyll is absorbed by the chloroplasts. In its passage through the leaf, radiation is progressively attenuated, so that the amount captured by successive cell layers falls off approximately exponentially. Not much infrared is absorbed in the region up to 2000 nm, but in the range of long-wavelength *heat radiation* above 7000 nm it is almost completely (97%) absorbed. Accordingly, the plant behaves like a black body with respect to heat radiation.

Fig. 2.5. The quality of the light at various heights in a tropical rain forest. From Johnson and Atwood (1970). Data for coniferous forests are given by Jarvis et al. (1976), for mixed forests by Alekseev (1975) and for stands of crops by Ross (1975)

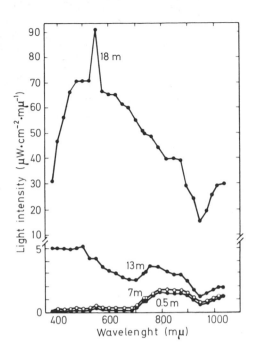

Transmission. Transmission by leaves depends on their structure and thickness. Soft, flexible leaves pass 10—20% of the sun's radiation, very thin leaves pass up to 40% and thick, solid leaves may not transmit any at all. Transmission is greatest at the wavelength ranges where reflection is great—that is, in the green and particularly in the near infrared. Radiation *filtered through foliage* is therefore particularly rich in wavelengths around 500 nm and over 800 nm. Beneath a canopy of leaves a red-green shade prevails, and in the depths of a forest only far red and infrared shade remains (Fig. 2.5).

2.1.3 Radiation and Plant Life

2.1.3.1 The Direct Effects of Radiation

For a plant, radiation is a source of energy (*photoenergetic effects*) and a stimulus regulating development (*photocybernetic effects*), but it can also cause injury (*photodestructive effects*). All these effects of radiation result from the capture of quanta. Because only those quanta that are absorbed can be photochemically active (the rule of Gotthuss-Draper), every radiation-dependent process is mediated by particular photoreceptors. Each of these is characterized by an *absorption spectrum* corresponding to the *action spectrum* of the associated photobiological events. Figure 2.6 shows the absorption spectra of a few important plant pigments; Table 2.1 indicates the involvement of various spectral regions in these photobiological phenomena.

In *photoenergetic* processes, the energy provided by absorption of radiation serves to drive metabolic reactions or causes chemical transformations, in a manner directly de-

Table 2.1. Effects of radiation on the plant. Expanded from Ross (1975)

Region of the spectrum	Wavelength (nm)	Percent of total solar radiation	Mode of Action			
			Photo-synthetic	Photo-morphogenetic	Photo-destructive	Thermal
Ultraviolet	290–380	0–4	Insignificant	Moderate	Significant	Insignificant
Photosynthetically active radiation	380–710	21–46[a]	Significant	Significant	Moderate	Significant
Near infrared radiation	710–4000	50–79	Insignificant	Significant	Insignificant	Significant
Long-wavelength radiation	3000–100,000	–	Insignificant	Insignificant	Insignificant	Significant

[a] The long-term average proportion of PhAR (400–700 nm) in the energy of the total incident short-wavelength radiation (300–3000 nm) in the northern hemisphere is 45–50% (Monteith, 1965; Szeicz, 1974); the mean is 47% (Yocum et al., 1964; Stanhill and Fuchs, 1977). This proportion changes greatly with the elevation of the sun and the degree of overcast. Further details and conversion factors for the PhAR range common in the USSR, 380–710 nm, are given by Ross (1975). Leaves absorb about 85% of the incident PhAR (Yocum et al., 1964; Ross 1975).

Fig. 2.6. Spectral distribution of energy in
(*1*) the sun's radiation outside the earth's
atmosphere, (*2*) the direct radiation from
the sun at sea level, (*3*) the radiation 1 m
under water in the coastal region of the
ocean, (*4*) the diffuse radiation from the
sky ("blue shade") and (*5*) the radiation
under a stand of plants ("red shade")
(Gessner, 1955; Gates, 1965; R. Schulze,
1970). *Lower graphs*, spectral absorption
by photosynthetically and photocybernetically effective pigments (Blinks, 1951;
French and Young, 1956; Siegelman and
Butler, 1965)

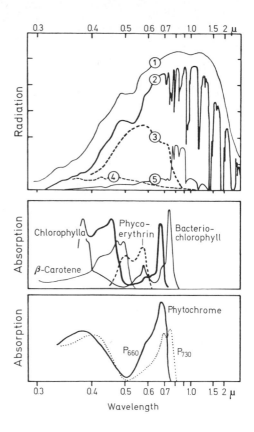

pendent on the amount of quanta absorbed. Energy-rich compounds can be con-
structed (*photosynthesis*), molecular structures can be altered (*photoconversion*), reactions can be accelerated (for example, the *photooxidation* of xanthophylls), or the
structure of a molecule can be destroyed.

The primary processes of photosynthesis (see Chap. 3.1.1.1) in green plants are driven
by radiation in the range of wavelengths between 380 and 710 nm. This *photosynthetically active radiation* (PhAR; often defined as the region 400–700 nm) is an
important quantity in plant ecology. The *photoreceptors* involved in photosynthesis are
the chlorophylls with absorption maxima in the red and blue, along with accessory
plastid pigments (carotin and xanthophylls) that absorb in the blue and UV regions. Be-
sides these, biliproteids (phycocyanin and phycoerythrin) are photosynthetic photoreceptors in blue algae, red algae, and cryptomonads. In place of chlorophyll,
photoautotrophic purple bacteria contain bacteriochlorophyll, a pigment with its main
absorption band in the far red.

Photodestructive effects occur with extremely high-intensity visible radiation or are
caused by UV. In both cases photoenergetic processes are involved. The plants most
sensitive to *strong light* are algae (especially red algae), mosses of shady habitats, some
ferns, and shade-adapted photolabile vascular plants—aquatic plants in particular. The
damage brought about by intense light consists primarily of photooxidation of chloro-

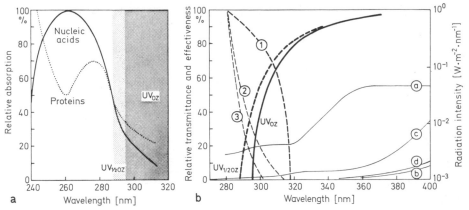

Fig. 2.7. a Absorption of ultraviolet by proteins and nucleic acids. The dark shading (UV_{oz}) marks the short-wavelength limit of solar radiation after passage through the ozonosphere; the light shading indicates the short-wavelength limit of radiation at sea level under the assumption that ozone density is reduced by half ($UV_{1/2oz}$). From Caldwell (1977). **b** Availability and effectiveness of ultraviolet radiation. UV_{oz}, short-wavelength limit of solar radiation with steep angle of incidence; $UV_{1/2oz}$, UV radiation at sea level if stratospheric ozone density were reduced by half. *1*, Cessation of protoplasmic streaming in epidermal cells of *Allium cepa*; *2*, mutagenic effect on liverwort spores and inhibition of photosynthesis of *Chlorella*; *3*, epidermal necrosis in *Oxyria digyna*. Spectral transmittance of: *a*, detached epidermal outer walls of *Sambucus caerulea*; *b*, living epidermis with flavonoids in the cell sap of *Sambucus*; *c*, cuticle of *Atriplex hastata*; *d*, cuticle of *Eryngium maritimum*. From various authors cited by Caldwell (1977) with data from Cappelletti (1961) and Robberecht and Caldwell (1978)

plast pigments. Some plants avoid high-irradiation damage by turning the leaves, or the chloroplasts in the cells, on edge (*phototaxis*) and by developing highly reflecting surfaces (e.g., with a covering of hairs) or surfaces that transmit little radiation (cork). In the presence of excess energy *glycolate metabolism* (cf. photorespiration, Chap. 3.1.2)—an oxygen-binding and energy-draining process in the cell—is most probably a crucial protective mechanism. Another important ability that helps a plant to tolerate strong light is the rapid *resynthesis* of decomposed plastid pigments.

UV below 300 nm causes not only photooxidation, but also photodestruction of nucleic acids and protein bodies, and acute damage to protoplasm (Fig. 2.7). In nature, UV-B occurs only at very low intensity—though we can expect an increase if the filtering capacity of the stratospheric ozone layer is much reduced by release into the atmosphere of nitrogen oxides and halogen hydrocarbons. *UV damage* to plants is evident in various symptoms: reduced capacity for photosynthesis, change in the activity of enzymes (peroxidase activity enhanced, cytochrome oxidase inhibited), disturbance of the process of growth (elongation, the growth of pollen tubes), the triggering of gene mutations and, finally, cell death. The molecular mechanism underlying injury to protoplasm by UV consists of the breaking of disulfide bonds in protein molecules and the dimerisation of thymine groups in DNA, which results in transcription defects. Because their epidermis strongly absorbs UV, higher plants are to a great extent protected from damage. By contrast, thallophytes and unicellular plants in particular are quite

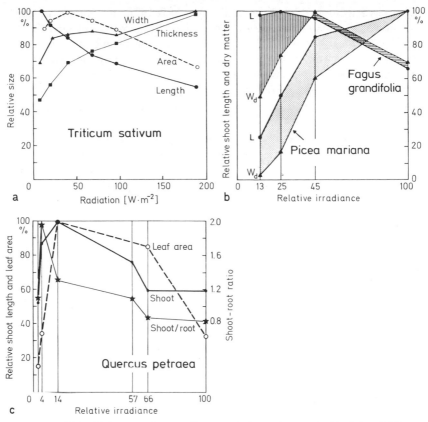

Fig. 2.8. a Length, width, thickness and area of leaves of wheat plants raised in light of different intensities. All data in percent of maximum. Friend et al. (1962). **b** Shoot length (*L*) and dry weight (*W_d*) of shoots of nine-year-old spruce and seven-year-old beech trees grown in full open-field light and in different degrees of shade. Young beeches rapidly grow taller in the shade, though an intermediate intensity is more favorable for optimal dry-matter production. The spruce develop best in all respects when maximally illuminated (Logan, 1969, 1973). **c** Structural features of two-year-old oaks grown under different levels of illumination. The average available light is given as a percentage of the open-field intensity; shoot length and leaf area are expressed as percentages of the maximum. On the *right-hand ordinate* is the ratio of shoot length to root length. From measurements by Trouchet, Gibondeau, and Goguely as cited by Roussel (1972)

vulnerable to the destructive action of short-wavelength radiation. They are killed by UV, which thus becomes a useful tool in sterilization procedures.

In *photocybernetic* processes receptor substances are chemically altered following uptake of radiation quanta, and this alteration affects the control of metabolism, growth, and development. Important factors here are the timing, duration, direction of incidence and spectral composition of the light; low intensities are sufficient. Radiation has a cybernetic action in the *photostimulation* of biosyntheses (e.g., the formation of chlorophyll from precursors and the synthesis of anthocyanin), in adjusting the direction of growth (*phototropism*), and in triggering germination and flower formation (*photoinduction*). Radiation affects the structure of the plant at the subcellular, cellular and whole-organism levels (*photomorphogenesis*; Fig. 2.8) and synchronizes development and the rhythmic events in the life cycle with the diurnal light/dark alternation and the seasons of the year (*photoperiodism*). Further details of these light-controlled processes of growth and development can be found in textbooks on plant physiology.

Light in two regions of the spectrum is effective *photocybernetically*—blue to ultraviolet radiation, and red to near infrared. Elongation and phototropism are regulated by *blue-light receptors* such as carotin and riboflavin, whereas most other processes depend on the *phytochrome system*. Phytochrome is a chromoproteid with a red-light-absorbing component related to the accessory pigments of algae (phycobilins). It is *photoconvertible*; that is, it occurs in two forms and can be changed from one to the other by appropriate illumination. The form that absorbs red, P-660, is converted by absorption in the spectral range 620—680 nm to the biochemically active far-red-absorbing form, P-730. Under far red light (700—800 nm) P-730 reverts to P-660. At the receptor sites (biomembranes) the proportions in which the two forms are present is determined by the proportions of red and far red in the available radiation; it is the ratio between the two states of phytochrome that regulates gene activity and thus the various eventual reactions.

Because of its importance in such processes, the *red/far red ratio* should always be included in descriptions of the light climate characterizing a plant habitat. It becomes apparent, for example, that below a plant canopy—where the seeds are germinating and the young plants growing—the amount of far red in the natural light can be 5 to 10 times that of red. There are many plant species with seeds that germinate only when red predominates; their germination is inhibited until the quality of the light changes, when the leaves fall or the upper strata become less dense. The photocybernetic postponement of germination serves to regulate the next generation—it has a population-ecological effect.

2.1.3.2 Adaptation of Plants to the Local Radiation Climate

In terms of their metabolism, development and morphology plants exhibit several kinds of adaptation—environmental (*modulative* and *modificative*) and genetic (*evolutive*)—to the prevailing quantity and quality of radiation.

Modulative adaptations occur rapidly and are *temporary*; when the original situation returns, the original behavior is soon resumed. Examples of photomodulation are leaf movements, by which the exposure of the leaf surface to the incident radiation is improved, and other phototropic, photonastic, and phototactic movements. The prime examples, however, are the functional adaptations of metabolism (chiefly photosynthesis) to fluctuations in light intensity (see Figs. 3.22 and 3.23).

Modificative responses adapt the plants to the *average* conditions of radiation during their period of growth; thereafter, the structural features of the plants are maintained. Plants adapted to shade develop extensive leaf surfaces and high concentrations of

Table 2.2. Differences between sun and shade leaves. Data from many authors; a classical study is that of Nordhausen (1903). A selection of recent papers: Boardman (1977), Lloyd (1976), Kozlowski (1971a), Mousseau (1977), Napp-Zinn (1973), Nobel (1976), Raven and Smith (1977), Tselniker (1978)

Characteristic	Sun leaves	Shade leaves
Structural features		
Area of leaf blade	−	+
Mesophyll thickness	+	−
Intercellular system (inner surface)	+	−
Cell number	+	−
Vein density	+	−
Thickness of epidermal outer wall and cuticle	+	−
Stomata density	+	−
Chloroplast number per unit area	+	−
Density of packing of the membrane systems in the chloroplasts	−	+
Chemical features		
Dry matter	+	−
Energy content of dry matter	+	−
Water content of fresh tissue	−	+
Cell-sap concentration	+	−
Starch	+	−
Cellulose	−	+
Lignin	+	−
Lipids	+	−
Acids	+	−
Anthocyanin, flavonoids	+	−
Ash	+	−
Ca/K	+	−
Chlorophyll a/b	+	−
Chlorophyll aI (P-700)	+	−
Photosystem II pigment complex	−	+
Chlorophyll/xanthophylls	−	+
Lutein/violaxanthin	+	−
Functional features		
Photosynthetic capacity	+	−
Respiratory intensity	+	−
Transpiration	+	−

+ High rates or large amounts, − low rates or small amounts.

chlorophyll and of accessory pigments in the chloroplasts. Plants exposed to stronger radiation develop an efficient axial system for water conduction; their leaves have several layers of mesophyll and cells with abundant chloroplasts (Table 2.2). As a result of structural adaptations and active metabolic processes, plants adapted to intense light produce greater amounts of dry matter, with a higher energy content, and their fertility is higher (frequency of flowering, setting of seed, yield of fruit). By contrast, plants adapted to dim light are distinguished by reduced dry-matter production, efficient synthesis of protein, and low respiration and water turnover. These characteristics enable them to thrive in locations where only modest amounts of energy are available.

Evolutive adaptations to the available radiation are based on *genotypic* changes; they determine the differences, sometimes quite conspicuous, that appear in the distribution ecology of various plant species and ecotypes. The classification of plants as dim-light plants, shade plants (sciophytes), sun plants (heliophytes) and strong-light plants (occupying open habitats in the high mountains, in deserts and on seacoasts) reflects genotypic differences in light requirements and in resistance to intense light. The response norm of a plant and its specific adaptation potential are *genetically* determined features. Thus sun plants can adapt to shade, but not to the same extent as shade plants, and shade plants exhibit analogous (but reversed) behavior.

Modulative, modificative and evolutive adaptations are not mutually exclusive, but *superimposed* so as to permit fine adjustments that guarantee the greatest possible utilization of the available radiation. When the space occupied by a stand of plants is densely filled, as is especially characteristic of tropical forests, it is because the light is completely exploited by a variety of life forms and plant species with light requirements genetically different but quite capable of adaptation. *Secondary effects* of radiation (heat, influences upon water balance) affect all adaptations to local light intensity. Sun plants, for example, must also be adapted to high temperatures, dry air, and a restricted supply of water.

Plants adapt not only to the intensity of light but to its *spectral composition* as well. Chromatic adaptation has been demonstrated primarily in Cyanophyceae and Rhodophyceae; the relative amounts of the specific accessory pigments change in accordance with the spectral shift in the light as it passes to deeper water. Land plants, however, are also able to change the composition of their chloroplast pigments, depending on the spectral composition of the light. Radiation with a high proportion of red appears to promote accumulation of chlorophyll a; when green and blue are present in greater amounts, chlorophyll b and carotenoids accumulate. The ability of plants to compensate for differences in the quality of the incident radiation by changing the amounts of pigment they contain is a great advantage—not only ecologically, but from the point of view of the grower, who can (for example) raise plants under artificial light.

2.2 Temperature

Plants are *poikilothermic organisms*—that is, their own temperatures tend to approach the temperature of their surroundings. However, this matching of temperatures is not precise. Because they *exchange energy with their surroundings* the temperature of the parts of plants above ground can depart considerably from that of the air. Thus the heat exchange of plants must always be considered with respect to the energy budget in the habitat.

2.2.1 The Energy Budget

2.2.1.1 Radiation Balance — the Source of Thermal Energy

Energy flows to earth in the form of radiation from the sun (insolation) which is absorbed by the plant cover, the soil, and the surfaces of the water. Opposing this gain in energy is a loss associated with the thermal radiation from terrestrial bodies. Thermal radiation, in correspondence to the relatively low surface temperature of the earth, is in the long-wavelength range (3—100 μm) and is strongly absorbed by dipole molecules in the atmosphere, particularly water vapor. As a result the air enveloping the earth is warmed, and most of the captured radiation is returned to earth as long-wavelength reradiation from the atmosphere. *The net radiation balance Q_I at a surface at a given time can be computed from the short-wavelength (0.3—3 μm) balance $\bar{I}_s$ and the long-wavelength thermal radiation balance $\bar{I}_l$.*

$$Q_I = \bar{I}_s + \bar{I}_l \qquad (2.3)$$

$$\bar{I}_s = I_d + I_i - I_r \qquad (2.4)$$

$$\bar{I}_l = I_a - I_g \qquad (2.5)$$

where I_d = direct solar radiation, I_i = diffuse skylight and cloudlight, I_r = reflected short-wave radiation, I_a = reradiation at thermal wavelengths from the atmosphere, I_g = long-wavelength thermal radiation from ground and plants.

The short-wavelength radiation balance received by the earth represents a gain in energy; the thermal radiation balance, as a rule, is a loss, so that the *overall radiation balance* is positive as long as the short-wavelength contribution predominates. This can apply of course only during the part of the day beginning shortly after sunrise and ending shortly before sunset. The balance becomes negative as soon as the daylight input is insufficient to compensate for the thermal radiant loss. The surplus energy is used in the biosphere for photosynthesis by plants, in warming the phytomass, soil, and air, and in phenomena associated with evaporation.

2.2.1.2 The Thermal Balance of the Plant Cover

Net radiation, energy consumption, and heat exchange are the main factors affecting the energy and hence the thermal balance of plants. The relationships are summarized in the *energy budget equation:*

$$Q_I + Q_M + Q_P + Q_{Soil} + Q_H + Q_E = 0. \qquad (2.6)$$

An example is given in Fig. 2.9, which illustrates the thermal balance of a tropical rain forest.

The terms in this equation, which must counterbalance one another in the overall energy budget of plants, are as follows:

1. The Radiation Balance Q_I.

2. The Energy Turnover in Metabolic Processes Q_M. Under insolation the capture of energy by photosynthesis predominates, while in the dark and in tissues without chloro-

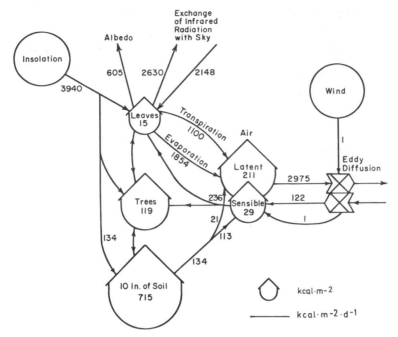

Fig. 2.9. Flow diagram of the radiation and thermal balance of a tropical rain forest in Puerto Rico. From Odum and Pigeon (1970). For the thermal balance of a temperate-zone deciduous mixed forest see Galoux (1971); energy balances for various plant stands are given in Monteith (1975, 1976)

phyll energy is liberated in respiration. Despite the eminent significance of basal metabolism to the organisms, the share of Q_M in the overall energy turnover is negligibly small, of the order of 1—2%.

3. Heat Storage by the Phytomass Q_P. Captured energy can be temporarily stored in the plants—for example, when more energy is taken up by the vegetation stratum than is given off to the atmosphere or to the soil. This happens when there is intense irradiation and little loss of heat by convection or evaporation. Under such circumstances leaves can warm up to 10° C, or in exceptional cases 15°—20° C, above the temperature of their surroundings. Massive plant organs with large heat capacity, such as succulent leaves and axial organs, fleshy fruits and tree trunks, sometimes *exceed the ambient temperature* by even greater amounts (Fig. 2.10).

4. Heat Storage in the Soil Q_{Soil}. In places free of vegetation, and where the plant growth is sparse (deserts, dunes, mountains, ruderal habitats, clear-cut woods, and newly plowed fields), a considerable part of the energy absorbed under insolation is conducted into deeper layers of the soil. Depending on the color, the content of water and air, the structure and the composition of the soil—and even more upon the slope and exposure of the surface—the ground is warmed to different degrees. Compact, wet soils conduct and store heat better than dry soils, so that the latter become hotter on the surface. Extremely high surface temperatures are produced in mountains at sites where the

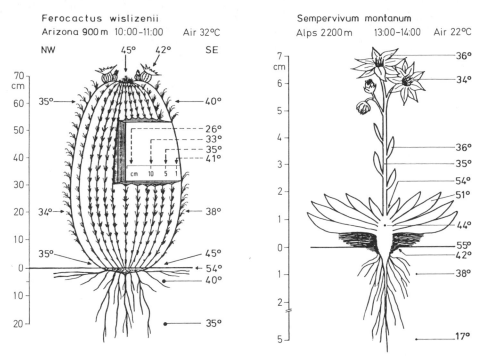

Fig. 2.10. The degrees to which succulent plants are warmed above air temperature under steeply incident radiation. The temperature in the center of the rosette of *Sempervivum montanum* can exceed that of the air by 32° C (unpublished measurements by W. Larcher). The barrel cactus *Ferocactus wislizenii* becomes warmest near the apex; when the sun is high the incident radiation tends to be tangential to the sides of the plant, which thus exceed the surrounding temperature by no more than 10° C (Monzigo and Comanor, 1975; K. Burian, pers. comm.). Further measurements of cactus temperatures are given by Lewis and Nobel (1977) and by Mooney et al. (1977)

sun's rays fall nearly perpendicularly onto exposed areas of dry dark raw-humus soils; there maximal temperatures exceeding 70° C have been measured. During the diurnal phase of net energy loss from the ground (that is, at night) the direction of heat transport in the soil reverses; the heat stored over the daylight hours is conducted back to the surface of the ground, which during the night has become steadily cooler. There are thus diurnal temperature fluctuations in the upper soil, down to depths of about 50 cm. In regions with a seasonal climate there is a superimposed annual fluctuation of temperature, which can be demonstrated even at depths of several meters (Fig. 2.11). Under a *closed cover* of vegetation, the soil is shielded from strong irradiation and radiant-energy loss. Even in the upper layers the diurnal temperature fluctuation is relatively small, and at depths below 30 cm it becomes insignificant (Fig. 2.12). Moreover, a thick covering of snow can have a similar effect. On the whole, the soil acts as a thermal buffer in the heat balance of a habitat, taking in considerable amounts of heat during the day only to give them up again at night.

SAINT PETER'S COLLEGE LIBRARY
JERSEY CITY, NEW JERSEY 07306

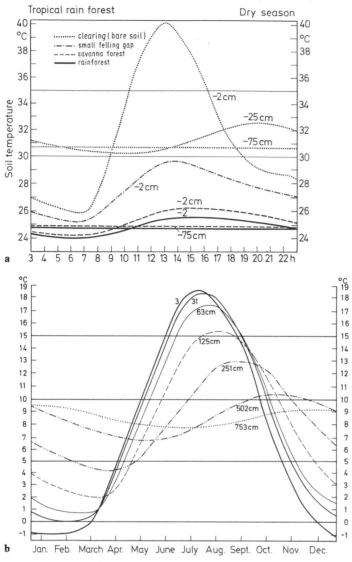

Fig. 2.11. a Diurnal fluctuation of soil temperature in a tropical rain forest and in areas free of vegetation during the dry season (Schulz, 1960). b Annual fluctuation of temperature at various depths in the soil in northern Europe (Schmidt and Leyst as cited by Geiger, 1961). Further examples are given by Rosenberg (1974)

5. Exchange of Energy with the Surroundings Q_H and Q_E. Exchange of heat between plants and their environment is also effected by *heat conduction and convection* (sensible heat exchange Q_H) *and by evaporation or condensation* (latent heat exchange Q_E). In neither the plant nor the community is conduction a significant factor; the redistribution of heat by convection and evaporation is mediated chiefly by mass flow of air and water.

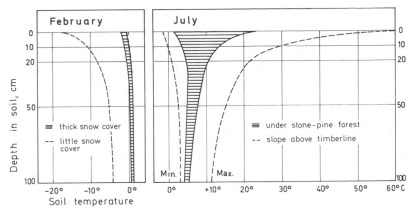

Fig. 2.12. Influence of snow cover and forestation upon the range of temperature fluctuation (curves show absolute maxima and minima) during the day at different depths in the soil; the habitats are in the region of the alpine timberline and above it (Aulitzky, 1961, 1962)

Convection Q_H

Under conditions of positive radiation balance, the direction of heat convection is usually away from the surface of the plants (Q_H is negative); on the other hand, if the plant surface is cooler than the air, heat is transferred to it from the environment (Q_H positive). Heat exchange with the surrounding air by convection is the more effective, the smaller and more subdivided the leaves and the higher the wind velocity (Fig. 2.13). Under strong insolation the plant is enclosed in a superheated envelope of air next to its surface. Wind sweeps away the boundary layer to within a few millimeters of the plant surface, thus increasing the rate of heat exchange. On the other hand, rosette plants and cushion plants, which grow close to the ground, lose heat less rapidly than erect plants—especially when they occupy sites protected from the wind (Fig. 2.14). In such a situation plants of the arctic and high mountains can maintain metabolic processes and growth in spite of the prevailing low air temperatures. In giant rosette plants of tropical high-mountain habitats, leaves that reflect light well owing to a dense covering of hairs surround the shoot apex and the floral primordia like a parabolic mirror; under strong irradiation these parts, particularly dependent on warmth, are thus brought to temperatures considerably higher than that of the surrounding air.

Evaporation Q_E

The thermal balance of plants is not determined by physical factors alone; because of the stomatal regulation of transpiration, it is also affected by physiological processes in the plant.

Q_E is negative when the plants transpire, and positive when dew or hoarfrost condenses on the leaves. The cooling effect of evaporation can be calculated from the thermal balance equation or by reference to the rate of transpiration. As water is given off, the amount of energy lost is that required to vaporize it. Q_E is the product of the rate of

23

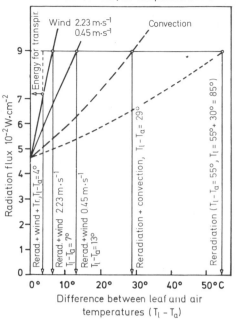

Computed thermal balance of an insolated leaf; air temperature 30°C

Fig. 2.13. Diagram of the energy turnover and the heat exchange of a leaf 1 cm in length under irradiation by $9 \cdot 10^2$ $W \cdot m^{-2}$ (1.3 $cal \cdot cm^{-2} \cdot min^{-1}$) at an air temperature of 30° C. The leaf temperature is determined by the energy uptake by absorption of radiation and the energy loss involved in reradiation, heat transfer by convection, air movement (two wind velocities shown), and transpiration (heat of vaporization of water). Here transpiration is considered to use energy at the rate of $1.6 \cdot 10^2$ $W \cdot m^{-2}$ (0.23 $cal \cdot cm^{-2} \cdot min^{-1}$, which corresponds to an average rate of transpiration of 40 $mg \cdot cm^{-2} \cdot min^{-1}$). The amount by which the temperature of the leaf exceeds that of the air is given by the intersection of the curve for each case with a horizontal line set at the energy level of the incident radiation. With irradiation amounting to about $4.5 \cdot 10^2$ $W \cdot m^{-2}$ (0.65 $cal \cdot cm^{-2} \cdot min^{-1}$) the leaf, at an air temperature of 30° C, would not be warmer than the air. After Gates (1965)

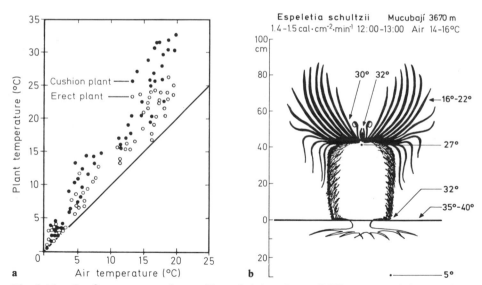

Fig. 2.14. a Leaf temperatures, in sunshine, of alpine plants of different growth forms. Plants growing close to the ground (*filled circles*) become distinctly warmer than those growing upright (*open circles*). From Salisbury and Spomer (1964). **b** The shoot apex of the giant rosette plant *Espeletia schultzii* in the Páramo level of the Venezuelan Andes becomes warmer than the air under full illumination by the zenith sun. Data of Pannier, Smith and Larcher as cited by Larcher (1975)

evaporation and the heat of vaporization of water (2.4–2.5 kJ · g^{-1} H$_2$O; see p. XVI).

$$Q_E = \lambda \cdot E. \tag{2.7}$$

E is the amount of water evaporated, and λ the "latent heat" of the water. Transpiration at a rate of 1 g H$_2$O · dm^{-2} · h^{-1} (0.1 mm H$_2$O · h^{-1}) corresponds to energy loss from the plant of 0.1 cal · cm^{-2} · h^{-1} (about 70 W · m^{-2}).

Table 2.3. Difference between the temperature of a *Canna* leaf and the air temperature, and the cooling effect of transpiration (Raschke, 1956)

	Heat exchange by convection and wind			
	Very low		Very high	
	T_{leaf}-T_{air}	Cooling by transpiration	T_{leaf}-T_{air}	Cooling by transpiration
Day (Radiation balance: +70 W · m^{-2})				
Tropical day 35° C, 90% *RH*	+26.4° C	−3.0° C	+1.9° C	−0.4° C
Steppe day 40° C, 7% *RH*	+20.9° C	−4.8° C	−1.1° C	−3.0° C
Temperate day 22° C, 61% *RH*	+15.9° C	−1.3° C	+0.7° C	−0.5° C
Night				
Tropical night (Radiation balance: −35 W · m^{-2}) 30° C, 94% *RH* Dew point: 29.0° C	−1.5° C	+0.9° C[a]	−0.2° C	−0.1° C
Steppe night (Radiation balance: −56 W · m^{-2}) 20° C, 23% *RH* Dew point: −1.9° C	−4.3° C	−0.1° C	−0.5° C	−0.2° C
Temperate night Radiation balance: −49 W · m^{-2}) 14° C, 95% *RH* Dew point: 13.2° C	−2.4° C	+1.3° C[a]	−0.3° C	0

[a] Heat transfer to leaf by condensation.

Example of the interpretation of the table: Under the conditions prevailing during a tropical day the leaf warms up by 26.4° C in still air, while with adequate heat exchange its temperature is only 1.9° C above that of the air; if it were not cooled by transpiration, the temperature of the leaf would be 3° C higher than that of the air in the first case, and 0.4° C higher in the second.

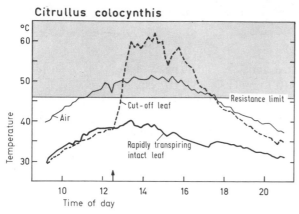

Fig. 2.15. Cooling effect of transpiration upon the leaves of a watered *Citrullus* plant under desert conditions. During rapid transpiration the leaves, despite intense insolation, are much cooler than the air. If a leaf is cut off (*arrow*) so as to make vigorous transpiration impossible, the leaf temperature rapidly rises above that of the air, becoming so high that signs of heat injury appear (the range of temperatures associated with heat injury is shown in gray). Plants like *Citrullus*, which ordinarily maintain a temperature lower than that of the air, can survive in hot habitats only if they are able to transpire at a high rate. After Lange (1959)

Cooling by evaporation is particularly effective when the air temperature is high, the humidity is low, and the plants are well supplied with water. From Table 2.3 it can be seen that under steppe conditions, one quarter of the heat lost by a leaf is accounted for by transpiration. If transpiration is accelerated by *wind*, it extracts so much heat that the leaves can become several degrees cooler than the air. The influence of transpiration upon thermal balance can be readily demonstrated by preventing the plant from giving off water. If one cuts through the petiole of a leaf in order to interrupt the supply of water, or treats the plant with antitranspirants (substances that inhibit transpiration), the leaf temperature rises within a few minutes (Fig. 2.15).

2.2.1.3 The Thermal Climate of Stands of Plants

In stands of plants, energy exchange occurs chiefly in a narrow zone near the upper surfaces (Fig. 2.16). In this active layer (the "*effective surface*" of the stand), the thermal individuality of the different plant parts is most clearly evident and the temporal variation of temperature is greatest. In this region the leaves and twigs warm up the most during the day and cool off most rapidly after sunset. On a hot summer day the temperature at the surface of temperate-zone forests was found to be ca. 4° above that of the air; the surface temperature of meadows was about 6° above air temperature. Under a canopy of leaves or in a dense stand of plants, the intensity of radiation is greatly reduced (cf. Fig. 2.2) and atmospheric reradiation is limited by the shielding leaves. The entire process of energy conversion is damped, so that within the stand more stable temperatures prevail. In dense tropical forests the temperature fluctuates by less than 10° C over the day and year; in the rainy season the amplitude of the daily temperature fluctuation is only 2–5° C.

Fig. 2.16. Radiation balance and air temperature (daily mean and daily fluctuation) over and within a dense young spruce forest on a typical day of fine weather in midsummer. The most pronounced turnover of energy takes place in the upper part of the treetop, where the daily air-temperature fluctuation is also greatest. Near the ground, where only a very small part of the energy is converted, the air temperature remains stable and cool (cf. also Fig. 2.2). After Baumgartner as cited by Geiger (1961). For tropical rainforests see Odum and Pigeon (1970) and Richards (1976)

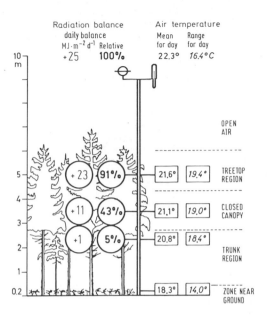

2.2.2 The Effects of Temperature upon the Vital Processes of Plants

2.2.2.1 Life-Supporting Range and Functional Range

Sufficient but not excessive *heat is a basic prerequisite for life*. Each vital process is restricted to a certain temperature range and has an optimal operating temperature, on either side of which performance declines. Thus for each *plant species*, and for each *stage of development*, one can determine characteristic "cardinal temperatures". These are not rigid constants but rather ranges about the genetically fixed norm; within these ranges, plants of a single species can develop into different metabolic and/or resistance *ecotypes*, and the optimal, minimal, and maximal temperatures for the plant can shift as the plant *adapts to environmental conditions*. Terrestrial vascular plants, as a rule, can thrive within a wide range of temperatures and are called *eurythermic*. For such plants, in an active state, the life-supporting temperature range usually extends from −5° C to about +55° C—a span of 60 degrees; but such plants are active in the production of dry matter and in growth (Fig 2.17) only between about 5° and 40° C. Among the aquatic plants, particularly the thallophytes, there are *stenothermic* species, specialized for life in very narrow—but sometimes extreme—temperature ranges. For example, snow and ice algae (e.g., *Chlamydomonas nivalis* on snow fields in the high mountains, and various green algae and diatoms in the polar ice) can thrive only near the freezing point. Stenotherms also include many parasitic bacteria and fungi that are adapted to the temperatures at which infection and spread through the host proceed optimally.

For ecological purposes it is necessary to know, for the different plant groups, the following thermal requirements and limits of tolerance:

The Survival Limits. These are the lowest and highest temperatures at which a plant can survive. A distinction is made between the activity limit and the lethal limit. When

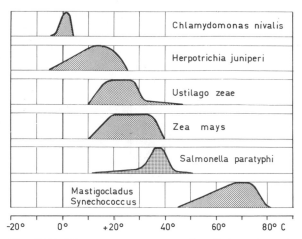

		Chlamydomonas nivalis
		Herpotrichia juniperi
		Ustilago zeae
		Zea mays
		Salmonella paratyphi
Mastigocladus Synechococcus		

-20°   0°   +20°   40°   60°   80° C

Fig. 2.17. The temperature ranges in which various plants thrive. Plants restricted to low temperatures are *Chlamydomonas nivalis* and other snow algae, as well as the mold *Herpotrichia juniperi*, which infests the snow-covered twigs of conifers. The mycelial growth of the corn smut *Ustilago zeae* is largely determined by the temperature range for activity of the host plant. Parasites of warm-blooded animals, like *Salmonella paratyphi*, display an especially narrow optimum for development. At extremely high temperatures the only organisms that can thrive are thermophilous bacteria and blue-green algae (for example, *Mastigocladus* and *Synechococcus* species from geysers). Data in Went (1957), Altman and Dittmer (1973), Raeuber (1968), Kol (1968), Brock (1967, 1978), De Wit (1970), Müller and Löffler (1971). Data for other plants can be found in the following publications: bacteria, Baross and Morita (1978); planktonic algae, Fogg (1975); soil fungi, Griffin (1972); forest trees, Lyr et al. (1967); crop plants, Brouwer and Kuiper (1972); tropical fodder and other crop plants, Sweeney and Hopkinson (1975)

the *activity limit* is exceeded the active vital processes are *reversibly* slowed to a minimal rate and the protoplasm enters a state of anabiosis (under excessive heat or cold). At the *lethal limit* permanent injury is sustained and life is extinguished (Tables 2.5 and 2.6). Resistance to great heat or cold is an advantage to any plant, but especially to those that must avoid competition; these cannot establish themselves under favorable temperature conditions and are found only in open, and therefore climatically extreme, habitats.

The Temperature Range for Dry Matter Production, Growth, and Development. Efficient metabolism and the synthesis of new tissue are the prerequisite for growth and development and thus are crucial determinants of the competitive ability of a species. The temperature range for the *photosynthetic acquisition of carbon* by most plants in an active state is only about 5° C narrower than that delimited by temperatures so cold as to injure, and so hot as to kill, the leaves. The range of optimal temperatures for photosynthesis and dry-matter production is, however, no wider, as a rule, than 10°—15° C; the temperature optimum is clearly related to the thermal climate of the region in which the species grows (see Chap. 3.2.3.2). The same relationships apply to growth of a plant. *Extension growth of shoots* of temperate-zone plants begins as soon

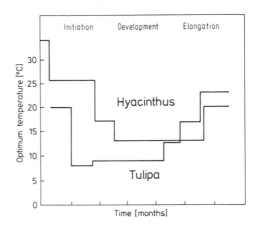

Fig. 2.18. Phase-dependent temperature optima for initiation and development of leaf and flower primordia and for leaf and shoot elongation, in tulips and hyacinths (Hartsema et al., 1930; Luyten et al., 1932)

as the temperature rises a few degrees above 0° C, whereas in tropical plants growth does not begin until 12°–15° C is reached. Plants in the arctic, mountain plants, and spring-blooming plants show signs of growth at 0° C. The temperature optimum, at which extension of shoots proceeds most rapidly, lies between 30° and 40° C for plants of the tropics and subtropics, and between 15° and 30° C for all other plants. The temperature range permitting elongation of the *roots* is usually very wide. The minimal limiting temperature for root growth in woody plants of the temperate zone is rather low, between 2° and 5° C. It is thus not surprising that roots begin to grow before the buds sprout and that they continue growing until late in the fall (cf. Fig. 2.38). Plants of warmer regions require higher temperatures. *Citrus* roots grow only above 10° C; within the natural geographic range of this genus, the soil temperature never falls below this point, even during the coldest season. But for orchards, this temperature limit can be critical, since it is in winter that the fruits of most *Citrus* species ripen—a process requiring a good supply of water and nutrients from the roots.

Active *growth by cell division* also occurs seasonally, being temperature-dependent in the same way as elongation growth. *Cell differentiation* can proceed at low temperatures, though it may be very slow. The differentiation of bud meristems of fruit trees and other perennial plants of the temperate zone is interrupted in winter only when the weather becomes especially cold. In bulbs of geophytic plants from the Mediterranean and Near Eastern steppes (for instance, tulips and hyacinths) leaf primordia are initiated at temperatures above 20° C, but low temperatures (around 10° C) such as prevail in the soil in winter, in regions where these plants naturally grow, are favorable to the final differentiation of the shoot apex in the bulb. Higher temperatures, again, are required when the shoot and inflorescence axis begin to grow out of the bulb (Fig. 2.18). It is generally true that the course of development of a plant is critically affected not by one particular optimal temperature range, but by a sequence of optimal temperatures.

The Temperature Limits and Heat Requirements for Reproductive Processes and Germination. If a population is to maintain itself, and a species to expand its range, it is not sufficient that the individual plants simply withstand extreme situations and perform their vegetative functions adequately; the requirements for flowering and the ripening and germination of seeds must be met. The range of temperatures suitable for the latter processes differs, as a rule, from that for growth and the development of the vegetative organs.

Flower formation is induced within certain temperature thresholds, and still different temperatures are effective in bringing about the development and unfolding of the flowers. Winter annuals and biennials, as well as the buds of certain woody plants (e.g., the peach) require a cold winter season in order to flower normally in the spring

Table 2.4. Minimal, optimal, and maximal temperatures (in °C) for the germination of seeds and spores. These data are representative, but in particular cases may be changed considerably by various factors, both external (light, soil moisture, thermoperiod) and internal (stage of maturation, age, and readiness of the seeds to germinate). From a number of authors cited by Abd El-Rahman and El-Monayeri (1967), Altman and Dittmer (1972), Bierhuizen (1973), Bierhuizen and Wagenvoort (1974), Bliss (1971), Chabot and Billings (1972), Griffin (1972), Kozlowski (1971a), Kramer and Kozlowski (1979), Mayer and Poljakoff-Mayber (1975), Valovich and Grif (1974), Vegis (1973), and Went (1949). For response to temperature fluctuations see Thompson et al. (1977)

Plant group	Minimum	Optimum	Maximum
Fungus spores			
Plant pathogens	0—5	15—30	30—40
Most soil fungi	ca. 5	ca. 25	ca. 35
Thermophilic soil fungi	ca. 25	45—55	ca. 60
Grasses			
Meadow grasses	3—4	ca. 25	ca. 30
Temperate-zone grain	(0) 2—5	20—25 (30)	30—37
Rice	10—12	30—37	(35) 40—42
C_4 grasses of tropics and subtropics	(8) 10—20	32—40	(40) 45—50
Herbaceous dicotyledons			
Plants of tundra and mountains	(3) 5—10	20—30	
Meadow herbs	(1) 2—5	20—30	35—45
Cultivated plants in the temperate zone	1—3 (6)	15—25 (30)	30—40
Cultivated plants in tropics and subtropics	10—20	30—40	45—50
Desert plants			
summer-germinating	(10)	20—30 (35)	
winter-germinating	(0)	10—20	ca. 30
cacti	10	15—30	
Temperate-zone trees			
Conifers	4—10	15—25	35—40
Broad-leaved trees	below 10 [a]	20—30	

[a] After cold-stratification.

("chilling requirement"). They do not become ready to flower until they have been exposed for weeks to temperatures between −3° and +13° C, ideally between +3° and +5° C. This acquisition or enhancement of the ability to flower by exposure to cold is called *vernalization*. If the cooling period is too short, comes at the wrong time, or is interrupted by warming above 15° C, the effect does not appear.

Fruits and seeds require more heat to ripen than is necessary for growth of the vegetative parts of the plant. In habitats with both shorter and cooler growing seasons, a plant species can better maintain itself if it has alternative means of propagation, such as the formation of runners, spread by rhizomes, or other *vegetative mechanisms*. The coldest regions, like the driest, are colonized almost exclusively by cryptogams, in which the elaboration of reproductive organs is minimal.

Germination is the process of greatest importance for distribution ecology. The cardinal temperatures for germination of spores and seeds must correspond to external conditions guaranteeing sufficiently rapid development of the young plants.

The temperature range for *initiation of germination* is wide in species that are broadly distributed and in those adapted to wide temperature fluctuations within their habitat (Table 2.4). *Tropical plants* germinate optimally (i.e., the percentage of seeds that germinate is greatest) between 15° and 30° C; the range for plants of the *temperate zone* is 8°−25° C, and for *alpine plants*, 5°−30° C.

The *rate of germination* rises with increasing temperature. In species germinating in summer (usually those of northern origin), as opposed to winter-germinating species (from regions with mild winters), germination proceeds extremely slowly at low temperatures; only after the seed-bed has warmed up to more than 10° C does the process accelerate, but then it soon makes up for lost time (Fig. 2.19). Thus synchronization is achieved with the season most favorable for development of the young plants.

In some plants there are complicated *mechanisms to prevent germination* at unfavorable times. The seeds of many woody plants in cold-winter regions (a number of forest trees and Rosaceae) and the seeds of mountain plants (e.g., *Silene acaulis* and many Primulaceae) germinate more readily if they have been exposed, while in a turgid state,

Fig. 2.19 Temperature dependence of rate of germination in various species of Caryophyllaceae. The northern species germinate in summer, and germination is greatly inhibited below 10° C; the southern species germinate in winter, and the process is rapid even at low temperatures. After Thompson (1968); cf. also Thompson (1970); for seed germination from a latitudinal series of populations of an arctic-alpine grass see Clebsch and Billings (1976)

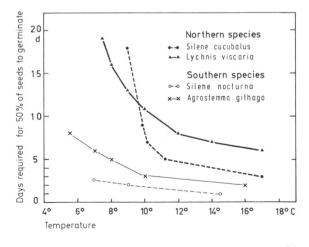

for a considerable time to low temperatures or mild frost (cold stratification at around +5° C). High temperatures trigger germination of some tropical plants; exposure of the dry seeds of rice and oil palm to 40° C rapidly brings dormancy to an end.

2.2.3 The Temperature Limits for Plant Life

2.2.3.1 Extremes of Temperature on the Earth

Only under water and deep in the ground is the temperature variation restricted to the range between about 0° C and 20°—25° C, where plants are in no danger and their vital functions are promoted. Near the soil surface on the continents, as well as in the tidal zones and in shallow waters, the temperature oscillates with the time of day (and, outside the equatorial zone, with the seasons) between limits that can be potentially lethal.

Particularly *high air temperatures* occur at latitudes near the northern and southern tropics; as an extreme case, maxima of +57° to +58° have been measured in Libya, Mexico and California. Over about 23% of the continental area of the earth, mean annual air temperature maxima of over 40° C are to be expected; in such areas plant temperatures of 50° C or more can occur in intense sunlight. Apart from these hot regions, *severe local overheating* to 60° and 70° C is possible, primarily on rocks and on open habitats sloping toward the sun. The hottest spots on earth inhabited by living organisms are geysers, in which the water coming to the surface is at 92°—95° C; near these, some bacteria colonize zones as hot as 90° C.

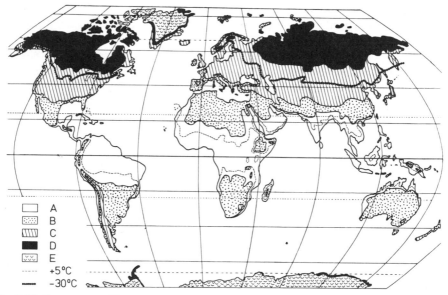

Fig. 2.20. Occurrence of low temperatures and frost on the Earth. *A*, frost-free zone; *B*, episodic frosts down to –10°C; *C*, average annual minimum temperature between –10° and –40°C; *D*, average annual minimum below –40°C; *E*, polar ice; --- +5°C minimum isotherm; ... –30°C minimum isotherm (Larcher and Bauer 1981). Frost zone map of the USA: USDA-ARS Publ. No. 814

The *coldest temperatures* measured on the earth have been in Antarctica (near −90° C); in Greenland and in eastern Siberia air-temperature minima between −66° and −68° C have been observed. Relatively *severe frost* (a mean annual air temperature minimum below −20° C) is to be expected over 42% of the earth's surface, and only a third of the land area never experiences freezing temperatures (Fig. 2.20). An important factor in the effects of periods of frost is whether they occur *regularly* in a region as the seasons alternate, or whether they are isolated episodes. *Episodic* frosts are usually more dangerous even though the temperatures are rarely very low, since they may catch the plants in a sensitive phase. In contrast, plants can "prepare" for the regular annual return of winter frosts, by allowing for a gradual hardening of their vegetative processes to low temperatures so that they suffer no damage.

2.2.3.2 The Limits for Existence and the Capacity to Survive

The probability that a plant species will survive is a function of its ability to come unscathed through extreme weather conditions and to maintain itself in endangered habitats. During a *heat spell*, the plants must cope not only with high temperature but also with the threat of desiccation; *winter* brings not only the direct effects of cold but also such dangers as frost drought, the pressure of snow, and avalanches.

In order to judge the degree to which a plant is *endangered by thermal stress*, one must have access to data about distribution, frequency, and probable time of occurrence of extreme temperatures. When suitable measurements of plant temperature are not available, temperature data from standard weather stations can provide rough estimates for large-scale comparisons involving plants of adequate size. Such inferences may be useful, but one must keep in mind that they are no substitute for a full, quantitative ecophysiological analysis.

The *probability of survival* of a species in a given habitat is higher, the greater the *resistance* of the most vulnerable vital part of the plant to all these factors, the sooner injuries are healed (the *recovery capacity*), and the less frequently the extreme conditions recur.

The *recovery capacity* of a plant is not easy to ascertain. The likelihood that damage can be repaired is more readily estimated if the resistance of the *perennating buds* and associated structures is known, and if the occurrence of unusually long and severe winters, late or early frosts, and abnormally hot spells is studied. In particular, it is advantageous if these climatic excursions can be employed as "field experiments", to test existing concepts in the natural habitat.

2.2.3.3 Temperature Stress Effects and Cell Death by Heat and Cold

Heat and low temperatures impair the vital functions and limit the distribution of a species, depending upon their intensity, duration, and variability; the state of activity and degree of hardening of the plants are even more important factors. Stress is the *exposure to extraordinarily unfavorable conditions*; they need not necessarily represent a threat to life, but they do trigger an "alarm" response (e.g., defensive and adaptive reactions) in the organism if it is not in a dormant state. Resting stages such as dry spores, and poikilohydric plants in a dry state, are insensitive and can survive undamaged any temperatures occurring naturally on the earth (Table 2.5).

Protoplasm responds to stress with an initial acceleration of metabolism. The increased respiration observable as a stress reaction is an expression of the effort being made to repair damage incurred and to make the adjustments in fine structure necessary for adaption to the new situation. The *stress reaction* amounts to a race between the adaptive mechanisms and the destructive processes in the protoplasm that would lead to death. When critical upper or lower temperature thresholds are exceeded, cell structures and functions may be damaged so abruptly that the protoplasm dies immediately. In nature *sudden destruction* often occurs during episodic frosts—for example, late spring frosts. Damage can, however, also come about *gradually*, as the equilibria of certain vital processes are affected and their operation is impaired; finally some *functions essential for life cease* and the cell dies.

Symptoms of Injury. Temperature-sensitivity varies with the vital process concerned (Fig. 2.21). The first effect to appear is the cessation of *protoplasmic streaming*, since this is directly dependent upon energy provided by respiration and upon the availability of high-energy phosphates. Next, the rate of photosynthesis is decreased. Damage to the chloroplasts is followed by a residual, sometimes permanent, inhibition of photosynthesis. In the terminal stage the semipermeability of the biological membranes is lost, so that the cell compartments, particularly the thylacoids of the plastids, collapse and the cell sap emerges into the intercellular spaces.

Causes of Death by Heat. Heat causes death by damaging membranes, and in particular by *denaturing proteins*. Even if only some especially thermolabile enzymes

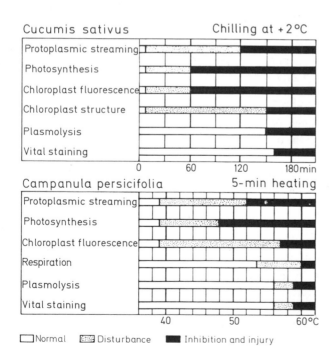

Fig. 2.21. The impairment of various cell functions by heat and cold. *Disturbance* indicates a reversible impairment, *inhibition and injury* are permanent damages (Alexandrov, 1964; Kislyuk, 1964)

are put out of action, so that nucleic-acid and protein metabolism becomes disorganized, the cells eventually die. Soluble nitrogen compounds then accumulate in such high concentration that they seep out of the cell; moreover, toxic decomposition products are formed and can no longer be made innocuous through metabolic processes.

Death by Chilling and Freezing. In cases of cold damage, there is a distinction between injury to protoplasm by the drop in temperature as such and injury by the process of freezing. Some plants of tropical origin suffer *chilling injury* even when the temperature is several degrees above freezing. Like death by heat, death by cold is the consequence of damage to biomembranes and the breakdown of metabolism, primarily that involving nucleic acids, proteins, and the provision of energy to the cell.

Plants resistant to low positive temperatures may be damaged by freezing—that is, by *ice formation in the tissues.* Protoplasts with a high water content that have been supercooled to low temperatures freeze *intracellularly*; ice crystals form very rapidly inside the cell and it perishes. Usually, though, ice is formed not in the protoplasts but rather in the intercellular spaces and between cell wall and protoplast. This sort of ice formation is called *extracellular.* As ice crystallizes out it has an effect like that of dry air, since the vapor pressure of ice is lower than that of a supercooled solution. Thus water is withdrawn from the protoplasts (Fig. 2.22), they shrink markedly (to $^2/_3$ of their volume), and the concentration of dissolved substances rises correspondingly. The redistribution between free and bound water and the ice phase continues until water potential equilibrium is reached between the ice and the water in the protoplasm. The water potential at which equilibrium is reached is temperature-dependent; at $-5°$ C it is about -60 bar and at $-10°$ C as low as -120 bar. *Low temperatures thus have the same effect upon protoplasm as desiccation.* The ability of a cell to withstand ex-

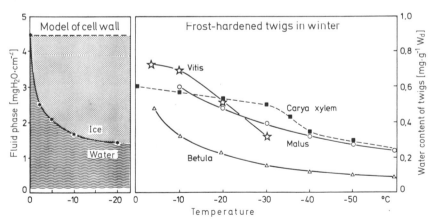

Fig. 2.22. Decrease in the amount of unfrozen water as ice is formed. *Left*, Filter-paper disk saturated with water, as a model of freezing cell walls. At temperatures as high as $-2°$ C half of the water is in the solid phase. From Olien (1974). *Right*, Twigs of woody plants in frost-hardened condition. The withdrawal of water, as the tissues gradually freeze during progressive cooling, affects the protoplasm in the same way as does desiccation (Krasavtsev, 1972; George and Burke, 1977)

tracellular ice formation is greater when a large part of the water remaining in the cell is osmotically bound or attached to protoplasmic components in the form of molecule clusters ("structured water"), and when the water that is not bound can easily pass through the biomembranes and out of the cell. When water is withdrawn from the protoplasm (whether by drought or freezing), enzyme systems associated with the membranes, which are crucially involved in ATP synthesis and phosphorylation, are inactivated (Heber and Santarius, 1973). This inactivation is caused by an excessive, effectively poisonous concentration of salt ions and organic acids in the solution remaining unfrozen. Sugar, sugar derivatives, certain amino acids and proteins, on the other hand, act to protect the membranes and enzymes from these effects (Maximov, 1929; Sakai, 1962). There are some indications that membrane damage can also result from the denaturation of proteins by freezing (Levitt, 1972).

2.2.3.4 Temperature Resistance

Temperature resistance can be defined as the net result of the capacity of a plant's protoplasm to survive extreme temperatures ("tolerance", according to J. Levitt, 1958) and the effectiveness of its mechanisms for delaying or preventing the onset of thermodynamic equilibrium ("avoidance"). Fig. 2.23 gives a diagrammatic survey of the components of temperature resistance.

Mechanisms for Avoidance of Injury

There are only a few mechanisms—and these not very effective—by which plants are able to protect protoplasm from extreme temperatures:

Insulation from overheating and frost is effective only for short exposures. For example, in the dense crowns of trees and in cushion plants the inner, more deeply located buds, leaves, and flowers are less endangered as the air temperature falls below freezing than are the outer parts of the plant; conifer species with particularly thick bark can better withstand fires in the undergrowth of the forest.

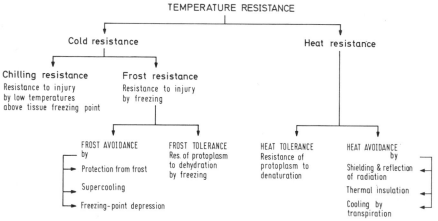

Fig. 2.23. Components of temperature resistance. After Levitt (1958)

36

Delayed Ice Formation in the Tissues. Dissolved substances and other water-binding (matric) forces *lower the freezing point*. Water in the vessels and cell walls is frozen only a few tenths of a degree below 0° C. Depending on the concentration of the solutes it contains, cell sap freezes between −1° and −5° C. Freezing-point depression provides moderate but reliable protection to tissues.

Moreover, the water in the cells can be *supercooled*; that is, it can actually be cooled to temperatures somewhat below the freezing point without freezing immediately. In large, water-filled parenchyma cells, and above all in the xylem vessels, the supercooled state is very *labile*. It is seldom maintained for more than a few hours and at most can assist the plant to survive a "radiation frost". In very small cells and cells with a subdivided vacuole (e.g., in seeds, bud meristems, or wood parenchyma) that are appropriately hardened, *persistant supercooling* to still lower temperatures is possible. This *"deep supercooling"* can actually prevent freezing in the buds and wood of various temperate-zone trees and shrubs (for example, rose bushes, apple trees, and many winter-deciduous forest trees) at temperatures around −30° C (−47° C at most). Supercooling and freezing can be detected by monitoring carefully the temperature drop in the plant. At first tissue temperature follows air cooling with little delay; then there is a sudden increase in temperature (the "exotherm") as heat of crystallization is liberated, which signals the onset of freezing (Fig. 2.24). As soon as equilibrium between fluid and solid phases has been reached, the temperature once again begins to fall.

Heat Reduction by the Reflection of Radiation and Cooling by Transpiration. Dangerously high plant temperatures—except of course in fires—occur only under intense

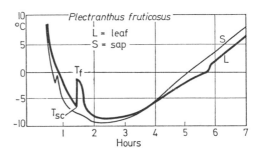

Fig. 2.24. The change of temperature in a leaf, and in the sap expressed from a leaf, of *Plectranthus* during a freezing experiment. At time 0 the samples were placed in a precooled chamber. There, their temperature rapidly fell to the supercooled level (T_{sc}). The leaf could be supercooled to a lower temperature (about −6° C) than the sap on a piece of filter paper (supercooling limit at about −2° C). As freezing began, the temperature in both leaf and sap suddenly rose as a result of the liberation of heat of crystallization, and the initial subsequent rate of fall was slow. The highest temperature reached in this exothermic process can be taken as the freezing point (tissue freezing point of the leaf, T_f, or freezing point of the sap/filter-paper system). During thawing, the rise of temperature in the leaf is delayed in the range of the freezing point because heat was used in melting the ice. After Ullrich and Mäde (1940). Similar curves were measured by Maximov (1914). More recent methods for quantitative analysis of the process of freezing are given by Olien (1967, 1978), Weiser (1970), Krasavtsev (1972) and Burke et al. (1976)

insolation. Increased *reflection* counteracts overheating, as does *turning the leaves* so that they intercept less sunlight. There are trees among the Caesalpiniaceae and Mimosaceae which, at air temperatures above 35° C, even though they are well supplied with water, fold the blades of their leaves together and thus cut their absorption of radiation in half.

Another protective measure against overheating is *cooling by transpiration*. As long as water can be supplied to the leaves of desert and steppe plants at a high rate, the leaves remain 4°–6° C, and in extreme cases as much as 10°–15° C, cooler than the air; without this heat-reducing mechanism, they would be injured (cf. Fig. 2.15).

Protoplasmic Tolerance

Protoplasmic tolerance is *genetically determined* and varies among species, varieties and ecotypes. Moreover, tolerance varies in the course of *development* of a single plant. Actively growing seedlings, sprouting shoots, and cultures of microorganisms in their phase of exponential growth are all extremely sensitive to high and low temperatures.

Freezing Tolerance and Hardening Against Frost. In regions with a seasonal climate, perennial plants become *hardened against frost*—that is, they acquire an *ability to survive considerable ice-formation in their tissues*. This ability follows an annual cycle in perennials growing outside the tropics; it is minimal during the growing season, develops gradually during the fall to a high point in winter, and in the spring is lost again (Fig. 2.25).

Only after extension *growth is completed* does the plant enter a state of *readiness for hardening*; the *hardening process* then advances in *phases*, with each stage preparing the way for the next. In the theory developed by Tumanov and his coworkers (1967), the process is induced in winter grain and fruit trees (these plants have been the most thoroughly investigated) by exposure for several days or weeks to temperatures just above zero. In this *pre-hardening* stage, sugar and other protective substances are accumulated in the protoplasm; the amount of water in the cells falls and the central vacuole divides into a number of smaller vacuoles. At this point the protoplasm is prepared for the next phase, which occurs when the temperature falls regularly to between −3° and −5° C. Now the biomembrane structure and enzymes are reorganized in such a way that the cells can withstand the removal of water by ice formation. Only then can the plants enter without danger the *terminal stage of hardening*, so that during prolonged freezing, with temperatures no higher than −5° to −15° C, the protoplasm achieves maximal frost tolerance. The critical temperature ranges vary from one species to another. Birch seedlings in readiness for hardening, which before the process was begun would have frozen to death at −15° to −20° C, at the end of the first phase can already endure −35° C; when they are completely hardened they can survive temperatures as low as −195° C. *Cold thus forces hardening to proceed*. Once the most severe cold spell is over, the protoplasm returns to the first stage of hardening, but tolerance can be returned to its highest level by renewed cold periods—as long as the plants remain in the winter state of development.

In the course of the winter, *induced adaptations* that promptly adjust the level of resistance to changes in the weather are superimposed on these seasonal variations.

38

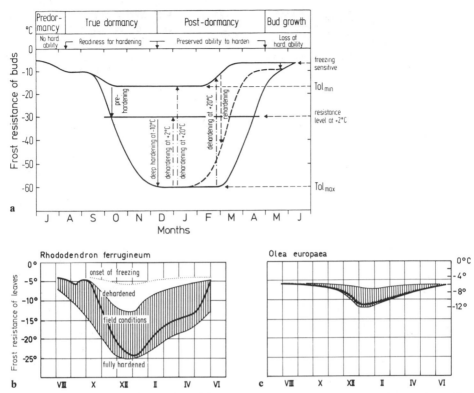

Fig. 2.25. a Stepwise frost hardening of the vegetative buds of Antonovka apple trees depending on dormancy and conditioning temperatures. *Upper curve*, lowest frost-tolerance level of frost-protected trees in a greenhouse. *Lower curve*, highest frost tolerance after progressive hardening and continuous freezing stress at −10° C. *Stippled curve*, reduced frost-tolerance in late winter and spring after dehardening and rehardening. After Tyurina et al. (1978). **b** Annual trends of frost resistance and the onset of freezing in the leaves of Rhododendron. **c** Olea, which does not become winter-dormant, demonstrates little adaptability. The shaded area indicates the range between minimal resistance after several warm days, and maximal resistance, achieved by stepwise hardening during severe frost. Between these limits lies the actual resistance, which depends upon the preceding weather conditions. After Pisek and Schiessl (1947)

Cold promotes hardening particularly in the early winter, when resistance can be brought to a peak within only a few days. *Above-freezing temperatures* toward the end of winter cause the plants rapidly to lose their tolerance; but even in midwinter considerable resistance can be lost after a few days at +10° to +20° C. The degree to which tolerance of freezing can be influenced by cold and heat—the *range for responsive adaptation of tolerance*—is characteristic of individual plant types, whereas the actual resistance level of a plant at any time depends on the degree of exposure to cold (Fig. 2.26).

As soon as a plant has emerged from *winter dormancy*, its resistance to cold, as well as its ability to develop such resistance, is *quickly lost*. In spring there is a close parallel be-

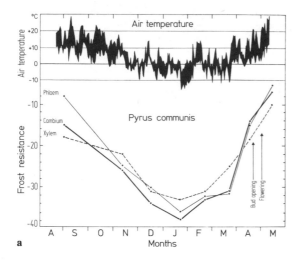

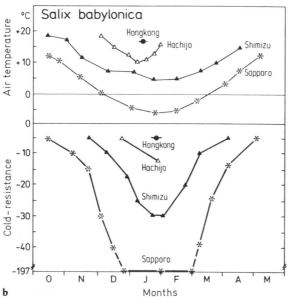

Fig. 2.26. a Seasonal and weather-dependent changes in frost resistance of the various tissues of pear twigs. After Larcher and Eggarter (1960). **b** Frost resistance of one-year-old twigs of weeping willows under tropical (Hong Kong, ca. 20° north latitude), subtropical (Hachijo Island, ca. 33° N), warm-temperate (Shimizu, ca. 35° N) and cool-temperate (Sapporo, ca. 43° N) climatic conditions. The air temperature indicated is the average for the month; the measure of frost resistance is the lowest temperature after exposure to which leaves and new shoots can still survive. All representatives of the genus *Salix*, which is distributed from the tropics far into the north, are capable of frost hardening, but they become freezing-tolerant only when they are actually exposed to cold (Sakai, 1970)

tween the plant's activity in putting out new growth and the lowering of resistance to cold (Fig. 2.27).

Heat Tolerance. Heat resistance, too, follows an annual cycle in many plants, though the changes are much less extensive than those of freezing tolerance. Seasonal changes in heat tolerance are controlled mainly by developmental processes but also by the actual temperature in the field (Fig. 2.28). All plants are very sensitive to heat during their *main growth period*. Numerous land plants in the temperate zone acquire their greatest *heat tolerance* during the *winter dormant period*. In addition to this developmentally dominated, ecologically paradoxical phenomenon, there are land plants which also (or

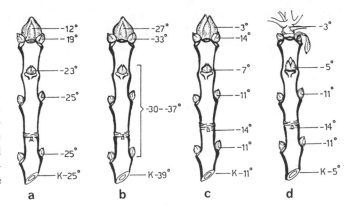

Fig. 2.27a–d. Changes in frost resistance of the buds of *Fraxinus ornus*, depending on winter dormancy and phenological development in the spring. **a** Autumn, prehardening phase; **b** winter, fully hardened; **c** spring, beginning of bud swelling; **d** spring, shooting. *K*, cambial zone (Mair, 1968)

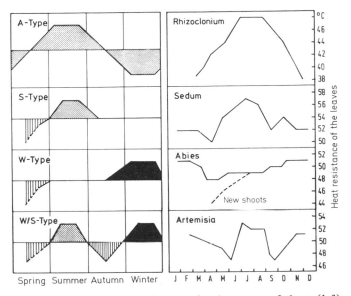

Fig. 2.28. Diagram of the seasonal variation in heat resistance of various types of plants (*left*) and the measured levels of heat resistance of the leaves of various species throughout the year (*right*). In the diagram, resistance changes occurring as an adaptation to the prevailing temperatures in the habitat are identified by *stippling*, the decline in resistance during the period of intensive growth is *hatched*, and the rise in resistance during winter dormancy is shown in *black*. *A-type*, species that continually adapt their heat resistance to the local temperature (example: the alga *Rhizoclonium* sp.); *S-type*, species with increased heat resistance during the warm season (example: *Sedum montanum*); *W-type*, species with increased resistance associated with protoplasmic alterations during winter dormancy (example: *Abies alba*); *S/W-type*, species with increased resistance in summer and winter and with two resistance minima associated with increased growth activity in spring and autumn (example: *Artemisia campestris*). Apart from these types there are plants with no detectable seasonal variation in heat resistance (e.g., *Asplenium ruta muraria*). Data from Schölm (1968), Biebl and Maier (1969), Bauer (1970), Gamper (1975). Diagram according to Lange (1967) expanded by Larcher (1973a)

only) show an increase of heat tolerance in the summer, and still others with no seasonal change in tolerance at all. The behavior of the *algae* makes sense from an ecological point of view; both fresh-water and marine algae adapt their heat tolerance to the temperature of the water. In late summer their tolerance is highest, and in winter lowest. The amplitude of this annnual oscillation is greater, the greater the annual water-temperature variation.

Plants respond to heat stress by rapid *adaptation*. Hardening to heat can take place within hours, so that resistance is higher in the afternoon of a hot day than in the morning. When the weather becomes cooler, the enhanced resistance disappears within a few days. The molecular mechanism of adaptation to heat is probably based chiefly on changes in the conformation of protein compounds and stabilization of the structure of macromolecules and biomembranes (Alexandrov, 1977).

2.2.3.5 Constitutional Types of Heat and Cold Resistance

Not all plants can survive low temperatures. Even among frost-tolerant plants not all are capable of passing through all the phases of hardening. Plant species also differ greatly in their sensitivity and ability to develop resistance to heat. Plants can be categorized according to the maximal resistance they can achieve and the roles of avoidance mechanisms and protoplasmic tolerance; such groups are of ecologic significance since specific temperature resistance is one factor limiting distribution (Tables 2.5 and 2.6). The *measure of resistance* is ordinarily taken as the temperature at which half of the plant samples are destroyed (the *temperature lethality* TL_{50}).

Constitutional Types with Respect to Cold

On the bases of the limits and the specific nature of cold resistance, one may distinguish three categories:

1. Chilling-Sensitive Plants. This group includes all plants seriously damaged even at temperatures above the freezing point: algae of warm oceans, some fungi and certain herbaceous and woody vascular plants in the tropics.

2. Freezing-Sensitive Plants. These can tolerate low temperatures, but they are damaged as soon as ice begins to form in the cells. Freezing-sensitive plants are protected from injury only by mechanisms that delay freezing. In the cooler seasons the concentration of osmotically effective substances in the cell sap and in the protoplasm is increased and the supercooling temperature is lowered, which causes the freezing point of the tissue to be lowered by a few °C. Plants sensitive all year round include the benthic algae of cold oceans and some fresh-water algae, tropical and subtropical woody plants, and various species from warm-temperate regions.

3. Freezing-Tolerant Plants. In the cold season, these plants survive extracellular freezing and the associated withdrawal of water from the cells. This category includes certain fresh-water algae, algae of the tidal zone and aerial algae, mosses of all climatic zones (even tropical), and perennial land plants in regions with cold winters. Some algae, many lichens, and a number of woody plants can become hardened to extreme cold; they remain unharmed even after exposure to long periods of hard frost and can even be cooled to the temperature of liquid nitrogen.

Table 2.5. Maximal temperature resistance of poikilohydric plants well supplied with water, as compared with the dry, anabiotic state. From the data of numerous authors; reference lists are given in Christophersen (1955), Biebl (1962), Altman and Dittmer (1973), Larcher (1973a), Zimmermann and Butin (1973)

Plant group	°C at which cold injury[a] occurs		°C at which heat injury[b] occurs	
	Wet	Dry	Wet	Dry
Rickettsias			50—70	
Bacteria				
Plant pathogens			45—56 (60)	
Animal pathogens	−15 to −196		50—70	
Saprophytic bacteria			up to 70	
Thermophilic bacteria			up to 95	
Bacteria spores		to −253	80—120	up to 160
Fungi				
Yeasts	to ca. −20			
Plant pathogens			45—65 (70)	
Saprophytic fungi	0 to <−10		40—60 (80)	75—110
Thermophilic fungi	ca. +5			
Fungus spores			50—60 (100)	over 100
Algae				
Marine algae				
sublittoral (tropics)	+16 to +5		32—35	
sublittoral (cold oceans)	ca. −2		22—26	
tidal zone	−10 to −40 (−70)		36—42	
Fresh-water algae	− 5 to −10 (−20)		40—45 (50)	
Aerial algae	−10 to −30	−196		
Thermophilic algae				
eukaryotic algae			45—50	
Cyanophyceae			70—75	
Lichens	−80 to −196	−196	35—45	70—100
Mosses				
Mosses of the forest floor	−15 to −25		40—50	80—95
Rock mosses	−30	−196		100—110
Ferns	−20	−196	47—50	60—100
Spermatophytes				
Ramonda myconi	− 9	−196	48	56
Myrothamnus flabellifolia		−196		80

[a] After at least 2 h exposure to cold. [b] After 0.5 h exposure to heat.

43

Table 2.6. Temperature resistance of the leaves of vascular plants from different climatic regions. Limiting temperatures are for 50% injury (TL_{50}) after exposure to cold for 2 h or more, or after exposure to heat for 0.5 h. The data were taken from many original publications. Sources are given by Altman and Dittmer (1973), Biebl (1962), Larcher (1973a), Larcher and Bauer (1981), and Kappen (1981)

Plants	°C for cold injury in the hardened state	°C for heat injury in the growing season
Tropics		
Trees	+ 5 to −2	45−55
Forest undergrowth	+ 5 to −2	45−48
Mountain plants	− 5 to −10	ca. 45
Subtropics		
Sclerophyllous woody plants	− 8 to −12	50−60
Subtropical palms	− 5 to −14	55−60
Succulents	− 5 to −10	58−65
C_4 grasses	− 1 to − 3 (−8)	60−64
Temperate zone		
Evergreen woody plants of coastal regions with mild winters	− 6 to −15 (−25)	50−55
Arcto-tertiary relict trees	−10 to −25 (−15 to −30)[a]	
Dwarf shrubs of Atlantic heaths	−20 to −30	45−50
Winter-deciduous trees and widely distributed shrubs	(−25 to −40)[a]	ca. 50
Herbs		
Sunny habitats	−10 to −20 (−30)	48−52
Shady habitats		40−45
Water plants	ca. −10	38−42
Cold-winter areas		
Evergreen conifers	−40 to −90	44−50
Boreal broad-leaved trees	(−196)[a]	42−45
Arctic and alpine dwarf shrubs	−30 to −70	48−54
Herbs of the high mountains and arctic	(−30 to −196)[a]	44−54

[a] Vegetative buds.

Constitutional Types with Respect to Heat

The effects of heat depend on the *duration of exposure*; that is, a slight excess of heat applied for a long time is as injurious as intensive heat applied for a short time. Therefore the heat-resistance data for plants must be standardized with respect to duration of exposure, and the generally agreed standard time is one-half hour. If the high temperature were instead maintained for an hour, the resistance limits would be about 1°−2° C lower. The categories of heat resistance are as follows:

1. Heat-Sensitive Species. This group includes all species that are injured even at 30°–40° C, or at the very most 45° C: eukaryotic algae and submersed vascular plants, lichens in a hydrated state (these, however, soon dry out in strong sunlight and then are completely heat-resistant), and most of the soft-leaved land plants. In addition, various bacteria pathogenic to plants, as well as viruses, are destroyed even at relatively low temperatures (for example, tomato wilt virus is killed at 40°–45° C). All these species can colonize only habitats in which they are not exposed to overheating—unless, of course, they are capable of keeping their own temperature down by means of transpiration (*Untertemperatur* species, in the terminology of O. L. Lange).

2. Relatively Heat-Tolerant Eukaryonts. The plants of sunny and dry habitats are, as a rule, capable of developing resistance to heat; they can survive heating for half an hour to 50°–60° C. Between 60° and 70° C there seems to be an absolute limit for survival of highly differentiated plant cells (i.e., those with nucleus and other organelles).

3. Heat-Tolerant Prokaryonts. Some thermophilic prokaryonts can endure exceedingly high temperatures: bacteria, as much as 90° C, and blue-green algae as much as 75° C. These, like the heat-stable viruses, are equipped with especially resistant nucleic acids and proteins.

Heat-resistance is a very *species- and even variety-specific* property; closely related species of the same genus can differ distinctly in this respect. The *evolution* of some of these characteristic adaptations has been associated with the environmental conditions affecting the life of the plant in its original or present range of distribution. For example, most tundra plants can resist heat-induced cessation of protoplasmic streaming at temperatures up to 42° C, while taiga plants are resistant up to 44° C, and desert plants up to 47° C (Fig. 2.29).

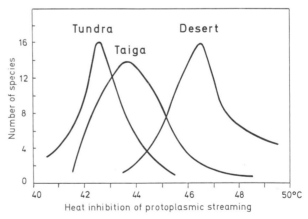

Fig. 2.29. Relationship between plant distribution and heat resistance. Here the measure of resistance is the thermostability of protoplasmic streaming after a 5-min exposure to the indicated temperatures. Plants of cooler regions are usually more sensitive to heat than those of dry habitats subjected to higher temperatures. Individual species, however, can deviate considerably from the statistical mode with respect to heat resistance. For example, among the tundra plants most resistant to heat protoplasmic streaming ceases only at temperatures high enough to produce the same effect in the most sensitive third of the hot-semidesert species that were studied (Kislyuk et al., 1977)

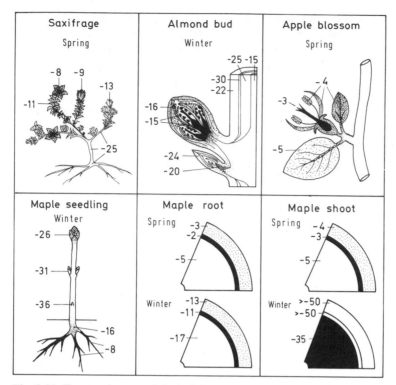

Fig. 2.30. Frost resistance of the individual organs and tissues of various plants. Resistance is given in °C. For sycamore-maple (*Acer pseudoplatanus*) seedlings the temperature given as the resistance limit is the lowest at which no injury is detectable (TL_0), but in all other examples it is the temperature for 50% damage (TL_{50}). The most sensitive tissue (or organ) is shown in black, and the most resistant, in white. After reproductive parts have been destroyed by freezing, the vegetative organs, as a rule, remain functional; and after extensive damage to the vegetative parts of a plant, particularly resistant tissues from which regeneration can occur usually still survive. If a plant is to maintain itself in habitats under year-round threat of frost (e.g., *Saxifraga oppositifolia* in the high mountains), it is advantageous for the regenerative parts to retain relatively high resistance even while the plant is actively developing. From measurements by G. Rehner as cited by Pisek (1958); after Harrasser (1969), Larcher (1970), and Kainmüller (1974)

Temperature Resistance and Organ Function

Various organs, and even tissues within an organ, differ greatly in temperature resistance (Figs. 2.30 and 2.31). *Reproductive organs* in general are especially cold-sensitive; among these are the floral primordia in the winter buds and the ovary within the flower. On the other hand, flowers are not necessarily more heat-sensitive than leaves. Dormant seeds are very resistant as a rule, but when they begin to germinate this resistance is soon lost; seedlings are usually highly sensitive. *Roots and underground shoot organs* are both cold- and heat-sensitive. *Geophytes* overwinter underground, so that cold-resistance of their bulbs, tubers and rhizomes is critical to the plant's survival.

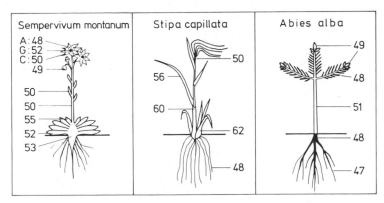

Fig. 2.31. Average summer heat resistance of an alpine rosette plant in flower (*Sempervivum montanum*), a steppe grass in a xerothermic habitat (*Stipa capillata*) and a young fir (*Abies alba*). All the temperatures indicated (in °C) are those for 50% injury. When hot weather persists for several days the resistance of the vegetative parts of *Sempervivum* and *Stipa* increases by 3°–6° C; dormant rosettes and grasses are always a few degrees more resistant than the flowering and shooting plants. *A*, anthers; *G*, gynoecium; *C*, corolla (Bauer, 1970; Pichler, 1975; Larcher, unpublished data)

In *woody plants* the resistance of the lignified parts of the root system (particularly the root collar) not infrequently limits the resistance of the whole plant to harsh winters; if these parts die off, the shoot must also eventually perish. The part of the *shoot above ground* is the least sensitive to both cold and heat. When hardened, its *cambium* has the greatest resistance of all tissues. The temperature resistance of the *perennating buds* is to some extent related to the degree of exposure (Fig. 2.32). Most buds not shielded from the winter cold become as well hardened as the axes that bear them, and in any case are more resistant than leaves. Buds that overwinter near the ground or under litter and snow develop only moderate resistance. By contrast, the basal buds are highly resistant to heat. Buds are very important to the survival of a plant; loss of leaves is no disaster if the buds remain healthy, but it is critical if many buds have been killed as well. However, many severely damaged plants are still able to sprout from more resistant reserve buds. Trees that are frequently forced to regenerate in this way become stunted and shrublike.

Resistance Differences Within Populations

Plants of different *ages* within a population are not uniformly resistant, nor are they uniformly endangered. When questions of plant distribution are concerned, particular attention should be paid to the flowers and young plants, for these *most sensitive stages* set the basic limits for maintenance and spread of a species (Thienemann's Rule). Intrapopulation variability is illustrated in Fig. 2.33, as exemplified by a stand of Mediterranean evergreen oaks. Lowering of the soil surface temperature to only −4° C can destroy all the potential seedlings from that year; if the temperature falls to between −8° and −10° C in several successive winters, a natural regeneration of the community becomes impossible, even though the mature shrubs and trees suffer not the slightest in-

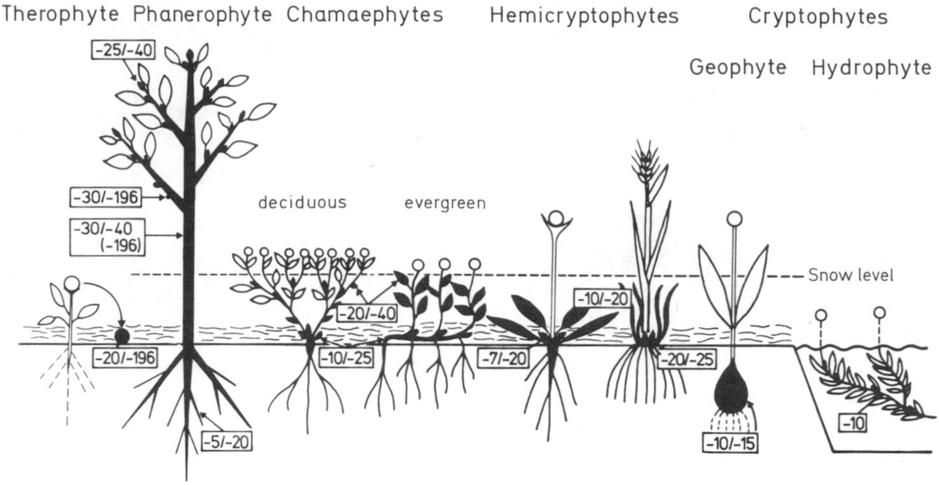

Therophyte Phanerophyte Chamaephytes   Hemicryptophytes   Cryptophytes

Geophyte Hydrophyte

-25/-40

-30/-196
-30/-40
(-196)   deciduous   evergreen

-20/-40   Snow level

-20/-196   -10/-25   -7/-20   -20/-25   -10

-5/-20   -10/-15

-10/-20

Fig. 2.32. Cold resistance of winter-survival ecomorphs of cold-winter regions. The degree of cold resistance is indicated by the range of temperatures below which the plants are damaged; the smaller number applies to sensitive species and the larger to those more resistant. Depending on their specific physiological constitution individual species may deviate widely from these limits. The terminology for winter-survival ecomorphs corresponds to the system of "life forms" of Raunkiaer (1934). The parts of the plants that overwinter are shown in *black*; those left *unshaded* die off at the end of the growing season. *Phanerophytes*, trees and shrubs with perennating buds above the snow; *chamaephytes*, small shrubs regularly protected by the winter snow cover; *hemicryptophytes*, perennial herbs with perennating buds just above the soil surface, under litter or enclosed in the remnants of dead leaves and leaf sheaths still attached to the plant; *cryptophytes*, perennial herbs with persistent organs under ground (geophytes) or under water (hydrophytes); *therophytes*, annual plants that complete their life cycles during the growing season and overwinter as seeds. These ecomorphs represent an evolutive adaptation not only to winter cold but also to winter drought (cf. Chap. 5.3.6.3) and other kinds of winter stress. The resistance data are taken from the measurements of many authors; for lists of data see Larcher and Bauer (1981)

jury at these temperatures. Only temperatures as low as −20° to −25° C are catastrophic for such a stand of oaks, and then only if the cold spells are long enough to freeze the thicker trunks. It is advantageous to the stand that the outer parts of the crowns, the parts most exposed to frost (radiation frost), develop a greater resistance to cold than the young stages growing up in the shade of the stand. The various components of a forest are adapted to the degree of cold normally encountered in their local subhabitat.

Not only the different age groups, but even the different *individuals* of a population vary with respect to the nature and degree of thermal resistance. This diversity can be the basis of the evolution of especially resistant *climate ecotypes*, which can advance into more exposed sites and can even survive when long-term changes in the climate (for example, glacial and postglacial fluctuations) might threaten the continued existence of the species in a particular region (Table 2.7).

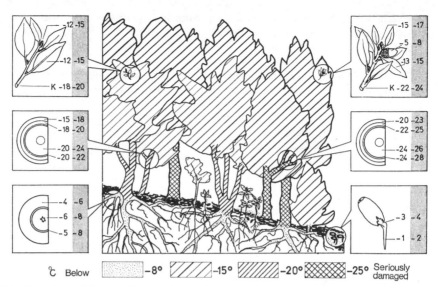

°C Below ░░░ −8° ▨▨▨ −15° ▨▨▨ −20° ▨▨▨ −25° Seriously damaged

Fig. 2.33. Zonation of frost-resistance levels in winter in a community of *Quercus ilex*. Temperatures indicated in the *unshaded part* of the marginal pictures represent the extremes below which first frost damage is to be expected; the data in the *shaded parts* refer to 50% damage. *K* in the upper pictures denotes the resistance of the cambium. Young plants, from one to three years old, are killed by temperatures between −10° and −15° C. From Larcher and Mair (1969)

Evolution of Frost-Hardening Processes

The appearance of ecotypes resistant to low temperatures, as well as the fine gradations in the resistance of plants from different climatic zones, indicates that the ability to become hardened to freezing has evidently been developed in a *stepwise* manner, along a resistance-ecological series of plant types (Larcher, 1971, 1981a). The first step in evolutionary cold-adaptation must have consisted in the adaptation of initially chilling-sensitive *enzyme patterns* and *protoplasmic fine structures* to temperatures between 0° and +10° C. The next step could have been adaptation to temperatures around and a little below 0° C, which the cells could survive by *supercooling* and lowering of the *tissue freezing point*. As a result, mild episodes of sub-zero temperatures could be endured without damage or marked disturbance of photosynthetic activity. The capacity of protoplasm to become freezing-tolerant in the course of a progressive hardening process could most readily have been developed by plants which had a genetically based alternation between periods of activity and dormancy (*endonomic activity rhythm*) in their life cycles—or at least were predisposed to periodic interruptions of growth activity.

The freezing resistance of flowering plants could have proceeded along two initially separate *evolutionary paths*; it is striking that there is an enhancement of the trend toward *improved avoidance of freezing with increasing altitude*, and of the trend toward *greater tolerance of freezing with increasing latitude*.

49

Table 2.7. Habitat-dependent ecotype differentiation in frost-sensitive and frost-tolerant plants. Further examples: Ivory and Whiteman (1978), Sakai and Wardle (1978), Larcher (1981a)

A. Seedlings of *Eucalyptus regnans* in the winter condition, collected at different altitudes in the southern Australian and Tasmanian mountains. In the experiment all were exposed to $-5.5°$ C for 12 h (Ashton, 1958)

Site of origin (m above sea level)	Extent of damage (% dead leaf area)
240	60
300—500	50
600—800	35
1100	30

B. Maximally frost-hardened twigs of conifers growing in mild-winter (Oregon) and cold-winter regions of North America. Frost-resistance is expressed as the lowest temperature (in °C) to which the plants can be brought without damage. From Sakai and Weiser (1973). For further examples see Flint (1972)

Plant, origin	Needles	Buds	Twigs
Tsuga heterophylla			
Oregon	-20	-20	-20
Alaska	-40	-35	-35
Rocky Mts. (Idaho, 1200 m)	-40	-40	-35
Pseudotsuga menziesii			
Oregon	-20	-20	-20
Rocky Mts. (Idaho)	-40	-30	-40
Rocky Mts. (Colorado)	-70	-50	-80

In the tropical highlands, which are considered to be a reservoir for early angiosperm evolution, the large daily temperature fluctuations characteristic of tropical mountains may have been influential in establishing and selecting for those *metabolic adaptations and mechanisms for protection* against freezing which represent the first phase in the hardening process. These remain effective throughout the year in mountain plants; evidently mechanisms protective against freezing can operate quite independently of transition into dormancy.

In regions with a subequatorial monsoon climate (alternation of dry and rainy periods), the *activity rhythm* that is a prerequisite for freezing tolerance could have been imposed on angiosperms as they expanded their range to the interior of the continents; in this case the onset of dormancy was coupled with an increase in *drought resistance*. If freezing tolerance of the protoplasm is understood as the capacity to survive the loss of water caused by extracellular ice formation in a tissue, it is a plausible inference that plants which can withstand severe desiccation during dormancy are also *resistant to water loss due to freezing*.

The prevailing opinion among phylogeneticists is that the evolution of vascular plants began in a humid tropical climate, later proceeding along mountain ranges into subtropical regions with alternating rainy and dry seasons and eventually into the temperate zone. Along the way, in the alternately wet and dry subtropics with episodic frosts, the evolution of preadapted plants in tropical mountains and highlands may have led to the perfection of freezing resistance that has enabled perennial plants to colonize cold-winter regions.

2.3 Periodicity

The rate of absorption of radiant energy at various places on the earth's surface depends upon their orientation with respect to the sun. This direction-dependence, because of the rotation and revolution of the earth, makes energy input a *periodically* varying environmental factor. It imposes periodicity upon all terrestrial phenomena, including the lives of organisms.

2.3.1 Climatic Rhythms

Diurnal Variation. The earth's rotation causes an alternation between day and night, between hours of light and of darkness; responses of plants to this alternation are termed *diurnal photoperiodism*. Responses to the corresponding temperature alternation reflect *diurnal thermoperiodism*. The changes in temperature, as a rule, lag somewhat behind those of light; the air temperature reaches a maximum not at the time the sun is highest but a little later, and the daily minimum temperature is reached near the end of the night.

At latitudes near the equator the photoperiod changes little in the course of the year. At the Tropics of Cancer and Capricorn the difference between the longest and shortest days is only 2 h. Thus there are no well-defined thermic seasons in the tropics, and the daily cycle is the important climatic rhythm (Fig. 2.34: Quito). Where there is strong insolation, on high plateaus and in the mountains, the daily cycle can be marked by very pronounced temperature oscillations. In tropical mountains the plants at high altitudes are exposed to night frosts (down to $-5°$ C) all year round, and during the day the temperature near the ground rises to $30°$ C or more.

Seasonal Variation. At higher latitudes the lengths of day and night change increasingly in the course of a year, and beyond the polar circles alternation between day and night ceases altogether in periods centered about the solstices (*seasonal photoperiodism*; Fig. 2.35). At an intermediate latitude such as that at Oxford (Fig. 2.34), the day lasts about 16 h in summer and only half as long in winter. This seasonal variation in photoperiod is accompanied by seasonal changes in average temperature (*seasonal thermoperiodism*). Toward winter the radiation balance becomes negative, primarily because of the steadily increasing duration of the nighttime phase in which reradiation predominates, but also because of the lowered intensity of insolation at the low sun elevations. The relatively dark season thus becomes a cold season.

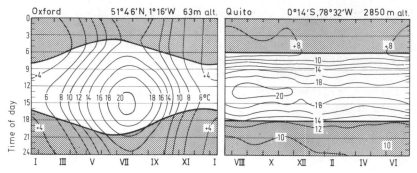

Fig. 2.34. Average air temperature, as a function of both time of day (*ordinate*) and time of the year (*abscissa*, Roman numerals for the months). Isotherms connect loci of constant temperature, and one can read the daily temperature variations by scanning vertically at any time of the year. Duration of daylight and nighttime (*shaded areas*) hours are also shown. *Left, Oxford*, England, is at an intermediate latitude and has a maritime climate (slight daily temperature fluctuations). *Right, Quito*, Ecuador, has an equatorial highland climate (marked daily fluctuation of temperature with almost no annual fluctuation). After Troll (1955)

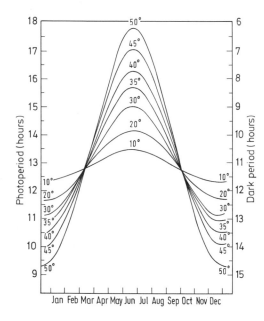

Fig. 2.35. Seasonal photoperiodism at various latitudes (Downs and Hellmers, 1975)

Vegetative processes may be suppressed not only by a season in which light is limited and temperature low, but also by periodic seasons of dryness (*seasonal hydroperiodism*). These are brought about by seasonally determined shifts in large-scale atmospheric circulation, and these in turn are caused by the differences between the energy balance in the strongly irradiated lower latitudes and that in less strongly irradiated northerly and southerly regions.

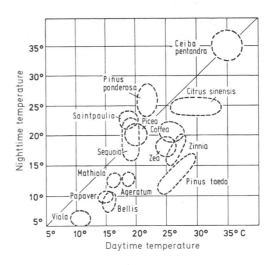

Fig. 2.36. Optimal temperature ranges for shoot growth of various plants. From Went (1957) with reference to the experimental results of Dormling et al., 1968 (*Picea abies*), Franco, 1958 (*Coffea arabica*; for further data see Maestri and Borros, 1977), Hellmers, 1966 (*Sequoia sempervirens*), Khairi and Hall, 1976 and Lenz, 1969 (*Citrus sinensis*), Kwakwa, 1964 (*Ceiba pentandra*), Kramer, 1958 (*Pinus taeda*), Laurie and Kiplinger, 1944 (*Viola odorata*) and Tinus, 1973 (*Pinus ponderosa*)

2.3.2 Activity Rhythms

2.3.2.1 The Influence of Diurnal Rhythmicity on Vital Processes

The diurnal *alternation between light and darkness* is reflected in many aspects of plant behavior. In some species photosynthetic capacity is impaired under continuous irradiation. Diurnal rhythmicity is evident in photosynthesis, respiratory activity, the permeability of protoplasm, and the cellular and organismic transport of materials; the processes of growth (e.g., photonastic and nyctinastic movements of leaves and flowers) are induced or controlled by diurnal photoperiodism.

A *temperature alternation between day and night* almost always promotes germination, the production of new tissue and the growth of shoots; such effects show a distinct evolutive adaptation to the amplitude of the diurnal temperature alternation in the habitat (Fig. 2.36). Plants growing in continental regions, where the temperature oscillates over a wide range during 24 h, develop best when the night is about 10°–15° C cooler than the day; for cacti and other desert plants an amplitude of 20° C is favorable. For most temperate-zone plants the optimal amplitude of the diurnal thermal cycle is 5°–10° C, but there are exceptions. Young spruce, for example, grow better with no temperature oscillation. Tropical plants, in accordance with the stable temperature regime in equatorial regions, are adjusted to low-amplitude oscillations (about 3° C). Finally, there are species with inverse thermoperiodism, such as young *Pinus ponderosa* and some varieties of African violet (*Saintpaulia ionantha*); growth of these plants is enhanced when the temperature is higher at night.

2.3.2.2 The Seasonality of Growth and Development

The Succession of Phases in a Life Cycle

The life of any organism begins with a reproductive process. This is followed by vegetative developmental processes such as growth and the formation of organs, which

53

in turn are followed by the reproductive processes leading to the next generation; the cycle is then complete. All these phases of development proceed according to a genetically set norm, coordinated by hormones and modified by environmental influences.

The ecological significance of this progression through phases in the life of an individual plant or a community has been emphasized in particular by Rabotnov (1950, 1978a, b). In each phase, a plant responds differently to environmental influences. The *seedling phase*, as a rule, is critical for survival and spread of a population. In the *juvenile phase* the plant is most capable of adaptation; during this period modificative adaptations to the amounts of radiation, mineral nutrients and water available are most pronounced. In the *adult phase* environmental factors that affect assimilative processes are reflected chiefly in frequency of flowering, the setting and development of fruit and the ability of the seeds to germinate—all properties that determine the future of the population in the next generation.

The Life Cycle of Continuously Growing Plants

In regions with a *pronounced seasonal climate* (summer-winter, wet—dry seasons) plants that grow continuously are *short-lived*: summer annuals in the temperate zone, winter annuals in winter-rain regions, and especially ephemerous plants in deserts. In all of these, the phases of the life cycle follow one another in uninterrupted sequence. Immediately after germination, the vegetative organs acquire their definitive form and multiply in number. When the intensive growth of the shoot subsides, the plants flower. Even while the fruits are ripening, signs of aging appear in the vegetative parts of the plant—the breakdown of proteins and the yellowing of the leaves. Finally the whole plant dies, leaving only the seeds; these remain in a dormant state until they are aroused by conditions favorable for germination.

In regions where conditions are *favorable to growth all year*, such as the permanently wet tropics and mild-winter regions of the warm temperate zone, there are *perennial* plants that grow continuously (Fig. 2.37); they can become as tall as trees, growing until flower formation absorbs all the new material synthesized. Up to that point the shoot elongates with no apparent interruption. Once the fruit has ripened, the entire plant dies (as happens to monocarp palms), although in some cases (e.g., the banana and pineapple plants) remnants of the vegetative part of the plant survive and produce new plants. In the tropics, continuous growth is exhibited by tree ferns, gymnosperms, and palms, as well as some large herbaceous and woody dicotyledons (for example, *Carica papaya*).

Plants with Intermittent Growth

Even in the *tropics*, most woody plants do not grow uniformly throughout the year. Although tropical forests are green all year, individual trees periodically replace old leaves with new and may even become temporarily bare. Usually new foliage is formed in the course of a few weeks, and shoot elongation also occurs during this period; apart from this time there is no increase in shoot length. In the permanently wet tropics these spurts of growth not uncommonly occur a *different times* in different species and even

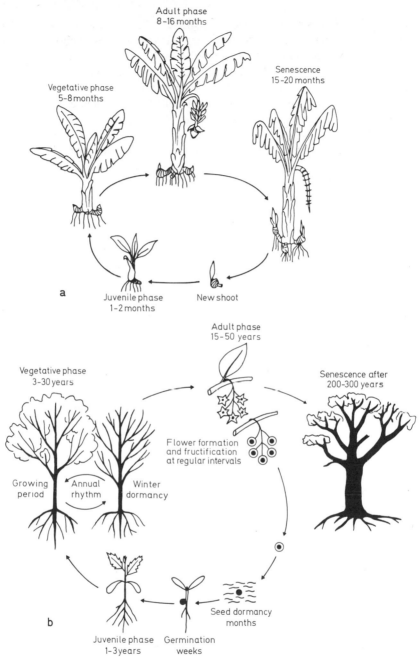

Fig. 2.37a and b. Life cycle of a tropical plant with continuous development (*Musa sapientum*) and a forest tree of the temperate zone with intermittent development. The time span given for each phase is a rough indication of its average duration; there may be considerable departures from these, depending on species, variety, and habitat. Modified from Hess (1975). Phase durations for the banana from V. Vareschi (pers. comm.), and for trees from Lyr et al. (1967), Kozlowski (1971a, b), and Wareing and Phillips (1978)

different individuals of the same species, and can occur more than once a year. Janzen (1975) has pointed out the ecological significance of this phenomenon: when new growth is produced in multiple, staggered sessions there is less danger that herbivores and parasites will strip the population of its foliage and multiply disastrously, in a climate that favors reproduction all the year round.

In regions with a *seasonal climate* the timing of development and resting periods is regulated by environmental factors, so that they occur more or less *simultaneously* in the different species of a region. Each stage of vegetative development begins with the growth of new shoots and unfolding of the leaves, and ends when the new growth has matured, primordia are initiated and the leaves fall (Fig. 2.37; in evergreens some foliage from previous years is lost).

The Alternation of Periods of Vegetative Activity and Dormancy

As a result of the seasonal periodicity of insolation, day length, temperature, and precipitation, there is a regular alternation between times when conditions are favorable to plant growth and those when they are not. Plants adjust to this climatic rhythmicity through rhythmic changes in the state (permeability, viscosity) of their protoplasm, in their metabolic activity, in their developmental processes, and in their resistance to environmental stress.

The regular alternation between developmental activity and dormant periods, in which growth is temporarily slowed or stopped, can be imposed on the plants by the recurrence of unfavorable environmental conditions. But the process can also, to a considerable degree, be preprogrammed and occur spontaneously. According to W. Pfeffer, it is useful to distinguish between *aitionomic* (imposed by the environment) interruptions in growth and *autonomic* (innate) rhythmicity having its origin in the genotype.

Aitionomic Alternation Between Activity and Suspended Growth. Plants with a continuous succession of phases have no provision for rest periods in their schedule of development. Their life cycles proceed without interruption unless they are forced to pause by unfavorable external factors. Many winter annuals (e.g., *Senecio, Cerastium, Capsella*) cease development only when the temperature falls below the freezing point, and resume growth when there is a thaw. The cryptogams, too, to the level of the ferns, suspend their growth only temporarily when forced to do so by external conditions; when the situation improves, they immediately begin to grow again. Similarly, the growth of roots and other underground organs is evidently only aitionomically controlled. The most decisive factor is soil temperature; lack of water affects growth only when the deficiency is very serious.

Autonomic Rhythmicity. Most flowering plants have an inherent, genetically determined tendency (hence one also speaks of "*endogenous*" or innate rhythmicity) to alternate between activity and rest. The defining property of autonomic rhythmicity is its existence and continuation *without external synchronizing factors*. In the permanently humid tropics there are trees in which leaf production and abscission occurs at different times in the individual branches. Grasses and geophytes retain their natural rhythm, evolved as an adaptation to the conditions of temperature and precipitation in the

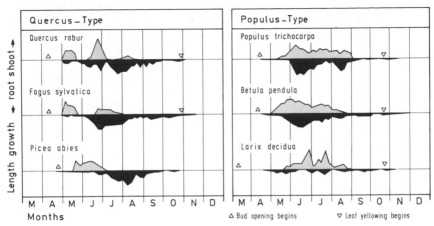

Fig. 2.38. Growth in length of shoot (*increasing upward and stippled*) and of roots (*increasing downward and black*) in several species of tree during the course of the year. In the *oak type* the growth of shoots is concluded early in the year; it proceeds in successive spurts separated by endogenously fixed growth pauses. Usually there are two such spurts per year, but the second is sometimes omitted; often there is an additional growth of shoots in late summer. This type includes pine and fir as well as the species shown in the figure. In the *poplar type*, shoot growth is synchronized by day length and the climatic temperature changes. This type also includes linden and black locust. The *growth of roots* in both types is regulated primarily by external factors; frequently it begins before shoot growth and continues until late in the autumn. After Hoffmann (1972); for further information see Kozlowski (1971a, b)

steppes and savannas, even when they have spread into regions where such behavior would not be necessary.

In regions with climatic rhythmicity, the genetically based "*physiological clock*" is adapted to the regular fluctuations in meteorological conditions (see, for example, the poplar type in Fig. 2.38). But even in cases of apparent synchronization, a close analysis of shoot growth discloses a pronounced autonomy of *some* species. An example of such endogenous periodicity is given by beeches and oaks, the shoots of which cease to elongate early in the summer—at a time when the days have not yet begun to shorten and other external factors also present no hindrance to growth. Some weeks later the plants become ready to resume development, and there may be a renewed shooting ("*lammas*" shoots). Not until autumn weather conditions prevail is growth brought to a final halt (Fig. 2.38, oak type).

The Winter Dormancy of Temperate-Zone Woody Plants

Toward the *end of the summer*, lateral buds are formed in the axils of the leaves, and the tips of the shoots become transformed into winter buds, or wither and die off. When the foliage begins to yellow, these buds are already dormant. Other parts of the plant also *enter dormancy* for the winter; the cambium, for example, and the other tissues of the shoot, can thus become hardened to frost and dehydration.

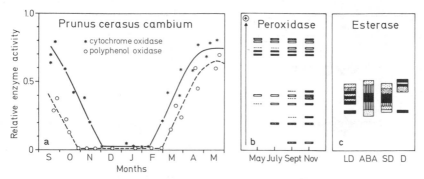

Fig. 2.39. a Relative activity of cytochrome oxidase and polyphenol oxidases in the cambium of cherry twigs in fall, winter, and spring. From histochemical findings by Meyer (1968). **b** Multiple forms of the peroxidase from spruce needles during the transition to winter dormancy. From Esterbauer et al. (1978). **c** Isoenzyme pattern of esterases from duckweed. *LD*, vegetative state under long-day conditions; *ABA*, after treatment with abscisic acid under long-day conditions; *SD*, short-day conditions; *D*, enzymes from dormant stages. From Perry (1971). For biochemical changes during the transition to dormancy see Leopold and Kriedemann (1975) and Wareing and Phillips (1978)

The onset of dormancy is marked by a number of physiological events. The *gibberellin* level falls, *growth inhibitors* such as abscisic acid begin to dominate, *gene activity* is selectively suppressed, translation processes are inhibited and mitotic activity in the meristems is much reduced or stopped altogether. The cell nuclei remain in the G1 phase of the cycle, in readiness for the reduplication of DNA that occurs toward the end of winter dormancy. The transition to dormancy is also evident in the *compartmentation and ultrastructure* of the cells—for example, the endoplasmic membrane systems and plasmodesmas are condensed (protoplast insulation), the mitochondria become smaller, the thylacoid structures in the chloroplasts of some plants and tissues (cortical chlorenchyma in particular) is reduced, and vacuoles disintegrate into smaller units. Basal *metabolic activity* declines, reserve substances such as starch, fats, and sometimes proteins are accumulated, various metabolites and minerals are shifted within the plant, and the enzyme patterns are altered. In various shoot tissues there is an *increase* in hydrolases, catalase, and peroxidases in the fall, in the cambial zone the activity of phenol oxidases, cytochrome oxidase and dehydrogenases is *reduced*, and certain enzymes are replaced by *isoenzymes* with the same action but different optimal temperatures (Fig. 2.39). This readjustment to the winter state is not an abrupt event, but occurs gradually, some alterations appearing earlier and others later; the transformation is not synchronized throughout the plant.

Distinct phases are observable in cases of endogenous winter dormancy. Three stages, with smooth transitions between them, can be distinguished: predormancy, true dormancy and postdormancy (Fig. 2.40).

1. Predormancy. This stage begins in the buds even before leaf abscision. In some woody plants, progressive decrease in day length induces the conclusion of the growth period and the transition to the dormant state; this is the case, for example, in poplar

58

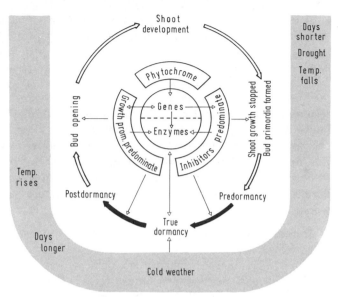

Fig. 2.40. Simplified diagram of environmental influences (shaded U), and hormonal interactions with cell activity and the developmental rhythm, in woody plants. The role of the phytohormones is to mediate between the genetically fixed predisposition and the environmental influences that affect function, and by coordinating a wide variety of vital functions—not only growth and development, but carbon acquisition and assimilative metabolism, water balance and resistance to climatic factors—to bring about the most harmonious adjustment possible between the life of the plant and the conditions under which it lives. Further details of regulatory processes and molecular mechanisms may be found in Wareing and Saunders (1971), Wareing and Phillips (1978), Perry (1971), Kozlowski (1971a, b), Villiers (1975), and Kwolek and Woolhouse (1981)

and willow, birch, hazel, beech, oak, maple, spruce and larch, most of them species with distribution extending far to the north. Very long days prevent the onset of winter dormancy in all these species, as well as in some species of juniper and many Atlantic plants. Long days can, moreover, also initiate the transition to bud dormancy—not in woody plants, but in geophytes such as lily-of-the-valley and various bulb plants. Species responding in this way evidently have *two* endogenous rest periods, one at the time of greatest winter cold and the other during the summer drought.

The second important regulating factor is *low night temperature*; critical levels are usually a little under 10° C, depending on the species. In all species not made dormant endogenously or by decreasing day length, falling temperature is the decisive signal. These are primarily genera native to the southern temperate zone, such as ash, horse chestnut, lilac, cherry, and many others. Falling temperatures can replace short days in experiments upon some photoperiodically controlled plants. In nature, seasonal photoperiod and seasonal thermoperiod are necessarily coupled, so it is not surprising that growth rhythms and many other vital processes are controlled by the joint action of these two variables. It is not uncommon to find that change in day length acts as the pacemaker inducing the transition in the plant, the process then being completed under the influence of changes in temperature.

2. True Dormancy. During predormancy, *inhibition of bud activity* increases steadily until finally, in November and December in the temperate latitudes, complete dormancy occurs. At this point the plants can *no longer be induced to sprout* by warming or by lengthening of the photoperiod. The inability of plants to emerge prematurely from the resting state once they have entered true dormancy is an important factor in their resistance, in view of the unpredictability of the weather. The plants of cold-winter regions must not respond to mild days in midwinter, or they would unavoidably suffer injury in the next cold wave. In species lacking a deep winter dormancy (*Citrus* species), this in fact happens time and again. Still lower temperatures accelerate the transition to postdormancy. Many woody plants, such as poplar, maple, linden, pine, some fruit trees, and the grapevine have a distinct *chilling requirement*; they can sprout normally only after a certain exposure to low temperatures—usually to *ca.* 0° C for 3—4 weeks.

3. Postdormancy. Various processes that begin when true dormancy is coming to an end are conspicuous characteristics of the postdormant period. The concentration of those *phytohormones* that enhance development (first gibberellins and cytokinins, then indole-acetic acid) rises, by *activation of genes and enzymes* the basal metabolism, the mobilization of reserves and biosynthesis are set in motion, and cell division proceeds at a gradually increasing rate. Once the plant has reached a state of activity, warmth and increasing day length trigger rapid development. At northern latitudes postdormancy ends during February; from this time on, the appearance of new shoots and leaves is determined entirely by weather conditions.

2.3.3 Synchronization of the Growth and Climatic Rhythms

The time courses of plant life cycles and vegetative activity are adjusted to the local duration of favorable conditions. The *vegetation period* in the tropics and subtropics is limited by increasing *water deficiency* when the dry periods begin; activity of plants in the temperate and cold climate zones is synchronized with the seasons by *seasonal photo- and thermoperiodicity*. North of 40° latitude the days are longer than the nights during the entire growing season, and above 50° the difference is quite pronounced (Fig. 2.41). Plant sibs with evolution centers at high and intermediate latitudes are adapted to this periodicity, mostly behaving as long-day plants with respect to the production of new shoots, leaves and flowers, and as short-day plants with respect to control of the duration of the growing season. In some cases *ecotype differences* have been demonstrated; in spruce of subarctic origin, formation of the terminal bud and thus the end of extension growth for a season is induced when the days become shorter than the critical duration 20 h (as opposed to 14 h in spruce originating in Central Europe). Photoperiodic regulation of the growing period of *Liquidambar styraciflua* in North America produces a shorter season in the northern than in the southern phenoecotypes. In strawberries the transition form the phase of asexual reproduction (by runners) to that of sexual reproduction (flowering) is controlled by *temperature and day length*; southern varieties cultivated in northern regions reproduce by runners for a long time, and only late in the season produce a few flowers, whereas northern varieties at low latitudes flower too early and send out only a few runners.

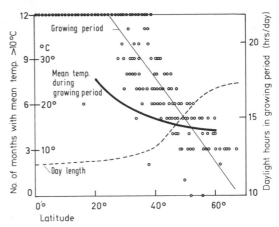

Fig. 2.41. Duration of the growing season (number of months with mean air temperature above 10° C) in locations at different latitudes (*circles*), and the dependence of day length and the mean temperature during the growing period upon latitude. At low latitudes the temperatures are favorable for growth and the days are short all or most of the year. At high latitudes the growing season is restricted to a short time in the summer, but then the days are long and the air temperature is on the average not much lower than during the growing season at intermediate latitudes. After Totsuka (1963). A computer map of the number of daylight hours during the growing season over the earth is given by Lieth (1974)

A plant species, a variety, or an ecotype is well acclimatized if the growing season is utilized to the full without risk of injury as the unfavorable season approaches. In general, this is ensured by the fact that the acquisition of resistance is coupled to developmental processes. Poorly adapted plants might sprout too late, continue to develop too slowly, and be damaged by the first winter frosts; conversely, the situation would be equally disadvantageous if they began to grow too early in the year (risking injury by late frosts) and stopped development too soon to make full use of the favorable season (Fig. 2.42). A lack of synchronization between periodic plant activity and the rhythmicity of the climate thus restricts the spread of a species; such maladaptations can be overcome, of course, in evolution by the differentiation of ecotypes.

The detection and analysis of photoperiodic and thermoperiodic phenomena is an important objective of research in controlled-climate greenhouses (*phytotrons*). Appropriate test programs can indicate the ecological requirements for the development of particular plants and the ecological barriers to their spread, and climate-ecotypes suitable for particular research purposes can be efficiently identified.

2.3.3.1 Phenology

The onset and duration of particular phases of development vary from year to year, depending on long-term characteristics of the weather. Phenology is the study of the cycle of sprouting, flowering, bearing of fruit, and senescence with respect to the timing of these events over the year. People close to nature began making and applying phenological observations early in history. Indeed, a phenological calendar existed in China

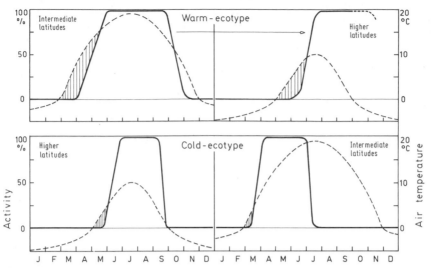

Fig. 2.42. Diagram illustrating the temporal coordination of the climatic rhythm and the rhythm of growth in trees. The *left half* of the figure shows a general synchronization between progression of temperature through the year (*dashed curve*) and physiological activity (*solid curve*); in the *right half* the consequences of lack of adaptation are indicated. Ecotypes adapted to warm conditions do not develop shoots until the temperature has risen considerably (cf. the areas *shaded with widely spaced lines*), whereas ecotypes from higher latitudes, adapted to cold, end their winter dormancy as soon as there is a slight rise in temperature (area with *closely spaced lines*). If warm-climate ecotypes are introduced to high latitudes (*upper right*), they retain their tendency to delay the onset of development; they sprout too late and the new growth has no time to mature. Thus they can be caught by frost before they have entered winter dormancy and may be damaged. Cold-climate ecotypes transplanted to a warmer region (*lower right*) sprout prematurely and therefore are endangered by spring frosts; moreover, they conclude their growth—which is adjusted to a short growing season—much too soon and fail to make use of a large part of the favorable season. From transplanting experiments by Langlet as cited by Bünning (1953). For pheno-ecotypes of spruce see Dormling et al. (1968); of liquidambar, Williams and McMillan (1971); of strawberry, Smeets (1968)

more than 2000 years ago. For centuries in Japan, the time of the first flowering of the cherry trees has been considered significant, and much of the traditional wisdom of farmers demonstrates a capacity for sharp observation and a deep insight into the relationship between the progression of meteorological phenomena and the development of the vegetation.

Phenological Dates. Phenology is based, even today, on the observation of externally visible changes (*phenophases*) in the course of a plant's life cycle. Phenological descriptions provide ecologically valuable information about the average duration of the growing and foliated periods of the plant species in an area (Figs. 2.43 and 2.44), and about local and weather-determined differences in the dates of onset of such phenomena (Fig. 2.45). Such descriptions can take into account large-scale events—snow thaw, leaf emergence, the ripening of grain, and the changing of leaf color

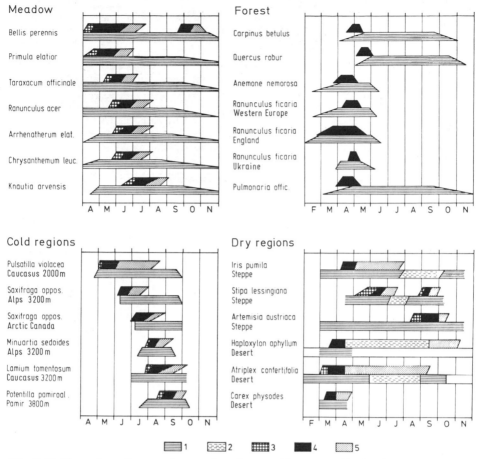

Fig. 2.43. Phenophase diagrams for meadow plants and for trees and herbs of an oak mixed forest of the temperate region (northwest Germany, with *Ranunculus ficaria* in England and the Ukraine for comparison), and plants from regions where the growing season is limited by cold (high mountains, arctic) and by dryness (steppe, desert). The strip diagrams of Schennikow (1932) are interpreted as follows: the lower strip represents vegetative development and the upper, reproductive, with *shading* to indicate: *1*, growing season, the foliated period in the case of trees; *2*, drought dormancy; *3*, flower buds visible; *4*, flowering period; *5*, ripe fruits and seed scattering. Data from Ackerman and Bamberg (cited by Lieth, 1974), Borissovaya (cited by Walter, 1968), Ellenberg (1939), Gamzemlidze (pers. comm.), Jankowska (cited by Lieth, 1970), Michelson and Togyzaev (cited by Voznesenskii, 1977), Moser et al. (1977), Nakhuzrishvili (1974), Salisbury (1916), Shalyt (cited by Beideman, 1974), Steshenko (1969) and Svoboda (1977)

over large expanses of the terrain—using aerial photographs and spectral reflection analysis by flying objects (*remote sensing*). As a science, phenology is not limited to the descriptive dating of events, but attempts to clarify the influence of climatic factors. *Analytic phenology* takes into account the developmental physiology of a plant, and applies such methods as analysis of variance and computer simulation. Nevertheless, it is

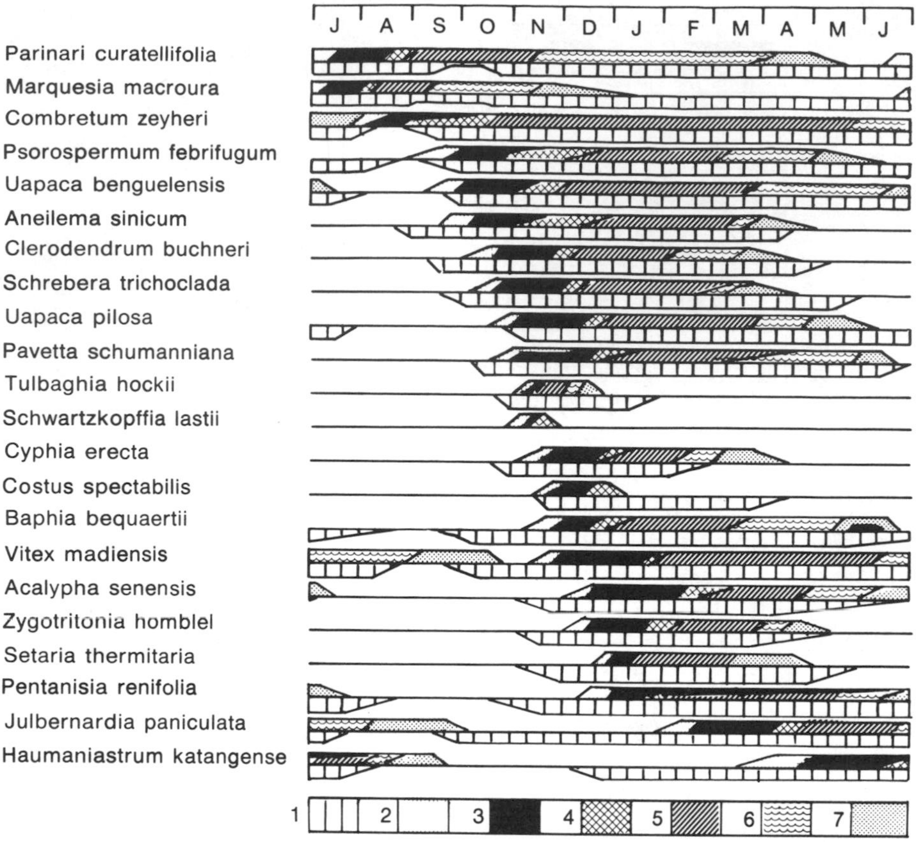

Fig. 2.44. Phenological spectrum for selected species of miombo woodland in Central Africa. *1*, foliation; *2*, flower and inflorescence buds; *3*, blossoming; *4*, faded flowers; *5*, unripe fruits; *6*, ripe fruits; *7*, scattering (Malaisse, 1974)

no simple task to discover the factors that trigger phenological phenomena, for the effective occurrence (the passing of a temperature threshold, for example) is long past by the time the associated phenophase becomes apparent (Fig. 2.46).

The Sequence of Phenophases in the Temperate Zone

The time of onset of *phenophases of the first half-year* depends primarily on the passing of certain *temperature thresholds*. This can be shown by comparing the temperature distribution over a certain terrain with the phenological dates (Fig. 2.47). The opening of the buds, sprouting, the onset of flowering in trees and shrubs and the germination of seeds are possible only after the temperatures of both air and soil exceed regularly a critical point characteristic of each stage. In general, the temperature threshold for the opening of the buds and flowering is 6—10° C, though in spring-flowering and mountain plants it is lower (around 0°—6° C) and in late-flowering plants, higher (in many ring-porous trees between 10° and 15° C, and about 15° C in grain). Poplars, birches

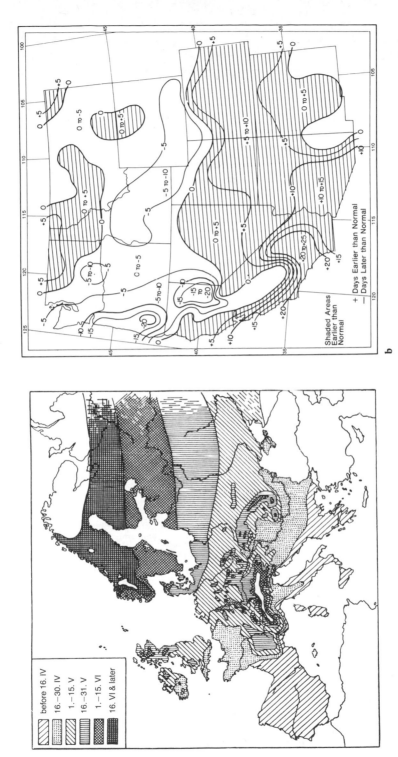

Fig. 2.45a and b. Phenological maps of the beginning of lilac flowering, as a signal of the onset of spring. **a** Average onset time in Europe. From Ihne as cited by Schnelle (1955). **b** Local departures from the average for the western USA (Caprio et al., 1974). Remote sensing of phenological events see Barrett and Curtis (1976) and Morain (1974)

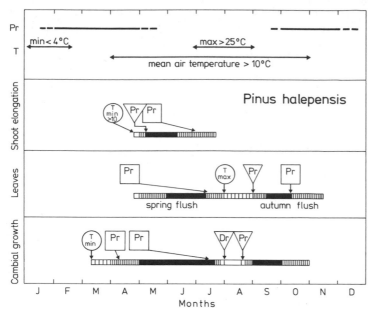

Fig. 2.46. Influence of climatic factors on growth processes in pines of the French Mediterranean coast. *T*, temperature; *Pr*, precipitation; *Dr*, drought. The *triangles* indicate events with triggering effects (e.g., thunderstorms) and the *squares*, amount-dependent effects (e.g., dependence of the duration of growth on the supply of water in the soil). *Black bars*, time of greatest activity. The chief factors determining the time course and amount of growth are the average minimum temperature in winter and spring, the quantity of rain during the main growth period, and the onset of summer heat and dryness. Effects of the previous year's weather conditions are not shown in the figure. Simplified from Serre (1976a, b)

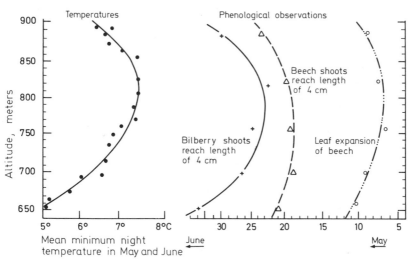

Fig. 2.47. Relationship between night temperature (*left*) and the phenological development of forest plants (*right*) at different altitudes on a mountain slope in southern Germany. After Frenzel and Fischer as cited by Geiger (1961)

Table 2.8. Influence of environmental factors upon leaf abscission in trees (Addicott, 1968)

Factor	Promotion	Retardation
Radiation		
Deficiency or surplus of PhAR		x
Longer photoperiod		x
Shorter photoperiod	x	
Temperature		
Moderate	x	
Light frost	x	
Heat or severe frost		x
Water		
Drought	x	
Flooding	x	
High humidity	x	
Minerals		
Deficiency of N, K, S, Ca, Mg, Zn, Fe, B	x	
Nitrogen fertilization		x
Excess Zn, Fe, Cl	x	
High soil salinity	x	x
Gases		
Oxygen deficiency in the air		x
Ethylene	x	
Noxious gases	x	

and some species of conifers sprout at just above 0° C. Sprouting and flowering, however, can be elicited by warmth only if the plants are already in a state of readiness to develop—i.e., if they have emerged from their winter dormant period.

Phenological dates falling in the *second half-year*, such as those for ripening of fruit, discoloration of leaves, leaf fall, and the times for harvest of crops, can be affected by all those environmental conditions that delay or accelerate the processes of maturation and aging. Once again, temperature is of greatest significance, but in this case its role is evident with respect to the enhancement of synthetic activity. Thus threshold temperatures are of less concern than the *heat sum*—i.e., the time integral of average daily temperatures (degree-days) exceeding specific threshold temperature levels. Other decisive factors are the supplies of nutrients and water and, above all, the influence of the diurnal *photoperiod* upon the times of onset of flowering, leaf fall, and winter dormancy (Table 2.8). In various species the plant is prepared for discoloration and leaf abscission by the shortening days. Then, as soon as the *temperatures* fall below threshold values between 5° and 10° C, these closing phases of the phenological calendar appear. Regional variations of these dates are also of interest; for example, in mountains the falling of leaves begins at high altitudes and progresses rapidly toward the valleys.

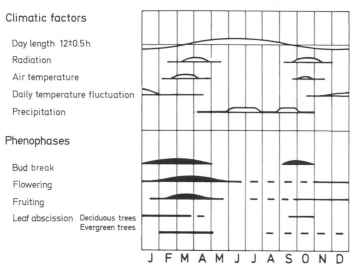

Climatic factors

Day length 12±0.5h
Radiation
Air temperature
Daily temperature fluctuation
Precipitation

Phenophases

Bud break
Flowering
Fruiting
Leaf abscission Deciduous trees
Evergreen trees

J F M A M J J A S O N D

Fig. 2.48. Seasonal occurrence of developmental processes in trees of evergreen tropical forests in Ghana (6° N), and the changes in climatic factors over the year. At the equinoxes there is a striking concentration of bud unfolding, abscission, flowering, and fruit formation (Longman and Jenik, 1974)

Phenological Processes in the Tropics

In those parts of the tropics that have *rainy and dry seasons* the phenophases are linked to this hydroperiodic alternation. The *rainy season is the main vegetation period*; in the *dry season* grasses and herbaceous plants dry up, deciduous trees shed their leaves, and in the first half of the dry season evergreen trees lose a large fraction of their older leaves. Most evergreen trees bloom at the beginning of the wet season, unfolding their preformed flowers after the first rainfall. Deciduous tropical and subtropical trees, by contrast, often flower in the middle of the dry season.

Even in those tropical regions with *abundant rain all year* there are phenological events, but these are not as conspicuous as those in regions with a marked seasonal rhythmicity. In the equatorial zone itself a number of *climatic factors* change in the course of a year (Fig. 2.48). In some tropical countries these variations occur at irregular times and to a slight extent, while in others they are both notable and predictable. One cannot conclude from the constancy of the mean monthly temperatures over the year that the air temperature is always the same. The slight variability in the monthly averages masks considerable short-term fluctuations. Within a few days, for example, the average temperature can fall by 5° C or more as a result of heavy rainfall. Even the very slight change in day length over the year affects the processes of development, for tropical plants respond to extremely weak photoperiodic stimuli; there are varieties of rice in which shoot growth is altered by changes in day length of as little as 10 min.

Phenological events in evergreen forests of the wet tropics must be described statistically, by noting when they occur in a greater than average number of species and in-

Fig. 2.49. Time courses of shooting, cambial growth and flowering of cocoa trees in Costa Rica (10° N), with the changes in air temperature and precipitation (the *bar at the top of the picture*; black regions indicate periods with more than 25 mm precipitation per week). Shoot growth is given as the percentage of branches with new shoots; the measure of cambial activity is the weekly increase in stem diameter. The time during which the trees are in flower is marked by *asterisks*. Shooting depends on the daily fluctuation in temperature, whereas cambial activity is promoted by high average temperatures; during the main elongation phase of the new shoots cambial activity is slowed. Precipitation is sufficient throughout the year and thus is not a controlling factor in this case (Alvim, 1964)

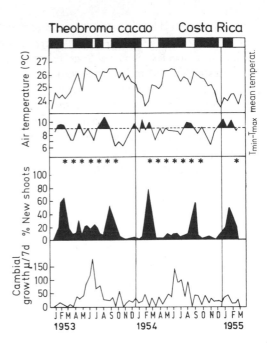

dividuals. The production of *new shoots*, unfolding of the leaves and *elongation growth* often culminate near the equinoxes; that is, in many tropical trees new shoots emerge and leaves unfold twice a year (see Fig. 2.48). The decisive environmental factor that triggers new growth is temperature in the case of cocoa (Fig. 2.49); the daily maxima must exceed 28° C, and the temperature must fluctuate by more than 9° C during 24 h. It is thought that opening of the buds of *Terminalia* and *Bosqueia* is elicited by an increase in water content of the stems. *Leaf fall* occurs throughout the year in tropical forests, but the amount of litter on the ground—two-thirds of which is made up of leaf material—is greater at certain times. Abscission in continuously growing plants occurs whenever the leaves reach a certain age, with no seasonal periodicity; abscission by intermittently growing evergreens follows the rhythm of leaf formation. Among the environmental factors involved in synchronizing leaf fall are day length and the distribution of precipitation, the effective signals being shortening of the light period and increasing dryness of the soil. The influence of day length becomes clear when individuals of a species growing north and south of the equator are compared. *Hevea brasiliensis, Manihot glaziovii, Erythrina velutina* and *Bombax malabaricum* in Ceylon (9° north) lose their leaves between December and March, and in Java (7° south) between June and August; in Singapore, on the equator, where day length stays about the same all year, *Hevea* and other species lose their leaves at irregular intervals. Many tropical shrubs and the continuously growing evergreen trees of the tropics—especially the bat-pollinated species—*flower* throughout the year. However, most tropical trees have a defined flowering period, which in some cases is very brief (e.g., *Tabebuia ochracea, Ceiba pentandra* and many lianas). Tropical plants can be classified by periodicity of flowering, into ever-blooming, occasionally blooming, gregariously blooming, and seasonally blooming species; a final category includes those species that flower at intervals of several years. *Ever-blooming* species such as those of *Hibiscus*, some *Ficus* species and *Carica papaya* produce flowers all year long. *Occasionally* blooming species, such as *Spathodea campanulata*, *Cassia fistula* and species of *Lagerstroemia*, form flowers at various times during the year, but not at particular seasons; flowering times vary among individuals and even among the branches of a single tree. In *gregariously* flowering species—for example, coffee and various epiphytic orchids, chiefly those of the genus *Dendrobium*—floral primordia develop continuously but remain dormant until an external stimulus triggers opening of the buds; the flowers then appear simultaneously over an extensive region. Unfolding of the flower buds of the coffee plant is inhibited as long as the soil is permanently wet. A few weeks of soil dryness removes this inhibition, so that 10 days after the onset of rains (or after irrigation is resumed in plantations) the plants come into flower. A sudden drop in temperature by about 3° C has the same effect. Temperature is also the triggering signal for gregariously blooming orchids. *Seasonal blooming* is found in regions with periodic dry seasons (*hydroperiodic* induction; P. de T. Alvim, 1964) or with a distinct variation in day length (*short-day induction* of the flowering process). Species of bamboo flower *at intervals of many years*. *Fruits* in tropical forests, like flowers, can be found in growing and ripening stages throughout the year, but as a rule fruit-bearing trees are more common in dry periods than when there is ample rainfall.

2.3.3.2 Phenometry

For a detailed analysis of the influence of external factors upon the course of development, simple phenological observations of dates of onset of the various phases are not enough. It is necessary to have phenometric records of growth, which quantitatively describe gradual processes like the elongation of the shoots and roots and the increase in thickness of the cambium. Phenometry also includes studies of the increase in mass and leaf area of individuals and stands of plants (Fig. 2.50). A good example of a problem in this field is that of cambial growth.

Phenometry of Annual Growth Rings. The level of mitotic activity in the cambium is closely related to the increase in length of the shoot. In annual plants cambial activity begins when elongation is to a great extent completed; in woody plants of the temperate zone it is initiated each year in connection with the sprouting of new shoots and the unfolding of the leaves. The first sign of beginning activity is a swelling of the cambium initials, in which the radial walls are stretched out and can be easily torn. Correspondingly, a phenological characteristic accompanying the beginning of cambial activity is that the twigs can be peeled. Soon thereafter the first divisions occur. In many trees of the tropics and subtropics, as well as in the fig, olive and the seedlings of some woody plants, growth in diameter is initiated, and stops, quite suddenly whenever external conditions happen to be appropriate. In the woody plants of the temperate zone just one growth ring is normally formed per year; hence the term *annual ring*. The ring is composed of histologically distinguishable zones corresponding to the spring and the late summer growth.

The duration of cambial activity and the type of wood formed—the differentiation into early or late wood—are affected by environmental factors. Precise measurements of the

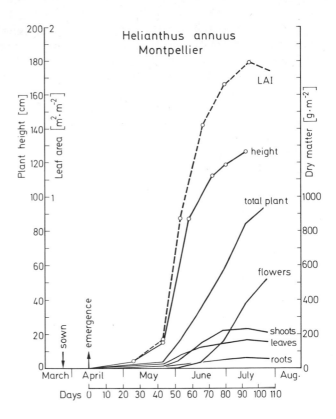

Fig. 2.50. Phenometric analysis of the growth of sunflowers in southern France. The expansion and die-off of the leaves is expressed quantitatively by the leaf area index (*LAI*), and the growth of shoot and roots, by the dry matter production. Note that the different organs do not all develop in parallel. After Eckardt et al. (1971). For a phenometric analysis of rice see Milthorpe and Moorby (1974)

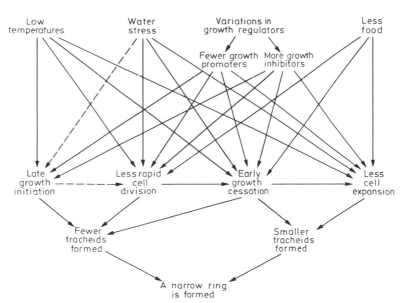

Fig. 2.51. Causal relationships among environmental conditions, endogenous factors and the growth of annual rings (Fritts, 1976)

increase in thickness of the stems during the year, using "dendrometers", and histological analyses of the new wood thus provide informative phenometric data. In general, the formation of spring wood is promoted by all those factors that favor the sprouting of buds and growth in length of the new shoots; all factors tending to slow shoot growth and accelerate the aging of the foliage lead to differentiation of late wood (Fig. 2.51). The thickness of the cell walls in the new wood depends upon the supply of carbohydrates, and thus is an indication of the yield of synthetic metabolism in that year. Other factors having direct or indirect influences on the thickess and appearance of the annual rings include radiation, temperature, availability of nutrients, water supply and duration of the photoperiod, as well as all kinds of harmful environmental influences; examples of the latter are attack by parasites, consumption by animals, excessive heat and frost, and absorption of pollutants. In cases where the particular influences to which the cambium is most subject are known in detail for a given species, the structure of the annual rings provides an important historical document of the growth-determining events of past years.

3 Carbon Utilization and Dry Matter Production

3.1 Carbon Metabolism in the Cell

3.1.1 Photosynthesis

In photosynthesis, radiant energy is absorbed and transformed into the energy of chemical bonds; for every gram-atomic weight of carbon taken up, potential energy amounting to 477 kJ (114 kcal) is obtained. Photosynthesis involves *photochemical* processes that occur in the presence of light, *enzymatic* processes not requiring light (the so-called dark reactions), and the processes of *diffusion* which bring about exchange of carbon dioxide and oxygen between the chloroplasts and the external air. Each of these subprocesses is influenced by internal and external factors and can limit the yield of the overall process. In the following discussion bioenergetic and biochemical aspects of photosynthesis are considered only insofar as they are of ecological importance. Detailed presentations of these can be found in textbooks of plant physiology, microbiology and biochemistry.

3.1.1.1 The Photochemical Process (Energy Conversion)

The primary requirement for the occurrence of photosynthesis is absorption of radiation by the chloroplasts. The degree to which radiation is utilized depends on the *chlorophyll concentration* —or, more precisely, on the concentration of photosynthetically active pigments. Particularly under intense light, this can be the factor that limits the photochemical process (Fig. 3.1). Chlorophyll deficiency, which is evident in the appearance of a plant ("chlorosis"), always considerably lowers the rate of photosynthesis. This condition occurs when the leaves first unfold and again in the fall, when they turn yellow; leaves also become chlorotic when mineral balance is disturbed, during drought, after infection and following exposure to noxious gases. Finally, chlorophyll deficiency can be genetically determined, as in mutants with mottled or yellow leaves.

The photochemical process is initiated when the chloroplasts capture photosynthetically utilizable radiation. Two pigment systems are involved in the light-driven reactions.

Photosystem I consists of pigment collectives with a particular structural organization; the predominant component of these is chlorophyll a (the ratio of chlorophyll a to chlorophyll b is about 6 : 1 to > 10 : 1). The reaction center of each is a chlorophyll-a-protein complex with an absorption peak at 700 nm—hence the alternative name for this complex, Pigment 700. The ratio of total chlorophyll to Pigment 700 is about

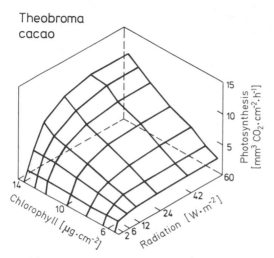

Theobroma cacao

Fig. 3.1. Relationship between potential photosynthesis under illumination of different intensities and the chlorophyll content of maturing cocoa leaves. Chlorophylls a and b are present in the same ratio over the entire range of variation. The measurements were done in CO_2-saturated conditions (2400 µl CO_2 per liter air) to ensure that the rate of photosynthesis was not limited by the secondary processes (Baker and Hardwick, 1976)

300 : 1 in herbaceous plants, about 450 : 1 in broad-leaved trees, and 600—1500 : 1 in evergreen conifers. *Photosystem II* contains a larger proportion of chlorophyll b (a/b ratio 1.2 : 1 to 2 : 1) and a chlorophyll-a-protein complex with maximal absorption at 680 nm. Both photosystems also include accessory pigments (carotenoids, and in algae phycobilins).

The fine structure of the chloroplasts and their pigment content are adapted to the *light climate*. Leaves that differentiated under high irradiation contain more chloroplasts per mesophyll cell, but the thylakoids in the chloroplasts are less densely packed than in shade leaves. Moreover, the chlorophyll a/b ratio in sun leaves shifts greatly in favor of chlorophyll a, and the proportion of Photosystem-I units is high (see Table 2.2).

After absorbing light quanta Pigment 700 releases electrons that are used for the reduction of $NADP^+$. The electrons required for the re-reduction of chlorophyll are provided by the photolysis of water or by another suitable electron donor—for example, H_2S in photoautotrophic bacteria (Hill reaction). During the photolysis of water oxygen is liberated, and enters the photosynthetic gas exchange process. Photosystem II pumps the hydrolytically acquired electrons to a higher energy level and supplies them to Pigment 700. In this *noncyclic electron transport* ATP is formed:

$$2H_2O + 2NADP^+ + 2ADP + 2P_i \xrightarrow[\text{P700 + P680}]{8\ h\nu} 2NADPH_2 + 2ATP + O_2 \,. \tag{3.1}$$

The electron released by Pigment 700 after absorption of radiation can also return to the oxidized chlorophyll molecule by way of several redox systems. This *cyclic electron transport* is also coupled with ATP formation (cyclic phosphorylation):

$$ADP + P_i \xrightarrow[\text{P700}]{2\ h\nu} 1\ ATP \,. \tag{3.2}$$

Quantum Yield of Photosynthesis. The yield of the photochemical reactions depends upon the energy of the light quanta, which in turn depends on the wavelength of the absorbed radiation. The *quantum yield Φ* of photosynthesis is the photochemical work performed by the absorption of light quanta. It is expressed in mol of oxygen liberated (occasionally in mol of CO_2 converted or of carbohydrate formed) per mole of absorbed photons:

$$\Phi = \frac{\text{Photochemical activity (mol } O_2)}{\text{Absorbed PhAR quanta (mol photons)}} \tag{3.3}$$

The quantum yield of photosynthesis depends primarily on the *spectral composition of the light* and the *photon flux*. As long as radiation is the sole rate-limiting factor, the intensity of photosynthesis increases in proportion to the photon flux. A steep slope of the light-dependence curve (see Chap. 3.2.3.1) thus reflects good utilization of quanta. Under favorable conditions plants can achieve quantum yields of $0.05-0.1$ mol O_2 per einstein, which corresponds to 33% utilization. However, on the average the quantum yield, and thus the effectiveness with which radiation is utilized is, lower; this is because, in strong light especially, the secondary processes of photosynthesis become rate-limiting.

3.1.1.2 Fixation and Reduction of Carbon Dioxide (Conversion of Matter)

The energy and "reducing power" gained in the primary reactions are used for the reduction of carbon dioxide to synthesize carbohydrates of higher energy value. This reaction occurs in the stroma of the chloroplasts. It is initiated by the binding of CO_2 to an *acceptor*, which after carboxylation decomposes into smaller molecules; these are finally reduced to trioses. The reaction is described by the formula

$$n CO_2 + n Acc + 2n NADPH_2 + 2n ADP \xrightarrow{\text{enzyme}} \tag{3.4}$$
$$(CH_2O)_n + 2n NADP^+ + 2n ADP + 2n P_i + n Acc + n H^2O .$$

The *rate of carboxylation*—that is, the speed at which the CO_2 taken up is processed—depends chiefly upon the CO_2 supply, the concentration of the acceptor, and the activity of the enzyme; the last of these depends upon the temperature, the cell water potential, the adequacy of the available minerals, and the state of development and activity of the plant.

The Pentose Phosphate Pathway for CO_2 Assimilation (Calvin-Benson Cycle)

In most plants a pentose phosphate, ribulose-1,5-bisphosphate (RuBP), is the CO_2 acceptor, and is decisive in determining the yield of the dark reaction of photosynthesis. Carboxylation is catalyzed by the enzyme RuBP carboxylase. The product of this reaction, a six-carbon molecule, decomposes immediately to produce two molecules of 3-phosphoglyceric acid (PGA). Each of these molecules contains three carbon atoms, and the process is therefore also called the C_3 *pathway* of CO_2 assimilation. PGA is reduced to glyceraldehyde-3-phosphate (GAP) over several steps involving ATP and $NADPH_2$ (Fig. 3.2). This is the final step in raising the CO_2 taken in to the energy level

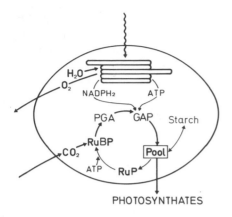

Fig. 3.2. Simplified diagram of CO_2 fixation and assimilation by way of the Calvin-Benson cycle in C_3 plants. *RuBP*, ribulose-1,5-bisphosphate; *PGA*, 3-phosphoglyceric acid; *GAP*, glyceraldehyde-3-phosphate; *Pool*, intermediary C_3 to C_7 compounds; *RuP*, ribulose-5-phosphate. The photosynthates are carbohydrates, carboxylic acids and amino acids. More detailed diagrams can be found in textbooks of plant physiology and biochemistry (see e.g., Bonner and Varner, 1976)

of a carbohydrate. GAP flows into a pool of carbohydrates of different carbon-chain lengths (C_3–C_7), which provides material for the synthesis of various substances (sugar, starches, carboxylic acids, amino acids) and for regeneration of the acceptor.

The Dicarboxylic Acid Pathway for CO_2 Assimilation (Hatch-Slack-Kortschak Pathway)

Comparative studies revealed that in certain plants the first product of CO_2 fixation was not a three-carbon molecule but rather oxaloacetic acid (OAA), a dicarboxylic acid with four carbon atoms. This form of *CO_2 uptake* was therefore called the C_4 pathway. *Reduction to carbohydrate* occurs by the C_3 pathway in C_4 plants as well. Uptake and subsequent processing of CO_2 takes place in two spatially separated and anatomically distinguishable tissues.

In the cells of the *mesophyll* CO_2 first binds to phosphoenolpyruvate (PEP). This process is catalyzed by the enzyme PEP carboxylase, which is very effective even at low CO_2 concentrations—so effective that C_4 plants can withdraw CO_2 from the outside air until its concentration falls below $10 \, \mu l \cdot l^{-1}$. In C_3 plants the limiting concentration for CO_2 uptake, the *CO_2 compensation concentration Γ*, is $30-70 \, \mu l \cdot l^{-1}$ (Table 3.1 and Fig. 3.3). For this reason C_4 plants, as a rule, achieve higher rates of photosynthesis than C_3 plants.

The carboxylation of PEP yields oxaloacetic acid (OAA), which in some species is converted to aspartic acid (aspartate) and in others is reduced to malic acid (malate) by a $NADPH_2$-dependent malate dehydrogenase. None of these dicarboxylic acids can be processed further in the mesophyll cells; they must be transported into *bundle-sheath cells* (Fig. 3.4). In the chloroplasts of the bundle-sheath cells the malate and aspartate are broken down into CO_2 and pyruvate (Py) by specific enzymes. The CO_2 released is captured by RuBP and enters the Calvin cycle, whereas the pyruvate returns to the mesophyll cells, where it can serve in the regeneration of PEP.

This apparently complicated combination of dicarboxylic acid synthesis and C_3 cycle gives the C_4 plants the advantage of *optimal CO_2 utilization*. The extremely high CO_2 affinity of PEP carboxylase and the special anatomy of the leaf enable the plant to re-

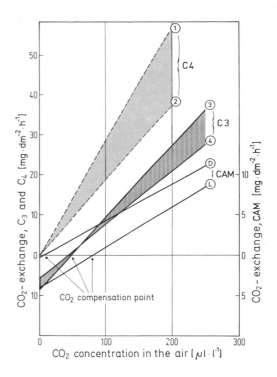

Fig. 3.3. Dependence of net photosynthesis on the CO_2 supply, and the position of the CO_2 compensation point in the C_4 plants *Atriplex spongiosa* (*1*) and *Atriplex rosea* (*2*); in the C_3 plants *Atriplex hastata* (*3*) and *Atriplex patula* (*4*), and in the CAM plant *Bryophyllum fedtschenkoi*, in the dark phase (*D*) and in the light phase (*L*) of a circadian rhythm in Γ in continuous light (Slatyer, 1970; Björkman, 1971; Jones and Mansfield, 1972)

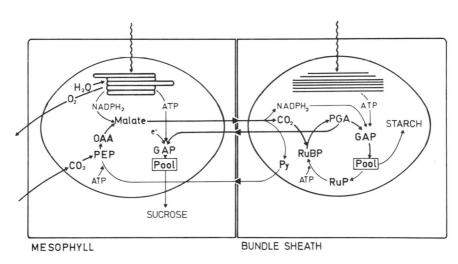

MESOPHYLL   BUNDLE SHEATH

Fig. 3.4. A much simplified diagram of CO_2 fixation via the Hatch-Slack-Kortschak pathway in C_4 plants. *PEP*, phosphoenolpyruvate; *OAA*, oxaloacetate; *PGA*, 3-phosphoglyceric acid; *GAP*, 3-phosphoglyceraldehyde; *RuP*, ribulose-5-phosphate; *RuBP*, ribulose-1,5-bisphosphate; *Py*, pyruvate. PGA is also produced by carboxylation of C_2 compounds which appear in the pool; the regeneration of PEP from PGA, in which water is given off, is not shown. For detailed diagrams see Hatch and Osmond (1976)

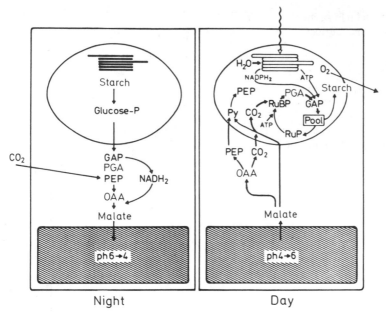

Fig. 3.5. Simplified diagram of the diurnal acid rhythm of CAM plants, and the photosynthetic utilization of the CO_2 released from malate. The way in which CO_2 produced during respiration is utilized is not shown. Labels as in Figures 3.2 and 3.4. Derived from Kluge (1971) and Osmond (1978)

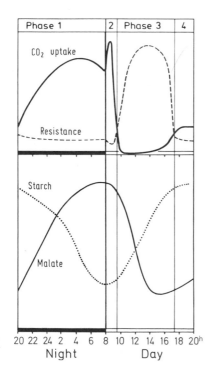

Fig. 3.6. Typical time courses of CO_2 uptake, diurnal acid rhythm, starch storage and stomatal opening in a CAM plant adequately supplied with water. *Phase 1*, primary CO_2 fixation and accumulation of malate in the dark with stomata open (diffusion resistance small), accompanied by starch breakdown; *Phase 2*, beginning of the light period, initial uptake of external CO_2 through the wide-open stomata; *Phase 3*, movement of malate out of the vacuole and into the cytoplasm and chloroplasts, malate decomposition and release of the CO_2 bound in it, stomatal closure, feedback inhibition of PEP carboxylase, photosynthetic assimilation of internal CO_2, starch synthesis; *Phase 4*, reactivation of PEP carboxylase following reduction of the high malate level. opening of the stomata and influx of external CO_2, and the beginning of renewed malate synthesis at the end of the light period. From experiments by various authors cited by Kluge (1977) and Osmond (1978). Details and variations in CAM behavior have been described by Neales (1975) and Kluge and Ting (1978), rhythms in enzyme capacities by Queiroz. For modulation by drought see Figure 3.48

use immediately the CO^2 liberated by photorespiration (see p. 81). C^4 plants are therefore capable of exploiting even the highest light intensities, providing high photosynthetic yields.

Thus far, CO_2 incorporation has been shown to occur by way of the C_4 pathway in almost 1000 angiosperm species, belonging to at least 18 plant families, and in algae (the blue-green alga *Anacystis nidulans*). Lists of species can be found in the publications of Downton (1975) and Raghavendra and Das (1978b). There are particularly large numbers of C_4 plants among the Poaceae (for example, sugar cane, maize, millet, and many savanna grasses), Portulacaceae, Amaranthaceae, Chenopodiaceae and Euphorbiaceae. In some genera (*Atriplex, Euphorbia, Cyperus, Panicum*) only certain species are of this type. Within species there can be C_3 and C_4 ecotypes–for example, in the grass *Alloteropsis semialata* and even in one and the same individual (e. g., of *Zea mays*) both C_3 and C_4 pathways may occur depending on leaf age and leaf position.

CO_2 Fixation in Succulent Plants (Crassulacean Acid Metabolism, CAM)

There are various plant species, mostly succulents, which take up large amounts of CO_2 *in the dark with stomata wide open*. Most of these are members of the families Agavaceae, Liliaceae, Bromeliaceae, Orchidaceae, Cactaceae, Crassulaceae, Aizoaceae, Euphorbiaceae, Geraniaceae, Portulacaceae, Asclepiadaceae, and Asteraceae, but the group also includes members of the Cucurbitaceae, Vitaceae, Lamiaceae, two epiphytic ferns, and even the submerged water plant *Isoetes howellii*. Lists of plants are given by Szarek and Ting (1977) and Kluge and Ting (1978). The CO_2 so acquired is captured by PEP in the cytoplasm, with the aid of PEP carboxylase (*CO_2 dark fixation*).

Oxaloacetic acid is produced and subsequently converted to malate by the $NADH_2$-dependent malate dehydrogenase (Fig. 3.5). The PEP comes from glycolysis, so that as malate is formed the starch content of the chloroplasts is correspondingly reduced (Fig. 3.6). Malate passes into the vacuole and is accumulated in the cell sap, which as a result becomes progressively more acid during the night. When it becomes light the next morning, the malate is transported back from the vacuole into the cytosol and the chloroplasts, where it is decarboxylated as in C^4 plants. The CO^2 liberated is taken up by RuBP and reduced to carbohydrate. During the daytime a normal C_3 cycle proceeds in the chloroplasts. The progressive breakdown of malate is accompanied by a rise in the pH of the cell sap (*diurnal acid rhythm*); the internal CO_2 concentration falls, the stomata begin to open, and new CO_2 can flow in from the outside and be fixed by PEP.

This 24-h sequence of events—CO_2 uptake, acid rhythm and assimilation—is demonstrable in principle in all CAM plants, but there are great differences in the amount and duration of CO_2 uptake in dark and light, depending on species, state of development, and environmental conditions such as water supply, temperature and day length. In addition to "*obligatory*" CAM plants, which always make full use of the opportunity to store malate, there are *inducible* species such as *Mesembryanthemum crystallinum* and *Sedum acre*, which only under the stress of drought (also high salinity, in the case of *Mesembryanthemum*) exhibit CAM behavior; in the absence of stress they respond like normal C_3 plants. Moreover, the CAM capacity can be restricted to *certain organs* of a plant. The deciduous *Frerea indica* (Asclepiadaceae)

79

Table 3.1. Distinguishing characteristics of plants with different types of CO_2 fixation (Black, 1973; Laetsch, 1974; Tieszen, 1975; Hatch and Osmond, 1976; Osmond, 1978; Raghavendra and Das, 1978a; Kluge and Ting, 1978). For functional significance of different photosynthetic pathways see Osmond et al. (1982)

Characteristic	C_3	C_4	CAM
Leaf structure	Laminar mesophyll, parenchymatic bundle sheaths	Mesophyll arranged radially around chlorenchymatic bundle sheaths („Kranz"-type anatomy)	Laminar mesophyll, large vacuole
Chloroplasts	Granal[a]	Mesophyll: granal; bundle-sheath cells: granal or agranal[b]	Granal
Chlorophyll a/b	ca. 3:1	ca. 4:1	$\leq 3:1$
CO_2-compensation concentration at optimal temperature	30—70 $\mu l \cdot l^{-1}$	<10 $\mu l \cdot l^{-1}$	In light: 0—200 $\mu l \cdot l$ in dark: <5 $\mu l \cdot l^{-1}$
Primary CO_2 acceptor	RuBP	PEP	In light: RuBP in dark: PEP
First product of photosynthesis	C_3 acids (PGA)	C_4 acids (malate, aspartate)	In light: PGA in dark: malate
Carbon-isotope ratio in photosynthates[c]	−20 to −40‰	−10 to −20‰	−10 to −35‰
Photosynthesis depression by O_2	Yes	No	Yes
CO_2 release in light (apparent photo-respiration)	Yes	No	No
Net photosynthetic capacity	Slight to high	High to very high	In light: slight in dark: medium
Light-saturation of photosynthesis	At intermediate intensities	No saturation at highest intensities	At intermediate to high intensities
Redistribution of assimilation products	Slow	Rapid	Variable
Dry-matter production	Medium	High	Low

[a] Thylacoids stacked up.   [b] Thylacoids lamellar.

[c] $\delta^{13}C = \left(\dfrac{^{13}C/^{12}C \text{ in sample}}{^{13}C/^{12}C \text{ in standard}} - 1 \right) \cdot 1000.$

PEP carboxylase discriminates between carbon isotopes present in the air, ^{12}C and ^{13}C, in a different way than does RuBP carboxylase.

binds CO_2 by the C_3 pathway in the leaves but by way of CAM in the succulent axial parts of the shoot. In so doing, this plant achieves high carbon yields while it is in leaf, during the wet season, and during the dry season when it is bare it can take up carbon in a way that conserves water (Lange and Zuber, 1977).

CAM requires that the chloroplast-containing cells have sufficiently *large vacuoles* to accommodate the nightly production of malate. The vacuole in CAM plants serves not only as a water reservoir, but also as a site of accumulation of carbon compounds which make photosynthetic activity temporarily independent of CO_2 exchange. For succulents, inhabitants of dry regions, it is ecologically advantageous to separate the elements of carbon assimilation in time, fixing CO_2 by night and processing it on the following day. In this way the supply of carbon is ensured in combination with optimal water conservation.

3.1.2 Photorespiration

Connected with photosynthesis, a metabolic process takes place in chloroplast-containing plant cells which, like respiration, takes up O_2 and releases CO_2 in the light, but which, contrary to respiration, ceases in the dark. This O_2/CO_2 gas exchange has been called light respiration or photorespiration. The substrate for the photorespiratory metabolism is again ribulose bisphosphate, which can be an acceptor not only for CO_2 but also for O_2. By taking up oxygen, RuBP is split into PGA and phosphoglycolate. The supply of O_2 and CO_2 regulates the relationship between acceptor oxidation (photorespiration) and acceptor carboxylation (photosynthesis) via the enzyme complex RuBP carboxylase/oxygenase. High partial pressure of O_2 favors photorespi-

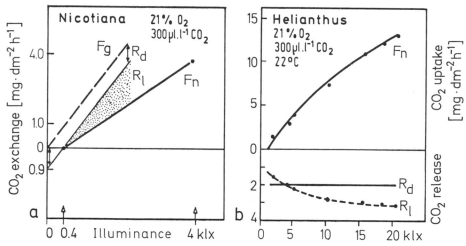

Fig. 3.7a and b. Influence of light intensity on photorespiration. **a** Diagram of the relationship between gross photosynthesis (F_g), net photosynthesis (F_n), photorespiration (R_l) and dark respiration (R_d), based on measurements of tobacco leaves. From Decker (1957). **b** Net photosynthesis, dark respiration and photorespiration of sunflower leaves at increasing intensities of illumination (Hew et al., 1969)

ration, a large supply of CO_2 favors photosynthesis. As the formation of phosphoglycolate is dependent on the supply of RuBP via the Calvin cycle the photorespiratory O_2 uptake and CO_2 release increase with increasing light intensity (Fig. 3.7).

The process of glycolate metabolism has not yet been understood in all details. A schematic review is given in Fig. 3.8. In the chloroplasts phosphoglycolate is split into glycolate and phosphate. The glycolate is transported out of the chloroplasts into peroxisomes, cell compartments about the size of mitochondria, which contain glycolate oxidase, catalase, and transaminases. In the peroxisomes, when O_2 is taken up, glycolate is oxidized to glyoxylate, and the peroxide thus produced is detoxified by catalase. Glycolate can either be completely reduced via oxalate by further O_2 uptake, or transformed to glycine by transamination. Glycine is transported from the peroxisomes into the mitochondria, where two molecules of glycine are coupled to form one molecule of serine with the release of CO_2. Serine is taken over by the amino acid metabolism or converted to glycerate after deamination by hydropyruvate. This can be phosphorylated in the chloroplasts and returned to the Calvin cycle or used elsewhere.

Under natural conditions (21% O_2 and 0.03% CO_2 in the air, strong irradiation, temperatures between 20° and 30° C) the C_3 *plants* immediately lose about 20%, or in the extreme case 50%, of the photosynthetically acquired CO_2 in the form of photorespiratory CO_2. Calculations of the utilization of energy [cf. Formula (3.18)] and the photosynthetic efficiency [cf. Formula (3.8)] for these plants can be based on a rate of photorespiration in intense light that is 1.5—3.5 times as large as dark respiration. C_4 *plants* in the light release no CO_2. Photorespiration in these plants occurs only in the cells of the bundle sheaths, and the CO_2 produced is refixed in the mesophyll cells be-

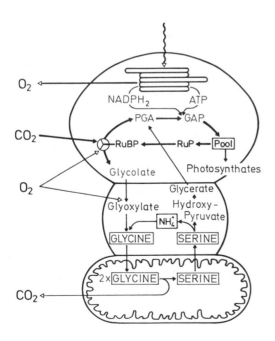

Fig. 3.8. Simplified diagram of glycolate metabolism in C_3 plants. Compartmentation of the reactions: *top*, chloroplast; *middle*, peroxisome; *bottom*, mitochondrion. For explanation see text; symbols as in Fig. 3.2. For detailed information see Tolbert (1971, 1979), Chollet and Ogren (1975), and Schnarrenberger and Fock (1976)

fore it can leave the leaf. This "CO_2 trap" prevents loss of matter during photo-respiration and makes possible the production of matter at a higher rate.

3.1.3 Catabolic Processes

Whereas anabolic processes bring about the synthesis of the substances that make up the plant, in catabolism substances are broken down to provide energy for the diverse metabolic functions of the cell. The substrate for these reactions is carbohydrate or fat; in the exergonic decomposition of these substances hydrogen is split off and energy re-leased in a stepwise manner. The hydrogen is incorporated into pyridine or flavin nucleotides. Most of the energy is obtained at the step in which hydrogen is transferred to the final hydrogen acceptor. In *respiration* this acceptor is atmospheric oxygen, which receives the hydrogen from an electron-transport chain by way of the sequence of respiratory reactions in the mitochondria. In the process, ATP is formed by respiratory-chain phosphorylation, and other energy-rich phosphates are produced by substrate-chain phosphorylation. In *fermentation* reducible organic compounds take up the hydrogen, and in *anaerobic respiration* the acceptors are inorganic ions such as nitrate and sulfate. Because the potential difference between hydrogen and oxygen is greater than those between hydrogen and other oxidizing agents, aerobic respiration provides much more energy than the other catabolic processes; it operates with an effi-ciency of 30—40%.

3.1.3.1 Glycolysis and Mitochondrial Respiration

The respiratory substrate glucose is decomposed in many sequential steps: glycolysis, decarboxylation of pyruvate, the citric-acid cycle (*Krebs-Martius cycle*), and oxidation of the terminal electron acceptor ($NADH_2$). Glycolysis takes place in the cytoplasm, while the citric-acid cycle and respiratory-chain phosphorylation occur in the mito-chondria. The entire process yields 36 mol of ATP, 2 mol of GTP, and a free enthalpy ΔG of -2.87 MJ ($= -686$ kcal) per mol of glucose.

3.1.3.2 Oxidative Pentose Phosphate Cycle

In addition to glycolysis and the citric-acid cycle, there is a third way in which glucose can be broken down, particularly in the highly differentiated cells of plants. This is the direct oxidation of glucose-6-phosphate, in which $NADP^+$ rather than NAD^+ is reduced. Decarboxylation in this pathway (also called the hexose monophosphate shunt) produces ribulose-5-phosphate, which is reconverted to glucose-6-phosphate by way of the oxidative pentose phosphate cycle. The significance of this metabolic pathway lies less in its energy yield than in the production of metabolites. The oxidative pentose phosphate cycle occurs in the cytosol and in the chloroplasts. It is closely relat-ed to the reductive pentose phosphate cycle (Calvin cycle), for the two have many reac-tions and enzymes in common and share in the $NADP^+$ pool.

3.2 CO_2 Exchange in Plants

3.2.1 The Exchange of Carbon Dioxide and Oxygen as a Diffusion Process

Carbon metabolism in the cell is linked to the external environment by gas exchange. In photosynthesis the chloroplasts use up CO_2, of which a supply must be maintained, and liberate oxygen. In parallel, by both day and night, the cells take up oxygen for respiration and give off carbon dioxide. In assimilating leaves, one or the other of these two opposed processes can predominate at a given time. The respiration occurring in the light comprises both photorespiration and mitochondrial respiration. During the day the rate of CO_2 uptake per unit of plant mass required for photosynthesis (gross photosynthesis, F_g) is greater, as a rule, than the rate at which CO_2 is freed by the total respiration in light (R_l), so that there is a net uptake of CO_2 into the leaf. Under these conditions one speaks of apparent or net photosynthesis (F_n).

$$F_n = F_g - R_l. \tag{3.5}$$

If the rate of photosynthesis decreases it can happen that the respiration occurring simultaneously just balances it (the "compensation point" is reached). If the rate of photosynthetic activity declines still further, respiration predominates, and in the dark, respiratory release of CO_2 alone prevails. The processes participating in respiration in the dark are mitochondrial respiration and the oxidative pentose phosphate cycle.

3.2.1.1 The Diffusion Rate

Gas exchange between the cells and the surroundings of the plant (the outside air or water) occurs by diffusion. The turnover is enormous; for every gram of glucose formed 1.47 g of CO_2 are required, and the volume of air from which this amount can be withdrawn amounts to about 2500 l. CO_2 and O_2 transport are described by Fick's law of diffusion:

$$\frac{dm}{dt} = -D \cdot A \frac{dC}{dx}. \tag{3.6}$$

The diffusion rate (quantity displaced, dm, in the time interval dt) depends upon the diffusion constant D and is greater the steeper the *concentration gradient* dC/dx in the direction of diffusion x and the greater the exchange area A. The diffusion constant depends both upon the substance considered and the medium in which diffusion takes place; in air, CO_2 and O_2 can diffuse about 10^4 times as fast as in water.
Fick's law can be applied to gas exchange in plants in the form derived by Gaastra (1959):

$$J = \frac{\Delta C}{\Sigma r}. \tag{3.7}$$

Here the flux of diffusing substance is expressed in molecules per unit area per unit time; ΔC is the *concentration difference* between the outside air and the site of the reaction in

84

the cell, and Σr is the sum of a number of terms representing *resistance to diffusion*. Σr thus takes into account the relevant diffusion constant, effects at phase interfaces, and the spatial dimensions involved in the situation.

3.2.1.2 The Concentration Gradient

If one assumes that when photosynthesis is proceeding to completion the CO_2 in the chloroplasts is used up, and that during respiration in the mitochondria the O_2 concentration there falls to zero, then the concentration gradients of these two gases are determined by their concentrations in the surroundings of the plant.

Carbon dioxide amounts to about 0.003% of the atmosphere by volume. It is readily soluble in water (saturation partial pressure at 10° C ca. 0.7%, at 20° C ca. 0.5%, and at 30° C ca. 0.4%), producing carbonic acid. The proportion of free CO_2 depends on the pH of the water, being high in the acid region; above pH 9 only hydrogen carbonate and carbonate ions are present, but these also represent CO_2 reserves for the plant. The used CO_2 is replenished under natural conditions by movement along a slight concentration gradient. Inadequate CO_2 supply is often a yield-limiting factor, for both land and water plants. When the CO_2 content of the air is artificially raised to 0.1—0.3% by volume, C_3 plants are able to bind 2—3 times, and C_4 plants 1.5 times, as much CO_2

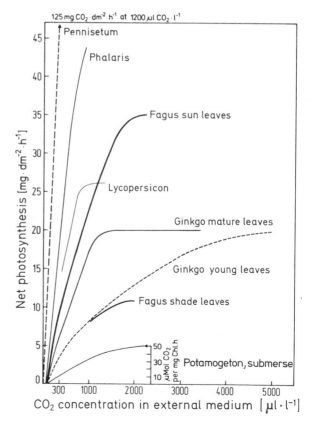

Fig. 3.9. Potential net photosynthesis of C_3 and C_4 plants as the supply of CO_2 is increased. C_3 plants: *Fagus sylvatica* (Koch, 1969), *Ginkgo biloba* (Ponomareva, 1960), *Lycopersicon esculentum* (Gaastra, 1959), *Phalaris arundinacea* (Gloser, 1976) and submerse leaves of the aquatic plant *Potamogeton amplifolius* (Lloyd et al., 1977); emerse *Potamogeton* leaves behave like *Lycopersicon*, becoming CO_2-saturated at 1100 $\mu l \cdot l^{-1}$. The tropical C_4 grass *Pennisetum purpureum* reaches CO_2 saturation at 1200—1500 $\mu l \cdot l^{-1}$ (Ludlow and Wilson, 1971a). For references see Strain (1978)

as under natural conditions ("potential photosynthesis", Fig. 3.9). The possibility of raising the yield of photosynthesis by increasing the CO_2 concentration (*CO_2 fertilization*) is sometimes applied by growers, particularly in greenhouses. By raising the plants in air containing 0.1% CO_2 (by volume) the growth of tomatoes, cucumbers, leafy vegetables, and tobacco could be doubled.

Oxygen is very much more abundant in the atmosphere than carbon dioxide. The shoots of *land plants* are always assured of a good supply. The oxygen concentration in bodies of *water* is lower than that in the atmosphere. The amount of dissolved O_2 depends on the temperature, and is at most 14.7 mg O_2 per liter of water (at $0°$ C) or 1% by volume; at $25°$ C only half as much dissolves. The oxygen concentration in the *soil* is also lower than in the open air, because oxygen is consumed by the respiration of the plant roots, soil animals and aerobic microorganisms. In compacted, waterlogged, and flooded soils O_2 diffuses so slowly that its concentration falls to a few percent or even to zero. Once the soil is free of oxygen, anaerobic microorganisms take over, producing a strongly reducing milieu within which Fe^{2+}, Mn^{2+}, H_2S, sulfides, lactic and butyric acids, and other compounds accumulate in toxic concentrations. Under these conditions nitrogen turnover in the soil is also severely disrupted.

For *mitochondrial respiration* to proceed normally an oxygen concentration of 1–3% suffices. When the concentration falls below this level, the first result is reduction of the rate of respiration, of the uptake of water and nutrient ions, and of root growth. There is increased production of abscisic acid and in particular of ethylene, so that shoot growth is disturbed and the leaves shed prematurely. In a state of anoxia—when there is no oxy-

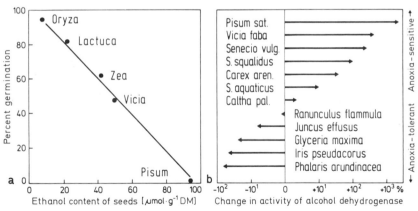

Fig. 3.10. a Ethanol content and percent germination (as a measure of vitality) of seeds differing in sensitivity to oxygen deficiency. The seeds were examined after being submerged under water for three days. From Crawford (1977). **b** Increase (*positive values*) and decrease (*negative values*) in alcohol dehydrogenase activity of the roots of plants differing in oxygen requirement, after growing for one month in waterlogged soil with inadequate oxygen supply. The activity of alcohol dehydrogenase is greatly enhanced in sensitive species, leading to ethanol accumulation and cell damage. In anoxia-tolerant species there is a shift to alternative metabolic pathways, and the activity of alcohol dehydrogenase actually decreases. After McManmon and Crawford (1971). For flooding injuries in trees see Gill (1970), Rowe and Beardsell (1973), Kozlowski (1976), Pereira and Kozlowski (1977)

gen at all—respiration gives way to *anaerobic catabolism*. Because the terminal oxidation is impossible acetaldehyde accumulates, and this induces and activates alcohol dehydrogenase. An *excessive ethanol content* is the characteristic symptom of cell anoxia (Fig. 3.10). Plants can germinate, take root and survive in soils poor in oxygen if they develop special metabolic or morphological adaptations. A *metabolic adaptation* to oxygen deficiency is the absence of induction of alcohol dehydrogenase by increased concentrations of acetaldehyde during anaerobiosis. In this situation the end product that accumulates is not ethanol, which is toxic, but rather lactic, shikimic, or malic acids. The activity of alcohol dehydrogenase during anoxia is a quantitative measure of the specific sensitivity of a plant species to oxygen deficiency. The *morphological adaptation* to an oxygen-poor environment consists in the development of tissues (*aerenchyma*) with a system of spacious intercellular channels through which oxygen diffuses easily from the leaves into the shoot axis and roots. The pore volume amounts to as much as 60% of the root parenchyma of swamp plants, whereas in plants on well-aerated soil it is less than 10%. Plants growing on particularly dense soils, where oxygen moves slowly, develop an extensive superficial root system, adventitious roots, and as an extreme specialization respiratory roots (in mangrove plants, swamp cypresses). Roots well supplied with oxygen are actually able to pass it into their surroundings and thus render reducing substrates harmless—for example, by the precipitation of Fe^{2+} as Fe-III oxides.

3.2.1.3 Diffusion Pathway and Transfer Resistances in the Leaf

In its course from the air to the chloroplasts, carbon dioxide encounters a series of transfer barriers. The CO_2 molecules enter the system of intercellular spaces through the stomata. In the cell walls, CO_2 moves from the gas phase, in which it diffuses relatively rapidly, to the liquid phase. The process of going into solution delays the CO_2 transport considerably. Within the cell, the dissolved CO_2 migrates slowly to the chloroplasts.

Figure 3.11 presents an overview of the pathways and resistances involved in transport near and within the leaf. During photosynthesis, the greatest CO_2 partial pressure is found outside a thin boundary layer of air near the leaf; the thickness of this layer depends upon the size and position of the leaves, the presence or absence of hair on the leaf surface, and above all on the degree of air movement. In still air the layer can be some millimeters thick, while a strong wind will sweep it entirely away. The thicker the boundary layer, the larger is the *boundary-layer resistance* r_a. If carbon dioxide is taken into the leaf more rapidly than it is replaced by diffusion, the film of air near the leaf is depleted of CO_2.

CO_2 enters the leaf through the stomata. Passage through the *cuticle* can occur with abnormally high outside concentrations, but under natural circumstances cuticular CO_2 uptake by land plants can be discounted. On the other hand, the CO_2 produced during respiration in the dark reaches such a high partial pressure that considerable amounts can escape through the cuticle and through peridermal covering tissues. The decisive constraint on CO_2 uptake into the leaf is *the stomatal resistance* r_s; when the pores are closed r_s is nearly infinite. The CO_2 concentration in the substomatal cavity and in the intercellular air C_i is lower than that in the outside air (C_a), but it is still appreciable. The

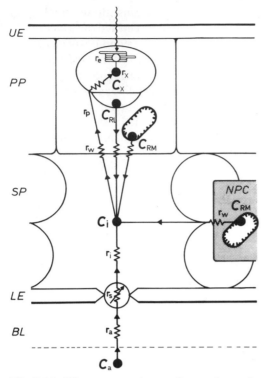

Fig. 3.11. CO_2 concentration gradient and transfer resistances in a leaf during photosynthesis. *UE*, upper epidermis; *PP*, palisade parenchyma; *SP*, spongy parenchyma; *LE*, lower epidermis; *NPC*, cells lacking chloroplasts and not photosynthetically active; *BL*, boundary layer (the film of air near the leaf). During photosynthesis a *gradient in the CO_2 concentration* is established from the outside air (C_a) via the intercellular air (C_i) to the minimal concentration at the site of carboxylation (C_x). In the intercellular system of the leaf, CO_2 arrives not only from outside but also from the cells, as a result of the respiratory activity of the mitochondria (C_{RM}) and photorespiration (C_{RL}). The *transport resistances* interposed are the boundary-layer resistance r_a, the physiologically regulatable stomatal resistance r_s, diffusion resistances in the intercellular system r_i, and resistances associated with the processes of dissolving and transport of CO_2 in the liquid phase of the cell wall (r_w) and in the protoplasm (r_p). r_x indicates the "carboxylation resistance", and r_e the "excitation resistance". For further details see Šesták et al. (1971) and Chartier and Bethenod (1977)

intercellular air not only serves as a source of CO_2, it also receives CO_2 as a result of respiratory processes in green and non-green cells. A "compensation point" in the gas-exchange equilibrium can thus be defined; this reflects the state in which C_i is equal to C_a. In this situation, there is no net gas exchange even though the stomata may be wide open.

The *diffusion resistance in the intercellular system* (r_i) depends upon the structure of the leaf, in particular the specific intercellular volume and the pathways available for movement of gases. In most species the intercellular volume amounts to 40—50% of that of the leaf; as a rule, the intercellular volume is greater in shade leaves than in sun

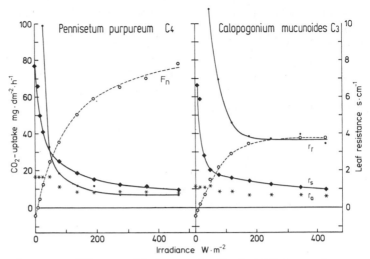

Fig. 3.12. Transfer resistances to CO_2 on and in leaves of a C_3 plant (*Calopogonium mucunoides*) and a C_4 plant (*Pennisetum purpureum*), and the intensity of net photosynthesis (F_n, *dashed curves*), in relation to incident radiation. Under intense irradiation the stomata are wide open, and the stomatal diffusion resistance r_s is only a little higher, or even lower, than the boundary-layer resistance r_a. In insufficient light the stomatal aperture is less and stomatal diffusion resistance increases. Before this stage is reached the light-driven primary processes become rate-limiting (r_e) for CO_2 uptake — that is, the residual resistance ($r_r = r_e + r_x + r_p + r_w$) is greatly increased. In C_3 plants under intense irradiation r_r is considerably larger than r_s, whereas in C_4 plants it is somewhat smaller. After Ludlow and Wilson (1971a). Examples for woody plants can be found in Holmgren et al. (1965; *Betula, Populus*), in Scholefield et al. (1977; *Persea*), and in Samsuddin and Impens (1978; *Hevea*)

leaves. Gas mobility is good in homobaric leaves with intercommunicating intercellular systems, and poor in septate heterobaric leaves. The *interface resistance r_w*, encountered when the gas enters the fluid phase at the cell walls and when it passes through the plasmalemma, is related to the specific intercellular surface area. A large internal diffusion area corresponds to a low interface resistance. In sun leaves the ratio of chlorenchyma area to leaf area is 15—30 : 1; in mesomorphic leaves it is 10—15 : 1, in shade leaves 5—10 : 1, and in mosses 2 : 1 (Nobel, 1977a, b).

The path of CO_2 transport ends in the chloroplasts with fixation by the acceptor. The speed with which CO_2 is processed also affects the steepness of the concentration gradient and thus the influx of CO_2. The rate of carboxylation, in comparison with the light reactions of photosynthesis, can be low and become a rate-limiting "bottleneck" in the overall process. This fact is sometimes expressed in the literature with reference to a "*carboxylation resistance*" r_x, though this resistance is not actually associated with the movement of CO_2. Because PEP carboxylase catalyzes CO_2-binding much more effectively than does RuDP carboxylase, the "carboxylation resistance" in C_4 plants is much smaller than in C_3 plants. Finally, photosynthetic performance and hence CO_2 influx are limited when energy and reduction equivalents are not supplied at a high enough rate by the primary reactions; this limitation can also be regarded as a transfer resistance, the "*excitation resistance*" r_e.

89

The diffusion resistances $r_a + r_s + r_i$ can be determined by way of the diffusion of water vapor whereas the remaining transfer resistances are customarily taken together as the "*mesophyll resistance*" r_m — or, better, as the "*residual resistance*" r_r — and calculated on the basis of net rate of photosynthesis. If the CO_2 compensation concentration Γ is taken as the terminal value for diffusion in the calculation, r_r represents only the diffusion resistances in the mesophyll and not the metabolism-related resistances r_e and r_x in the chloroplasts. The orders of magnitude of these resistances are given in Figure 3.12.

3.2.1.4 The Regulation of Gas Exchange by the Stomata

(The stomata are the most important regulators of the diffusion process. By varying the width of the stomatal pores the plant simultaneously controls CO_2 entry into the leaf) and release of water vapor.

Opening of the Stomata

Number, distribution, size, shape, and mobility of the stomata are species-specific characteristics, though they vary according to habitat and even among individuals (Table 3.2). The critical anatomical dimension determining stomatal resistance is the *pore width*. Stomatal diffusion resistance increases exponentially with decreasing pore width, following a hyperbolic curve. Its reciprocal $1/r_s$, the *stomatal conductance*, is thus directly proportional to pore width. The maximal width to which the pore of a stoma can be opened, which depends upon the shape and the properties of the walls of the guard cells, limits the rate at which gas can flow through. The opening capacity is greatest in the leaves of herbaceous dicotyledons, in the foliage of deciduous trees with open crowns, and in trees of tropical forests; it is particularly low in woody plants with thick, stiff leaves.

The proportion of the shoot surface over which stomatal diffusion is possible is called the *pore area*. The pore area is the product of the pore density (number of stomata per mm^2 of leaf surface) and the maximal pore width (ostiole area). In most plants the pore area amounts to $0.5-1.2\%$ of the leaf surface, though in tropical forest plants it can be as much as 3%. The pore area is especially small in succulents and scleromorphic leaves; on the assimilation organs of succulents the pore density is low, and the leaves of sclerophyllous plants and evergreen dwarf shrubs have stomata capable of opening only slightly.

The Physiological Mechanism of Stomatal Movement

Opening and closing of the stomata is brought about by a *turgor difference* (see Chap. 5.2.2) between the guard cells and the adjacent epidermal cells (subsidiary cells). If the turgor of the guard cells becomes greater than that of the subsidiary cells the stomata open; when not under tension, they are closed. Increase in turgor is an *osmoregulatory* process associated with active transport of K^+ from the adjacent cells into the guard cells. Charge imbalance is prevented by the movement of inorganic anions (e.g., Cl^- in grasses), the formation of organic anions (e.g., malate), or release of protons. Transport

Table 3.2. Stomatal density, pore width and pore area on the abaxial (lower) side of leaves, with the minimal stomatal diffusion resistance for CO_2 (computed for one side of the leaf). The data for stomata were taken from summaries by Stocker (1929, 1971, 1972), Meidner and Mansfield (1968), Cintron (1970), Pisek et al. (1970), Parcevaux (1972, 1973), Napp-Zinn (1973) and from original publications. The diffusion resistances for CO_2 are from the measurements of numerous authors (sources cited by Körner et al., 1979)

Plants	Stomatal density (number per mm^2 leaf area)	Pore length (μm)	Maximal pore width (μm)	Pore area (% of leaf area)	Minimal stomatal diffusion resistance to CO_2 (s · cm^{-1})
Herbaceous plants of sunny habitats	100—200 (300)	10—20	4—5	0.8—1	0.6—2.6
Herbaceous plants of shady habitats	40—100 (150)	15—20	5—6	0.8—1.2	2.2—6.5
Grasses	(30) 50—100	20—30	ca. 3	0.5—0.7	0.9—5.2
Palms	150—180	15—24	2—5	0.3	ca. 6
Tropical forest trees	200—600 (900)	12—24	3—8	1.5—3	
Winter deciduous trees	100—500	7—15	1—6	0.5—1.2	1.7—6.5
Sclerophyllous plants	100—500 (1000)	10—15	1—2	0.2—0.5	1.7—5.2
Conifers	40—120	15—20	—	0.3—1	2.0—6.5
Desert shrubs	150—300	10—15	—	0.3—0.5	1.6—7.8
Succulents	15—50 (100)	ca. 10	ca. 10	0.1—0.4	3.1—6.5

by the ion pumps requires a supply of energy (ATP) and is influenced by endogenous effectors (phytohormones, photocybernetic sensitizers). Because of these factors, the readiness of the stomata to open and close varies during the course of a day and in different states of activity, development, stress, and adaptation. An example of *modulation of reactivity* of the stomata is given by the effect of phytohormones. (+) abscisic acid blocks the ion pumps, reducing the degree to which the pores can open; cytokinins enhance their readiness to open. Stomatal function can also be impaired by *external influences*, such as toxic excretions of plant parasites (the stomata may be locked in the open position by the wilting toxin of *Fusicoccum amygdali* or paralyzed by a phytotoxin of *Helminthosporium mayalis*) and environmental poisons like SO_2, PAN, ozone and the fungicide phenylmercury acetate.

Control of Stomatal Movement

The physiological mechanisms underlying stomatal movement have not yet been adequately explained, and still less is known about the way this movement is regulated.

Although the stomata respond to many influences, their movement appears to be controlled chiefly by two control circuits, one involving CO_2 and the other H_2O.

The *CO_2 control circuit* operates in relation to the partial pressure of carbon dioxide in the intercellular space. Its influence is most apparent in the dark, when the epidermis comes into contact with air of varied CO_2 concentrations: With an air CO_2 content of more than $150-250 \, \mu l \cdot l^{-1}$ the stomata are closed, and they open when the CO_2 concentration is reduced. In the *light* the CO_2 pressure in the intercellular space of C_3 and C_4 plants falls, as CO_2 is used up during photosynthesis. Pore opening in the light is thus essentially caused by reduction in CO_2. The effect of CO_2 concentration on the degree of pore opening is particularly clear in *CAM plants*. These open their stomata during the night, when the CO_2 partial pressure in the intercellular space falls due to intensive malate formation, and close them in the light, when the CO_2 released during malate breakdown accumulates in the intercellular space before it can be further metabolized (Fig. 3.6).

The *H_2O control circuit* comes into play during water deficiency; abscisic acid is formed in the leaves and hampers osmoregulation of the guard cells. As a result pore opening is progressively impaired, so that when severe water deficiency threatens, the stomata remain closed and are not affected by external factors.

Factors Influencing Pore Width

The degree of stomatal opening, and thus the stomatal diffusion resistance, can be adjusted in response to changes in the environment and within the plant.

The *water potential of the plant* (see Chap. 5.2.3) is the fundamental factor on which all regulation of pore width is based. The stomata can open widely only when the turgor potential is high; if it should fall below specific thresholds, as a result of water

Table 3.3. Limiting values of leaf water potential for the onset of pore-width reduction and for complete closure. From the measurements of many authors. Data collections are given by Ritchie and Hinckley (1975), Boyer (1976), Ludlow (1976), Richter (1976), Turner and Begg (1976), and Poole and Miller (1978)

Plant group	Onset of closure	Complete closure
Herbaceous crops		
Dicotyledons	− 6 to −10 bar	−10 to −30 bar
Grain and fodder grasses	− 6 to −10 (−20)	−20 to −30 (−50)
Winter-deciduous trees	(−2) − 6 to −10 (−20)	−10 to −20 (−30)
of the temperate zone		
Evergreen conifers	− 5 to −10	−10 to −20
Ericaceous dwarf shrubs	−10 to −20	−20 to −30
Sclerophyllous plants	−10 to −30 (−57)[a]	−30 to −50 (> −60)[a]
Desert shrubs	from −20	−35 to −80
Succulents	ca. −5	

[a] *Acacia harpophylla* (Van den Driessche et al., 1971; Tunstall and Connor, 1975).

deficiency, all other factors that promote opening become ineffective. Depending on the species of a plant and the structural peculiarities, age, insertion, and adaptation to the habitat of its leaves, there are limiting values of water potential for the beginning of closure and for complete closure; some of these are given in Table 3.3.

Low *humidity* causes stomatal closure in many species, especially when the water potential is beginning to fall. Air movement brings about a steeper vapor-pressure gradient near the leaf surface, so that the direct influence of humidity on pore opening is particularly strong when there is *wind*.

In the *light*, when water is plentiful, the stomata open wider the higher the intensity of illumination. In the dark the pores are narrow, except in CAM plants, but not necessarily closed.

The *temperature* primarily affects the *rate* of opening, which depends upon the energy available to the movement mechanism (ion transport). The *degree* of opening is indirectly affected by temperature, by way of its influence on photosynthesis (CO_2 control circuit) and transpiration (H_2O control circuit). At low temperatures (about $+5°$ to $0°$ C) the stomata open slowly and incompletely; between $0°$ and about $-5°$ C stomatal gas transport ceases. Great heat evidently disturbs the regulatory processes in some species (for example, *Phoenix dactylifera* at $50°$ C) so that the stomata open wider than before. If the water lost can be replaced rapidly enough, the result is a high rate of transpiration and a corresponding cooling (see p. 25).

The *interplay among the external factors* usually results in an intermediate pore width (Fig. 3.13). The stomata are completely open only rarely and briefly, because it is uncommon for all those conditions that favor opening to coincide. On the other hand, ex-

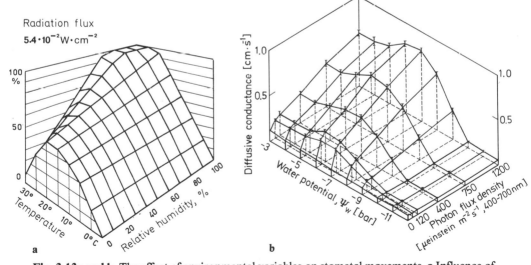

Fig. 3.13a and b. The effect of environmental variables on stomatal movements. **a** Influence of temperature and humidity upon stomatal aperture in the leaves of *Ligustrum japonicum* under low-intensity radiation. From Wilson (1948). **b** Effect of irradiation and water potential of the leaves of *Phaseolus vulgaris* on the epidermal diffusive conductance. From Pospišilová and So-lárová (1978). Further diagrams are given by Hinckley et al. (1975), Lösch (1977), Oechel and Lawrence (1979), and Schulze and Hall (1982)

treme situations that force prolonged complete closure occur regularly in dry regions, the arctic and the high mountains.

Of all the internal factors, the *phytohormones* are the most significant. By way of abscisic acid, phaseic acid, cytokinins and gibberellins in particular, the stomatal regulatory system can be adjusted to the processes of *growth and development* and to the seasonal *activity rhythm* of the plant. Opening behavior varies with the age and state of development of a plant (for example, at the transition into the reproductive phase). In many plants that stay green all winter the readiness of the stomata to open falls to a minimum during winter dormancy, when the abscisic-acid level is high.

3.2.2 Photosynthetic Capacity and Specific Respiratory Activity

3.2.2.1 Photosynthetic Capacity

The maximum rate of net photosynthesis by a plant at a given state of development and activity, under natural conditions of atmospheric CO_2 content and optimal conditions with respect to all other external factors, in the terminology of A. Pisek and W. Larcher, is called its *photosynthetic capacity*.

Photosynthetic capacity is measured under standard conditions; it can be used to characterize certain physiological types of plants as well as plant species, ecotypes and even individual varieties. Plants can differ profoundly in this respect. A survey is given in Table 3.4. The greatest photosynthetic capacity (108 mg $CO_2 \cdot dm^{-2} \cdot h^{-1}$) that has been found (Ludlow and Wilson, 1971a) is that of *Cenchrus ciliaris*, a tropical C_4 fodder grass. Values around 80 mg $\cdot dm_1^{-2} \cdot h^{-1}$ have been measured in tropical C_4 plants by several workers. Of the C_3 plants, only one species is so far known to achieve un-

Table 3.4. Average maximum values for net photosynthesis under conditions of natural CO_2 availability (0.03% by volume), saturating light intensity, optimal temperature and adequate water supply. Summary drawn from the data in original publications by many authors

Plant group	CO_2 uptake	
	$mg \cdot dm^{-2} \cdot h^{-1}$ [a]	$mg \cdot g^{-1} W_d \cdot h^{-1}$ [b]
Land plants		
Phanerogams		
Herbaceous plants		
C_4 plants	30—80 (108)	60—140
C_3 plants		
Crop plants	20—45 (60)	30—60
Plants of sunny habitats (heliophytes)	20—40 (94)	30—60
Shade plants (sciophytes)	4—20	10—30
Plants of dry habitats (xerophytes)	20—45	15—33
Grain and fodder grasses	15—35 (40)	
Wild grasses and sedges	8—20 (25)	8—35

Table 3.4 (continued)

Plant group	CO_2 uptake	
	$mg \cdot dm^{-2} \cdot h^{-1}$ [a]	$mg \cdot g^{-1} W_d \cdot h^{-1}$ [b]
CAM plants		
In the light	3—20	0.3—2
In the dark	10—15	1—1.5
Woody plants		
Tropical and subtropical trees		
Fruit trees	18—22	10—25
Forest canopy trees	12—24	
Understory trees	5—10	
Broad-leaved evergreens of the subtropics and warm-temperate regions		
Sun leaves	10—18	
Shade leaves	3— 6	
Seasonally deciduous trees		
Sun leaves	15—25 (35)	
Shade leaves	5—10	
Conifers		
Winter-deciduous		10—40
Evergreen	5—18	4—18
Mangrove trees	6—12 (20)	
Sclerophylls of periodically dry regions	5—15	3—10
Bamboos	5—10	
Palms	6—10 (12)	
Desert shrubs	(4) 6—20 (30)	(2) 5—15 (35)
Dwarf shrubs of heath and tundra		
Winter- deciduous	10—25	15—30
Evergreen	5—10 (15)	2—10
Cryptogams		
Ferns	3—5	
Mosses	up to 3	0.6—3.5
Lichens	0.5—2 (6)	0.3—2.5 (4)
Aquatic plants		
Swamp plants, emerse hydrophytes	20—40 (50)	
Submerse cormophytes	2—6	5—25
Seaweeds	3—10	1—20 (30)
Planktonic algae		2—3

[a] To allow comparison of photosynthetic capacity of different plant types, the photosynthetic rates are normalized per unit surface area. The surface area is that area receiving radiation, not the total area of upper and lower surface.

[b] The photosynthetic rate per unit dry weight of leaf; this number can be used to calculate the length of time required for a leaf to acquire the carbon necessary to form another leaf of a given mass.

usually high rates of photosynthesis (94 mg $CO_2 \cdot dm_1^{-2} \cdot h^{-1}$), the winter-annual desert plant *Camissonia claviformis* of Death Valley (Mooney et al., 1976); this plant contains extraordinarily large concentrations of RuBP carboxylase, and the stomatal diffusion resistance with stomata open is especially low. Most C_3 plants photosynthesize at no more than half that rate. In this group, *agricultural crops* take the lead; the high rates of photosynthesis in these plants are due in great degree to successful breeding. Among these crops rice, wheat, potatoes, and the sunflower show particularly high performance. The *thallophytes* are at the bottom of the list. The C_4 plants take up 30 times as much CO_2 as do mosses, lichens, and algae, and they assimilate CO_2 with almost double the yield of most of the agricultural plants. Among the vascular plants, herbs outperform woody plants, and in both the herbaceous and woody plants those forms adapted to shade manage only half to a third of the carbon utilization of plants growing in sunny locations. Species with assimilation organs of small surface area, like grasses with rolled leaves, dwarf ericaceous shrubs with grooved leaves, the needles of conifers, shrubs with assimilating shoots, and succulents, capture little of the incident light and therefore photosynthesize at only moderate rates. Leafless, suberized *shoot axes* can assimilate CO_2 in the cortical chlorenchyma and even in chloroplast-containing woody elements. The yield of such assimilation is very low, and plays a role at most in compensating for some of the respiratory losses of the twigs. However, in plants that are without leaves for considerable periods — for example, winter-deciduous trees and shrubs of the temperate zone and seasonally deciduous trees and shrubs of semi-arid regions — this form of photosynthesis may have ecological significance.

The *aquatic plants* form a distinct group with a surprisingly small capacity to bind CO_2. Even if one disregards the values for planktonic algae, which are difficult to apply comparatively, a clear deficiency in photosynthetic capacity remains — even with respect to the submerse vascular plants. A primary reason for this is thought to be that sessile aquatic plants are not as well supplied with CO_2 as land plants. It is true that fresh water contains about 160 times as much CO_2 as the air; but the rate of diffusion of CO_2 in water is only about 10^{-4} that in air, so that the CO_2 supply underwater at the surface of the leaves is restricted.

The differences in photosynthetic capacity among species and varieties are of the greatest importance as a basis for the selective breeding and cultivation of plants valuable in agriculture, gardening and forestry. The causes of the specific differences in net photosynthetic capacity are found in the effectiveness of the enzymes and the amount of photorespiration, as well as in anatomical peculiarities of the leaf structure, the ease with which air passes through the intercellular system, and the shape and distribution of the stomata.

3.2.2.2 The Specific Activity of Mitochondrial Respiration

The respiratory activity in a plant species differs from one organ to another, changing with the availability of respirable substrate, with the state of development and activity, and with the temperature. To facilitate comparison, the *specific respiratory activity* is expressed as the rate of respiration measured in the dark at a *standard temperature*, usually 20° or 25° C (Table 3.5). Herbaceous species, especially those with a rapid

Table 3.5. Respiration of mature leaves in the dark in summer at 20° C. From the measurements of numerous authors

Plant group	CO_2 release $mg \cdot (g W_d \cdot h)^{-1}$
Crop plants	3—8
Wild herbs	
Heliophytes	5—8
Sciophytes	2—5
Winter-deciduous foliage trees	
Sun leaves	3—4
Shade leaves	1—2
Evergreen foliage trees	
Sun leaves	ca. 0.7
Shade leaves	ca. 0.3
Evergreen conifers	
Sun needles	ca. 1
Shade needles	ca. 0.2
Ericaceous dwarf shrubs	
Winter-deciduous	2—3
Evergreen	0.5—1.5
Desert shrubs	1—3

growth rate, respire twice as rapidly as the foliage of deciduous trees under the same conditions, and the latter in turn respire on the average at five times the rate of the assimilation organs of evergreens. Within a given group, heliophytes respire distinctly more rapidly than sciophytes at 20° C, and plants of the arctic and high mountains respire more strongly than those of warmer regions and valleys. The respiratory activity of submerse cormophytes is in the range 1.4—3 g $CO_2 \cdot g$ $W_d \cdot h^{-1}$, and that of seaweeds is 0.5—5 g $CO_2 \cdot g$ $W_d \cdot h^{-1}$. Different species of plants exhibit characteristic differences in respiratory activity; ratios may be of the order of 1 : 10 to 1 : 20. In a sin-

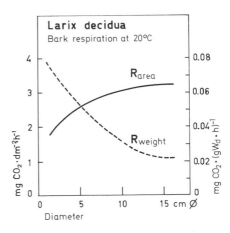

Fig. 3.14. Respiration of larch branches and trunks of different thicknesses. As the diameter increases the proportion of nonrespiring woody material rises, so that the value for respiration per unit dry weight (R_{weight}) decreases with increasing diameter; the CO_2 given off per unit surface area of shoot (R_{area}) increases only slightly from an intermediate thickness on. After Tranquillini and Schütz (1970); see also Müller and Nielsen (1965) and Yoda (1967)

gle plant, *flowers* and unripe *fruits* respire at a greater rate than leaves, and roots more rapidly than the axial parts of the shoots. In *branches and tree trunks* the primary respiratory regions are the bark, the cambium, and the outermost cell layers of the woody tissue. When expressed with respect to the periderm area (gas exchange per unit surface area), the respiratory activity measured at a given temperature in twigs, branches and tree trunks of different thicknesses increases with diameter (Fig. 3.14); when expressed with respect to weight (exchange per unit mass) it decreases, because the woody axes of shoots contain much nonrespiring material, the proportion of which steadily increases with increasing diameter. In studies of plant organs with a large and highly variable content of dead supporting material, it is advisable to express respiratory activity with respect to the content of protein or protein nitrogen.

3.2.2.3 The Photosynthetic Efficiency Coefficient

The net yield of photosynthesis is greatest when high photosynthetic capacity is paired with moderate respiration. This situation can be quantified in terms of the coefficient

$$k_F = \frac{F_g}{R} \doteq \frac{F_n + R}{R} \tag{3.8}$$

where, as before, F_g = gross photosynthesis, F_n = net photosynthesis, and R is total respiration. The *photosynthetic efficiency coefficient* k_F indicates how much the leaf organs must divert from the overall photosynthetic intake for their own respiration, under a given condition. Since F_g is not directly measurable, it is customary to approximate it (as in Eq. 3.8) by the sum of net photosynthesis and respiration (at a corresponding temperature).

k_F is useful in comparing the yield obtained in the gas exchange process in different plants. The leaves of vascular plants can bind at most 10—20 times as much CO_2 as they release by *simultaneous dark respiration,* and algae bind up to 25 times as much. However, when *photorespiration* is considered, it becomes evident that C_3 plants are actually much less efficient; k_F falls to 3—5. The C_4 plants, which have no apparent photorespiration, thus again perform best. Photosynthetic performance is determined not only by the yield of the process of photosynthesis, but also by the consumption of these products during respiration. The situation is particularly unfavorable in conifer needles and lichens, both of which contain in their assimilation organs a large proportion of tissue that respires but is photosynthetically unproductive.

3.2.2.4 Influence of Developmental Stage and Activity State upon Respiration and Photosynthetic Capacity

Photosynthetic capacity and respiratory activity are characteristic of a plant species, but they are not constants. Within a given plant, gas-exchange behaviour changes in the course of development and with seasonal and even diurnal fluctuations in activity.

Respiration. Younger plants respire more rapidly than older plants. In the growing parts of plants, respiration is particularly prominent; for the extensive synthesis of new

98

Fig. 3.15. Respiration of leaves of deciduous trees (*dashed line*) and evergreen woody plants (*stippled area*) as a function of the state of development of the vegetation. Based on the data of Eberhardt (1955), Pisek and Winkler (1958), Neuwirth (1959), Larcher (1961), Negisi (1966), E. D. Schulze (1970), and Ledig et al. (1976). Examples of development-dependent changes in respiratory activity of grasses are given by André et al. (1978), Koh et al. (1978a), and Jones et al. (1978)

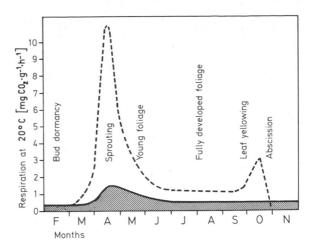

tissue, ATP is hydrolyzed at a rate far exceeding normal operating requirements. A feedback mechanism enables the cell to speed up respiratory ATP formation according to demand. In seedlings, at the tips of roots, during leafing-out and in developing fruits, *construction respiration* (the respiration that supports synthesis) amounts to between three and ten times the *maintenance respiration*. The intensity of construction respiration is proportional to the rate of growth. With increasing differentiation and maturation of the tissues, respiratory activity returns to a much lower level (Fig. 3.15). The onset of the breakdown processes of aging in the leaves, and especially in the fruits of some plant species, may be presaged by a *transient sharp increase* ("climacteric rise") in respiration. This is a sign of altered metabolism, also recognizable in the discoloration of the foliage and the release of gaseous metabolic products (for example, ethylene) by fruits.

Photosynthetic capacity also changes continually in the course of development. In the *sprouting phase* photosynthetic activity is initially so slight that it is exceeded by the very intensive, simultaneous respiration associated with synthesis of new tissue. When the new shoots of woody plants are being extended, therefore, one can measure a net CO_2 release all day long in the light; the young shoots cannot maintain themselves and must be supplied with carbohydrates from older parts of the plant (Fig. 3.16 and 3.19). As the leaves unfold, however, the capacity for intensive CO_2 fixation develops rapidly. Young, fully developed foliage is at a peak in this respect, and only days or weeks later photosynthetic capacity begins to decline, falling steadily as the plant grows older. This *decline in performance* occurs more rapidly in assimilation organs that are functional for only one season, whereas in foliage that lasts for several years it is a slower, interrupted process. In herbaceous plants that develop rapidly the individual leaves of the shoot may differ considerably, depending on their level of development and state of differentiation; the same is true of leaves produced at different times of the year (Fig. 3.17). In evergreen plants from regions with a cold season, photosynthetic capacity falls off after every winter and each time new leaves are formed (Fig. 3.16). If the formation of new leaves is omitted, or if the new growth is removed, the life span of the old

99

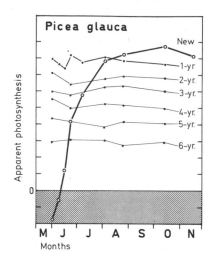

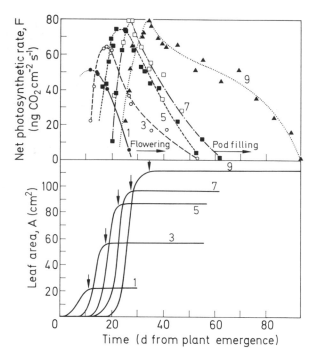

Fig. 3.16. Dependence of photosynthetic capacity of spruce upon the age of the needles. The newly-sprouted needles develop their full photosynthetic capacity only gradually; during the expansion phase the rate of photosynthesis is low and respiration is much increased, and as a result they give off CO_2 even in the light (stippled area = CO_2 release). In subsequent years the photosynthetic capacity of the needles sinks a little lower after each winter that the needle survives. After Clark (1961). Photosynthesis parameters of developing poplar leaves: Ceulemans and Impens (1979)

Fig. 3.17. Net photosynthesis and growth in area of the first, third, fifth, seventh and ninth leaf on the main stem of soybean plants (*Glycine max*), in relation to the age of the plant. After Woodward (1976). For the photosynthetic gradient along a vertical sequence of leaves of *Phragmites* see Gloser (1977); of *Miscanthus*, Sawada and Iwaki (1978)

leaves is extended and the process of aging, with respect to photosynthetic capacity, is slowed.

The dependence of photosynthetic capacity on the stage of development has several causes. Very young leaves have not yet acquired their full surface area and capture correspondingly less light; furthermore, they are usually low in chlorophyll and respire at a high rate. The chief effect, however, is due to changes in the activity level of the enzymes during development (Fig. 3.18) and to shifts in the distribution of photosynthates,

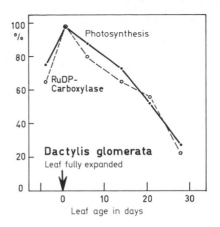

Fig. 3.18. Decline of photosynthetic capacity and activity of RuBP carboxylase with increasing age of the leaves in orchard grass. After Treharne and Eagles (1970). Examples of development-dependent changes in photosynthetic capacity of rice are given by Tanaka and Kawano (1966) and Lerch (1976); of wheat, by Koh et al. (1978a); of maize, by André et al. (1978); of potato, by Mokronosov and Nekrasova (1977); of understory herbs, by Goryshina (1969). Reviews: Šesták (1977, 1978), Zima and Šesták (1979)

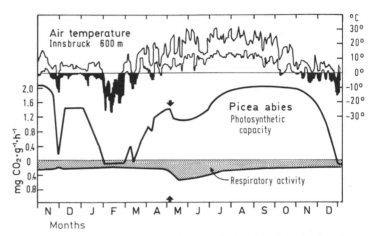

Fig. 3.19. Photosynthetic capacity and respiratory activity of the needle-bearing tips of twigs of a mature spruce. The rate of respiration is shown as the *stippled band*, increasing downward. The twigs were gathered outdoors and investigated in the laboratory under standard conditions (photosynthesis at 12° C and 10 klx, respiration in the dark at 20° C). The effect of outdoor air temperature persists in the laboratory; photosynthetic capacity is reduced as soon as the daily minima of the air temperature fall regularly below 0° C, and if the temperature maxima also remain below freezing, net photosynthesis is completely suspended. When the new shoots appear (*arrows*) there is a transitory decline in net photosynthetic capacity as a result of the increased respiration of the growing tips of the twigs. After Pisek and Winkler (1958). For further examples see Parker (1961), McGregor and Kramer (1963), Neilson et al. (1972), and Strain et al. (1976). For late-season photosynthesis of intensively cultured *Populus* clones see Nelson et al. (1982)

both of which in turn are controlled by phytohormones. For this reason photosynthetic capacity is often particularly high in the generative period of life — during the *flowering phase* and the *formation of fruit*. Shortly before the plant dies back or the leaves fall, photosynthetic capacity collapses entirely for the additional reason that chlorophyll is broken down or, in grasses for example, the stomata stiffen.

Photosynthetic capacity and respiratory activity also change with the alternation between the active and dormant periods of plants. During the period of *winter dormancy* the photosynthetic capacity can fall to zero for weeks, with a simultaneous decrease in respiratory activity — an indication of the relative suspension of overall metabolism of the plant (Fig. 3.19). In regions with a mild winter the reduction of photosynthesis is not so dramatic, but it is distinctly depressed.

3.2.3 The Effect of External Factors on CO_2 Exchange

CO_2 exchange is influenced by a number of external factors. As a photochemical process, photosynthesis is of course directly dependent upon the availability of radiation. The dark reactions of photosynthesis and respiration are purely biochemical processes, limited especially by temperature and the supply of CO_2.

3.2.3.1 The Dependence of Net Photosynthesis on Light

The Light-Dependence Curve

If leaves or suspensions of algae are exposed to increasing intensities of illumination, the CO_2 uptake increases at first in proportion to light intensity and then more slowly to a maximum value. That is, the relationship between net photosynthesis and radiation is represented by a *saturating curve*. This light-dependence curve in dim light reflects a net release of CO_2, since more CO_2 is given off by respiration than is fixed by photosynthesis (Fig. 3.20). At somewhat greater intensities the light compensation point is reached. At the *compensation light intensity* I_K photosynthesis fixes exactly as much CO_2 as is set free by respiration. Plants that respire rapidly thus require more light for compensation than do those with slower respiration rates. Once the compensation point has been passed, CO_2 uptake increases rapidly. In the lower range of this increase there is a strict *proportionality* between the yield of photosynthesis and the available radiation. The speed of the light reactions is the limiting factor for the overall process in this range; the greater the quantum yield Φ, the steeper the rise of the light-dependence curve in the range of proportionality. With very high light intensity the yield of photosynthesis continues to increase only slightly or not at all; the reaction is *light-saturated* at this point (I_S) and the rate of CO_2 uptake is now limited not by photochemical but rather by enzymatic processes, and by the supply of CO_2.
A comparison of the light-dependence curves of different plant species (Fig. 3.21) shows that the species that fix CO_2 via the dicarboxylic acid pathway stand out. C_4 *plants* such as millet and maize are not light-saturated even at the highest intensities, and even at intermediate irradiance they operate more efficiently than the C_3 plants. Evidently PEP carboxylase, even at the strongest light intensities applied here, is capable of keeping pace with the light reaction. The Calvin cycle of the C_3 *plants* is much less efficient, and thus the light-dependence curves saturate at lower intensities. There are also plants in which photosynthetic performance falls off under excessive illumination, so that the curve shows an intensity optimum. Most of these are cryptogams, but tree seedlings, plants growing in the underbrush of dense woods, and cultivated

Fig. 3.20. Dependence of CO_2 exchange upon light intensity, in sun and shade leaves of the beech. Measurements were made at 30° C. The region of weak light (the section enclosed by the *box in the upper drawing*) is shown in the lower drawing with the abscissa expanded. Leaves adapted to shade respire at a lower rate than those adapted to light; they reach the light compensation point (I_K) at a lower intensity. Moreover, in the region of the curve between I_K and I_S, they utilize the light more efficiently, but I_S is at lower intensity. After Retter (1965); see also the classical study of Boysen-Jensen and Müller (1929) and the review by Boardman (1977) and Björkman (1981)

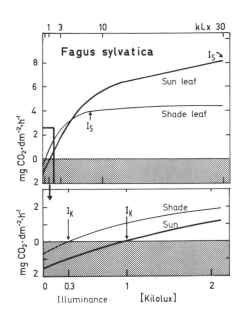

Fig. 3.21. Light-dependence of net photosynthesis in various plants at optimal temperature and with the natural supply of CO_2. After Stålfelt (1937), Böhning and Burnside (1956), Gessner (1959), Retter (1965), Stoy (1965), Hesketh and Baker (1967), Ludlow and Wilson (1971 a, b, c), Dawes et al. (1978) and the results of numerous other authors

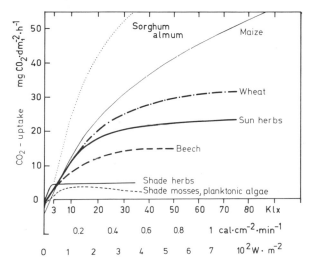

plants sensitive to intense light (for example, coffee, cocoa, and various house plants) also exhibit this behavior.

The positions of the cardinal points I_K and I_S reflect the light conditions in the natural habitat of the plants and characterize the different kinds of plants (Table 3.6 and Fig. 3.20). Leaves adapted to shade respire less than sun leaves and therefore compensate at considerably lower light intensities. In general, the I_K of shade plants lies at 0.5—1% of full sunlight. Shade leaves furthermore utilize weak light better than do the sun leaves and reach their light-saturation point at very low intensities, near 10 klx (about 100

103

Table 3.6. Light-dependence of net photosynthesis of single leaves, under conditions of natural CO_2 availability and optimal temperature. From the measurements of numerous authors

Plant group	Compensation light intensity I_K, in klx	Light saturation I_S, in klx
A. Land plants		
1. Herbaceous plants		
C$_4$ plants	1—3	over 80
Agricultural C$_3$ plants	1—2	30—80
Herbaceous heliophytes	1—2	50—80
Herbaceous sciophytes	0.2—0.5	5—10
2. Woody plants		
Winter-deciduous foliage trees and shrubs		
Sun leaves	1—1.5	25—50
Shade leaves	0.3—0.6	10—15
Evergreen foliage trees and conifers		
Sun leaves	0.5—1.5	20—50
Shade leaves	0.1—0.3	5—10
3. Understory ferns	0.1—0.5	2—10
4. Mosses and lichens	0.4—2	10—20
B. Water plants		
Planktonic algae		(7) 15—20
Tidal-zone seaweeds	1—2	10—20
Deep-water algae		1—2
Phanerogams	<1—2	(5) 10—30

$W \cdot m^{-2}$). Heliophytes, inferior to sciophytes in twilight, make better use of bright light and thus produce a significantly higher photosynthetic yield. Agriculturally important plants, which should produce as large yields as possible, must therefore be heliophytes.

Short-term adaptation to changes in habitat light level, such as occur when previously shaded understory plants are exposed by opening of the tree canopy (when trees are logged off, for example, or blown over), can occur within one to several weeks by *modulative* adaptation of photosynthesis and respiration (Fig. 3.22). However, adaptation is primarily *modificative*, in accordance with the level of available radiation while the assimilation organs are being initiated and differentiated (Fig. 3.23). In this process there arise the features of plant morphology, cellular and subcellular structure, and biochemical activity that determine the characteristic properties of CO_2 exchange in strong-light and dim-light forms. The extent and rate of all these processes of adaptation depend on the *stage of development* (young plants are especially capable of adapting to shade) and are *genetically* programmed. Species and ecotypes evolutively adapted to habitats in full sun develop the photosynthetic capacity and activity typical of strong-light plants only under high relative irradiance. Sciophytes are capable of developing forms adapted to extremely low light levels; on the other hand, under high irra-

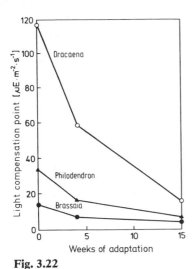

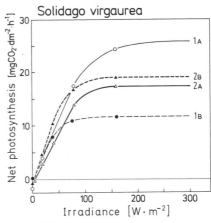

Fig. 3.22   Fig. 3.23

Fig. 3.22. Shifting of the light compensation point in house plants of tropical origin during adaptation to shade (27 $\mu E \cdot m^{-2} \cdot s^{-1}$, or 1600 lux). After 15 days of adaptation dark respiration is reduced by 50%—70% and there is somewhat better photosynthetic utilization of the little light available. After Fonteno and McWilliams (1978). Adaptive shifts in the light compensation point of gametophytes and sporophytes of a tropical tree fern are described by Friend (1975)

Fig. 3.23. Light response curves for net photosynthesis of *Solidago aurea* ecotypes from exposed (*1*) and shaded (*2*) habitats grown under high (*A*, 150 W $\cdot$ m^{-2}) and low (*B*, 30 W $\cdot$ m^{-2}) irradiance. After Björkman and Holmgren (1963). Response of *Solanum dulcamara* ecotypes to varying light intensities connected with water shortage: Gauhl (1979). Experiments on light adaptation of photosynthesis in crop plants have been done, for example, by Burnside and Böhning (1957), Hiroi and Monsi (1966), and Louwerse and Zweerde (1977). See also the classical studies of Harder (1923) and Myers (1946)

diation their utilization of the available light is poor. However, shade plants that would undergo a depression of photosynthesis when suddenly exposed to strong light can adapt to gradually increasing light, at least to the extent that they are not damaged by fairly high light intensities.

Light-Dependence in the Natural Habitat

The variability in light dependence of photosynthesis among different types of plants can also be seen under the conditions prevailing in the natural habitat. If gas exchange is not restricted by other local environmental factors such as the water supply and the temperature, net photosynthesis parallels light availability up to the saturation region. In the case of C_4 plants, this means that they can make full use of the light at noon on a clear day (Figs. 3.24 and 3.25). In C_3 plants the increase of photosynthetic activity ceases at the time of day when the irradiation exceeds I_s. Bright passing clouds have little effect on the rate of photosynthesis of heliophytes, but there is an effect of the more

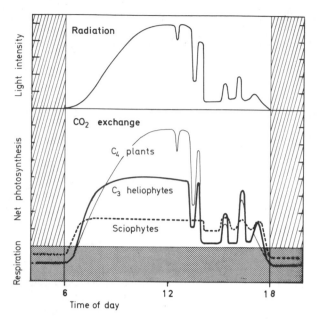

Fig. 3.24. Schematic diagram of the daily fluctuation in CO_2 exchange as a function of the available radiation. *C_4 plants* can utilize even the most intense illumination for photosynthesis, and their CO_2 uptake follows closely the changes in radiation intensity. In *C_3 plants* photosynthesis becomes light-saturated sooner, so that strong irradiation is not completely utilized. *Sciophytes*, adapted to utilization of dim light, take up more CO_2 in the early morning and late evening, as well as during periods when the sun is obscured, than do the heliophytes; but the former do not utilize bright light as efficiently

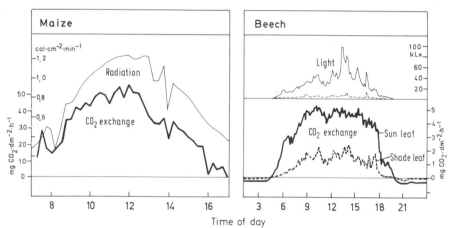

Fig. 3.25. Variations in net photosynthesis measured throughout the day as a function of the available radiation. In maize (a C_4 plant), photosynthesis follows the daily fluctuation in radiation. In sun leaves of the beech, F_n follows the changing illumination only up to about 50 klx, and in shade leaves, F_n undergoes short-term fluctuations associated with the brief fluctuations in brightness. After Hesketh and Baker (1967) and E. D. Schulze (1970). Examples of light-dependent diurnal CO_2 uptake by understory plants in tropical forests are given by Lemée (1956), Odum and Pigeon (1970), and Björkman et al. (1972)

marked fluctuations in illumination caused by variable cloud cover. The sciophytes on the woodland floor and the shade leaves in the interior of tree crowns are affected by the variations in the sunlight penetrating the foliage only if illumination stays below about 10 klx. However, little is lost by their inability to follow at higher intensities, for beneath

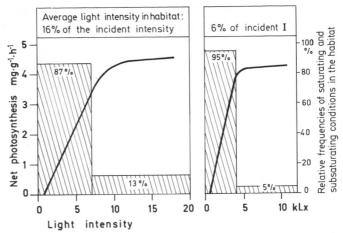

Fig. 3.26. Light-dependence of net photosynthesis of oak saplings in the undergrowth of a wood. *Left*, plants that have developed at an average relative intensity of 16%. *Right*, plants adapted to deeper shade, having grown at an average relative intensity of 6%. The graphs indicate that under the tree canopy the undergrowth is provided with saturating light during only 5% or 13%, respectively, of all daylight hours. During the remaining hours of the day the light intensity is in the range of proportionality of the light-dependence curve; in this region the available radiation is fully utilized. After Malkina et al. (1970); cf. also Tselniker (1978, 1979)

a moderately dense tree canopy (average available radiation in the herbaceous layer is 6−16%) the light intensity is above their saturation point for only about $^1/_{10}$ of the time during the daylight hours (Fig. 3.26).

Light-Dependence of Photosynthesis Within a Stand of Plants

Observations of single leaves could lead to the mistaken inference that there tends to be a surplus of light. However, for the plant as a whole and for stands of plants this is not so. It is true that the individual leaves of a plant are often arranged so as to favor interception of the strongest average light, but leaf orientation is seldom perpendicular to the direct incident radiation from the sun. In the course of the day the leaves of a plant are struck by light at many different angles, and only rarely are they exposed to the full incident radiation; moreover, the leaves shade one another.

In a stand of plants, the contribution of the various layers of foliage to the overall photosynthetic yield is quite different, depending on the arrangement and the amount of shading of one layer by another (Fig. 3.27). In the morning, in a field of alfalfa 30 cm high, the light compensation point is exceeded two hours later in the lowest layer of leaves than in the top layer of leaves, and even under strong sunlight the bottom layer displays only 3% of the photosynthetic activity of fully illuminated leaves. Even the layer of leaves at half height (10−20 cm above the ground) attains little more than 10% of full performance (Fig. 3.28). The situation is similar within the dense crowns of trees and beneath the tree canopy of a forest. There, too, light saturation of photosynthesis is possible only in the outermost and uppermost regions, while within the crown net

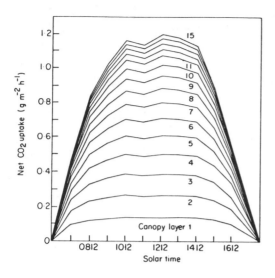

Fig. 3.27. Simulated net photosynthesis at 60% diffusive radiation for each of the 15 layers of a *Goethalsia* forest in Costa Rica. Layer *1* is at the *top* of the canopy and Layer *15*, at the *bottom*. The area between each pair of curves represents the simulated photosynthesis of the indicated (*upper*) layer (Allen and Lemon, 1976). For stratified measurements of photosynthesis in crowns and stands of trees see Neuwirth (1968), Helms (1965, 1970), Woodman (1971), Schulze et al. (1977), and Malkina (1978)

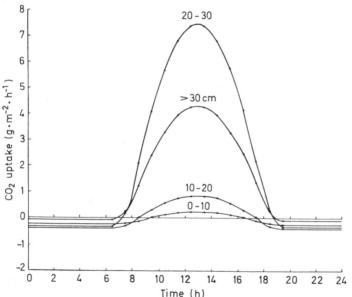

Fig. 3.28. Computed net photosynthesis of various leaf layers in a stand of the alfalfa *Stylosanthes humilis* on a clear day. *Numbers* indicate height above soil (Begg and Jarvis, 1968)

photosynthesis at noon falls to 15% or less of the possible value. The compensation point in such shaded regions is reached only in late morning, and the illumination falls below it again several hours before sunset. The photosynthetic performance of a stand of plants as a whole rises gradually with increasing irradiation, and continues to increase even after the light-saturation points of the outer leaves have been exceeded (Fig. 3.29).

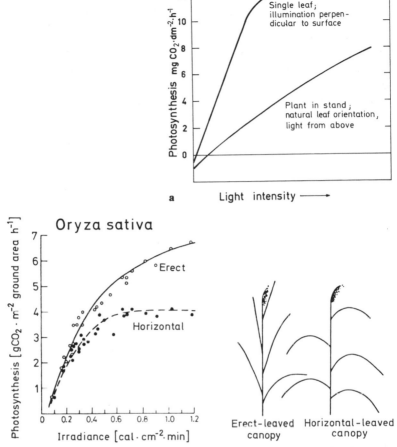

Fig. 3.29a and b. Net photosynthesis of grain plants in a stand. **a** Comparison of the photosynthesis of oat leaves receiving radiation incident perpendicular to the surface with those at the natural angle to the sunlight, under increasing irradiation. Within a stand, because of the angle at which light strikes the leaf surface and because of the shading of the leaves by one another, light-saturation is not reached even under strong irradiation. After Boysen-Jensen (1932). Data on single leaf and canopy photosynthesis in a ryegrass sward are given by Woledge and Leafe (1976). **b** Comparison of the rates of photosynthesis (per m² covered by the stand) of rice plants with erect and with horizontal leaves. After Tanaka as cited by Monsi et al. (1973); for photosynthesis of plant stands with different foliage angles see Kuroiwa (1978)

Photosynthesis by Planktonic Algae

In bodies of water, within the euphotic layer colonized by phytoplankton, there is a characteristic gradual variation of photosynthetic activity with depth — even though the illumination gradient is approximately exponential (Fig. 3.30). The photosynthetic

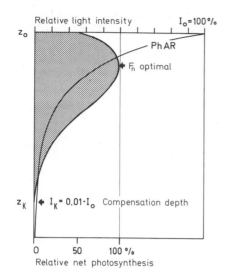

Relative light intensity $I_0 = 100\%$

PhAR

F_n optimal

$I_K = 0.01 \cdot I_0$ Compensation depth

0   50   100 %

Relative net photosynthesis

Fig. 3.30. Photosynthetic activity of phytoplankton and relative light intensity, as a function of depth under water. At the water surface (z_0) photosynthesis is supersaturated with respect to light when the sun is high, and therefore is somewhat reduced; a few centimeters to decimeters beneath the surface the illumination becomes optimal; and at the compensation depth z_K, where the light intensity is only 1% of that at z_0, net photosynthesis falls to zero. After Talling (1970). The depth gradient of the photosynthetic efficiency coefficient k_F in tropical-ocean algae is given by Buesa (1977)

depth profile is associated with the shape of the light-dependence curve for algae, which instead of saturating passes through an optimum and then falls (cf. Fig. 3.21). The highest rates of photosynthesis when the sun is high are thus measured not directly below the surface of the water but somewhat deeper, between 2 and 15 m (2—5 m in lakes, 10—15 m in the open ocean); the effect of course depends on the intensity of the light and the turbidity of the water. On overcast days and during the seasons of diminished radiation the light does not reach above-optimal intensities, and the region of maximal photosynthesis is shifted up to the water surface. At greater depths, photosynthetic activity declines until eventually it is just able to compensate for respiration. Compensation as a rule is found at a depth (the *compensation depth*) at which not more than 1% of the surface radiation penetrates.

3.2.3.2 Temperature-Dependence of Photosynthesis and Respiration

The Influence of Temperature on Reaction Rate

Temperature affects metabolic processes by way of its influence on the reaction kinetics of chemical events and on the effectiveness of the various enzymes involved. In general Van't Hoff's *reaction-rate/temperature rule* holds, according to which reaction rate k rises exponentially with temperature. The increase in reaction rate that results from a temperature increase of 10° C is expressed by the temperature coefficient Q_{10}, given approximately by

$$\ln Q_{10} = \frac{10}{T_2 - T_1} \ln \frac{k_2}{k_1} \tag{3.9}$$

where T_2 and T_1 are two absolute temperatures, and k_2 and k_1 are the associated reaction rates. Because of the exponential relationship between k and T, the ratio k_2/k_1, and thus the temperature coefficient, is fairly constant over a small range of tem-

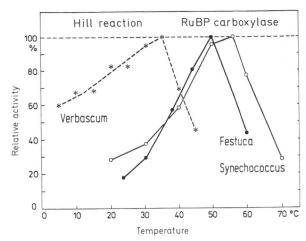

Fig. 3.31. The influence of temperature on noncyclic electron transport (Hill reaction) in chloroplasts of *Verbascum thapsus* (Williams III et al., 1975) and on the activity of the RuBP carboxylase of *Festuca arundinacea* (Treharne and Nelson, 1975), and of a high-temperature clone of the thermophilic blue-green alga *Synechococcus lividus* (Sheridan and Ulik, 1976). The light-driven reactions (photolysis, electron transport, phosphorylation) reach a maximum at lower temperatures than does the activity of RuBP carboxylase, the key enzyme in the secondary processes of photosynthesis, and they exhibit heat injury at lower temperatures than the enzyme

peratures: 1.4–2.0 for most enzyme reactions and 1.03–1.3 for physical processes. When the influence of temperature over a larger range is of interest, it should be kept in mind that the Q_{10} of metabolic processes varies with temperature. At low temperatures it is large, for as a rule enzyme reactions are rate-limiting, whereas at high temperatures it becomes smaller because physical processes such as diffusion velocity become limiting.

Temperature-Dependence of Carbon Incorporation

Temperature affects photosynthesis by way of the secondary processes (Fig. 3.31); the photochemical process is nearly independent of temperature. The fixation and reduction of carbon dioxide occur with increasing speed as the temperature rises, until a maximum value is reached; this rate is then maintained over a broad range of temperatures. Not until it becomes quite hot, so that membrane-bound light reactions are slowed and the interaction among the different reactions is disturbed, does photosynthesis rapidly break down. The CO_2 fixation and malate formation that occur by night in CAM plants also follow a curve with a maximum (Fig. 3.35).

Temperature-Dependence of Respiration

As the temperature rises dark *respiration* increases exponentially. Below 5° C the energy of activation for the various processes involved in respiration is large, and the Q_{10} is high. In tropical plants the Q_{10} is around 3, or even more, at temperatures below

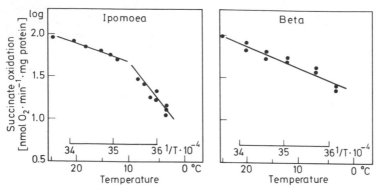

Fig. 3.32. Arrhenius plots for the metabolic activity of mitochondria of chilling-sensitive *Ipomaea* tubers and chilling-resistant beet roots (Lyons and Raison, 1970)

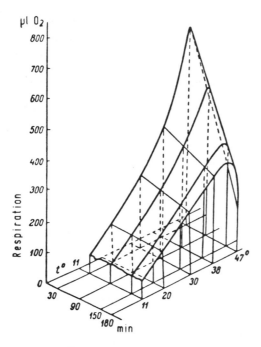

Fig. 3.33. Temperature-dependence of dark respiration in the leaves of *Podophyllum peltatum*. After Semikhatova (1974)

10° C (Fig. 3.32). Above 25°–30° C the temperature coefficient for respiration in most plants falls to 1.5 or less. At still higher temperatures biochemical processes occur so rapidly that the supply of substrate and metabolites (e.g., ADP) cannot keep pace with the turnover of matter and energy. The rate of respiration thus declines in a short time (Fig. 3.33). At temperatures between 50° and 60° C enzymes and functionally important membrane structures are damaged by heat, and respiration ceases.

Few studies have been done on the temperature-dependence of *photorespiration*. In principle it is to be expected that photorespiration, which relies directly upon photosynthesis for the provision of substrate, depends on temperature in the same way as those processes.

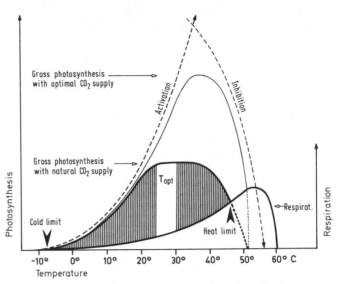

Fig. 3.34. Diagram of the temperature dependence of photosynthesis and respiration. Gross photosynthesis increases as a result of thermal activation of the enzymes involved, until inhibitory effects lead to a decline in photosynthetic activity. The difference between gross photosynthesis and respiration gives the net photosynthesis (*cross-hatched area*); note the corresponding positions of the cold limit, the temperature optimum (T_{opt}), and the heat limit for net photosynthesis (original). For simulation models see Tenhunen and Westrin (1979)

Limiting Temperatures and the Temperature Optimum for Net Photosynthesis

The temperature-dependence of net gas exchange results from the difference between the rate of photosynthetic CO_2 incorporation and the rate of respiration prevailing at a given temperature (Fig. 3.34). Net photosynthesis is measured over a range in which increasing temperature has a stimulative effect, and over another in which the effect is inhibitory. These regions are defined by three *cardinal points*: the cold limit or the temperature minimum (T_{min}) for net photosynthesis, the temperature optimum (T_{opt}), and the heat limit or the temperature maximum (T_{max}) for net photosynthesis (Figs. 3.34 and 3.35).

Optimal Range of Temperature. That range of temperature in which net photosynthesis is more than 90% of the maximum obtainable can be regarded as optimal. The temperature optimum for *net* photosynthesis is narrower than the optimal temperature span for CO_2 fixation; that is, while *gross* photosynthesis is still operating at top speed, the rate of respiration steadily increases, diminishing the net photosynthetic yield.

For many C_4 *plants* the optimum temperatures lie between 30° and 40° C, and in some cases are as high as 50° C (Table 3.7). The C_4 pathway of carbon assimilation, then, is the genotypic prerequisite and an ecological advantage for colonization of hot habitats. There are, however, also C_4 plants with low temperature optima for net photosynthesis — for example, varieties of maize grown in the temperate zone and certain cool-adapted

113

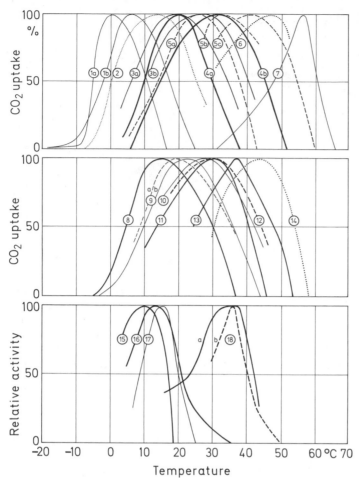

Fig. 3.35. Examples of the temperature-dependence of net photosynthesis in thallophytes and herbaceous tracheophytes (*top graph*) and in woody plants (*middle*), and of the dark fixation of CO_2 and the photosynthetic primary processes in CAM plants (*bottom*).

1, Lichens and mosses; *a, Lecanora melanophthalma*, an Antarctic lichen (Lange and Kappen, 1972); *b, Cetraria nivalis*, a subarctic-alpine lichen (Kallio and Heinonen, 1971); the subarctic tundra moss *Rhacomitrium lanuginosum* is similar to *b* but with the cold limit at −12° C (Kallio and Heinonen, 1973). *2, Oxyria digyna*, an arctic-alpine tracheophyte found in the central Alps at 2500 m (Pisek et al., 1969); arctic ecotypes have a somewhat higher temperature optimum (Billings et al., 1971). *3*, C_3 grain crops: *a*, wheat (Sawada and Miyachi, 1974; Vong and Murata, 1977); *b*, rice (Vong and Murata, 1977). *4*, dicotyledonous crops: *a*, potato (Lundegårdh, 1927; Winkler, 1961); *b*, soybean (Ludlow and Wilson, 1971a, b). *5*, C_4 grain crops: *a* maize varieties for cool climates (Winkler et al., 1975); *b*, maize varieties for warmer regions (Moss, 1963); *c, Sorghum vulgare*, Arizona (El-Sharkawy and Hesketh, 1964). *6, Tidestromia oblongifolia*, a summer-annual C_4 desert plant, Death Valley (Pearcy et al., 1971). *7, Synechococcus lividus*, a thermophilic blue-green alga, high-temperature clone (Sheridan and Ulik, 1976).

Woody plants: *8, Pinus cembra*, central Alps, tree line 1850 m (Pisek and Winkler, 1959); *9, Fagus sylvatica*, central Europe, *a*, shade leaves, *b*, sun leaves (Retter, 1965); *10, Cassia fistula*,

Table 3.7. Temperature dependence of net photosynthesis under conditions of natural CO_2 availability and light saturation. From the original publications of many authors

Plant group	Low-temperature limit for CO_2 uptake (°C)	Temperature optimum of F_n (°C)	High-temperature limit for CO_2 uptake (°C)
Herbaceous flowering plants			
C_4 plants of hot habitats	+5 to 7	35—45 (50)	50—60
Agricultural C_3 plants	−2 to ca. 0	20—30 (40)	40—50
Heliophytes (temperate zone)	−2 to 0	20—30	40—50
Sciophytes (temperate zone)	−2 to 0	10—20	ca. 40
Desert plants	−5 to 5	20—35 (45)	45—50 (56)
CAM plants (CO_2 fixation at night)	−2 to 0	5—15	25—30
Winter annuals, spring-flowering and alpine plants	−7 to −2	10—20	30—40
Woody plants			
Evergreen trees of the tropics and subtropics	0 to 5	25—30	45—50
Sclerophyllous trees and shrubs of dry regions	−5 to −1	15—35	42—55
Winter-deciduous trees of the temperate zone	−3 to −1	15—25	40—45
Evergreen conifers	−5 to −3	10—25	35—42
Dwarf shrubs of heath and tundra	ca. −3	15—25	40—45
Mangrove trees		25—30	ca. 40
Thallophytes			
Arctic and subarctic mosses	ca. −8	ca. 5	ca. 30
Lichens of cold regions	(−25) −15 to −10	5—15	20—30
Desert lichens	ca. −10	18—20	38—40
Tropical lichens	−2 to 0	ca. 20	
Snow algae	ca. −5	0—10	30
Thermophilic algae	20 to 30	45—55	65—70

drought-deciduous forest, Java (Stocker, 1935a); *11, Camellia sinensis*, tea plantation in Assam (Hadfield, 1975); *12, Ficus retusa*, rain forest in southern Japan (Kusumoto, 1957); *13, Acacia craspedocarpa*, arid bush in western Australia, in summer (Hellmuth, 1967); *14, Hammada scoparia*, a C_4-shrub with assimilating stems, Negev Desert, in summer (Schulze et al., 1976).

CAM plants, which accumulate malate at night (dark fixation of CO_2); *15, Echinocereus fendleri*, cactus steppe, Arizona, 1260 m above sea level (Dinger and Patten, 1972); *16, Ferocactus acanthodes*, Colorado desert (Nobel, 1977a); *17, Tillandsia recurvata*, epiphyte, Venezuela (Medina et al., 1977); *18*, isolated chloroplasts of *Opuntia polyacantha*, Colorado (Gerwick et al., 1977); *a*, temperature dependence of photosynthetic electron transport, measured by O_2 release; *b*, temperature dependence of photophosphorylation, measured by ATP formation. The optimal temperature for dark fixation of CO_2 in cacti is about 20° C lower than that for photosynthetic processing in the chloroplasts of the CO_2 liberated during the day

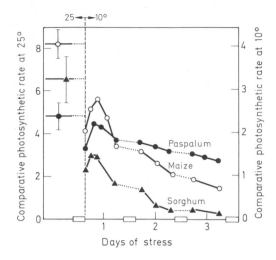

Fig. 3.36. Progressive depression of net photosynthesis in chilling-sensitive C_4 grasses kept for several days at 10° C. The symbols at the *left* indicate the initial levels at 25° C. After Taylor and Rowley (1971)

species of *Spartina* and *Atriplex*. Among the C_3 plants the optimum (like the other cardinal points) can be set anywhere within a wide range, depending on the ecological constitution of the plant. Sciophytes, which only occasionally encounter direct radiation and are warmed less than the plants in sunny locations, function optimally between 10° and 20° C; so too do spring-blooming and high alpine plants, which grow during a season or in a locality characterized by low average air temperature. Herbs adapted to sunny habitats and trees of warm climates, on the other hand, achieve their highest photosynthetic yields at 25°–30° C, and for desert shrubs the optimum is 35°–45° C. In *CAM plants*, temperature during the light phase appears to have little effect on photosynthesis over a broad range, whereas CO_2 fixation in the dark clearly is adjusted to the lower nighttime temperatures; the optimum range for the dark fixation of CO_2 usually lies between 5° and 15° C, and above 25°–30° C the process comes to a halt. It will come as no surprise that the *lichens,* which occupy the extreme outposts of plant life both in the high mountains and in the polar regions, are adapted to the cold climate of the places where they grow. But foliose and crustaceous lichens from warmer countries, and even those from hot deserts, also have optima at lower temperatures. This adaptation is also ecologically reasonable, for really productive assimilation occurs in these lichens only when they are well supplied with water — when they are wet with rain, dew or fog and when the humidity is high. These conditions normally prevail only when the sky is overcast or in the early morning.

The Low Temperature Limit for Net Photosynthesis. Tropical plants function productively only at temperatures of 5°–7° C or higher (Fig. 3.36), whereas plants of the temperate zones and cold regions assimilate CO_2 even at temperatures below 0° C. In the higher plants CO_2 uptake is blocked as soon as the assimilation organs begin to freeze; in most vascular plants during the growing season this occurs at temperatures around −1° to −3° C, though winter annuals, spring geophytes and alpine plants do not freeze until the temperature has reached −5° to −7° C — a range in which evergreen leaves and needles also freeze in winter. Many lichens behave differently. They can take

116

Fig. 3.37. After-effect of night frost on the time course and yield of net photosynthesis of pine twigs during the following day. After Polster and Fuchs (1963). For recovery of conifers after frost see Tranquillini (1957), Pisek and Kemnitzer (1968), Neilson et al. (1972), of wheat Koh et al. (1978b), of lichens Lange (1962), and Kallio and Heinonen (1971). Cold stress effects on CO_2-uptake are reviewed by Larcher (1981b) and Larcher and Bauer (1981)

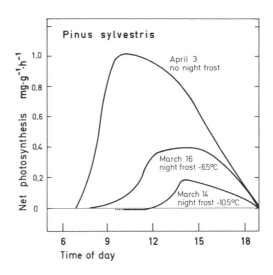

up CO_2 even when their thalli are partially frozen; some species have been shown to assimilate CO_2 down to $-25°$ C.

The High Temperature Limit for Net Photosynthesis. With very high temperatures, the rate of photosynthesis falls off sharply, and at the same time the intensified rate of respiration frees larger amounts of CO_2. The "temperature maximum for net photosynthesis" is the highest temperature at which all the CO_2 given off by respiration is reassimilated; if the temperature rises further, CO_2 begins to escape. One can therefore consider the high temperature limit for net photosynthesis as a compensation point (the heat compensation point), reached sooner the more heat-sensitive the photosynthesis and the more rapid the enhancement of respiration. Plants of high photosynthetic performance and those with slow respiration thus have an advantage; in C_4 plants of extremely hot habitats, heat compensation points have been measured at $58°$ and $60°$ C — just under the temperature causing death of the leaves.

After-Effects of Frost and Heat

As long as the temperature is extremely low or high, CO_2 uptake is entirely suppressed. Subsequently, as more favorable conditions reappear, it is only in rare cases that the plants recover immediately.

After freezing, some plants resume assimilation without delay, but this is by no means always the case. Photosynthesis begins slowly after the plants have thawed out. Not until several to many hours have passed is CO_2 uptake resumed. The lower the temperature and the longer the period of exposure, the more severe and prolonged is the setback. Repeated freezing has the same effect as more severe cold. The daily rhythm of *net photosynthesis after night frosts* is distinguished by a slower increase and lower peak value of CO_2 uptake, the lower the temperature in the preceding night (Fig. 3.37). A series of night frosts progressively restricts the period during the day that can be used for CO_2 uptake, and thus diminishes considerably the CO_2 uptake of the plants.

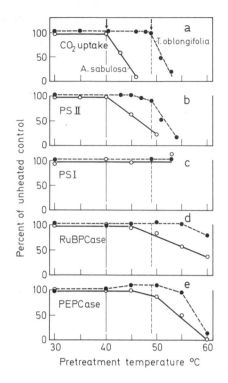

Fig. 3.38. a Depression of net photosynthesis in the C_4 plants *Atriplex sabulosa* (cold-adapted) and *Tidestromia oblongifolia* (heat-adapted) following exposure to heat. **b–e** Heat stability and impairment of subprocesses in photosynthesis. Note the differences in heat sensitivity of the chloroplast membrane reactions: Photosystem II (**b**), Photosystem I (**c**), RuBP carboxylase (**d**), and PEP carboxylase (**e**) (Björkman et al., 1976)

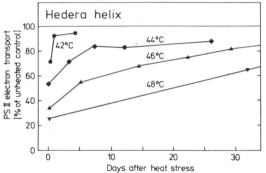

Fig. 3.39. Depression and recovery of heat-sensitive primary photosynthetic processes (electron transport through Photosystem II) in ivy leaves heated to various temperatures for 30 min. After Bauer and Senser (1979). A survey of studies on the after-effects of extreme temperatures on CO_2 exchange is given by Bauer et al. (1975) and Berry and Raison (1981)

The impairment of photosynthesis by *heat* also may last beyond the actual period of exposure. When the particularly heat-sensitive thylacoid structures are damaged, noncyclic phosphorylation and the function of Photosystem II are impeded until the structures are restored and their original performance recovered (Fig. 3.38). Depending on the species and the degree of heat exposure, this recovery can take from a few days to several weeks (Fig. 3.39).

Temperature Adaptations

The rates of photosynthesis and respiration adapt to the temperature prevailing at a given time.

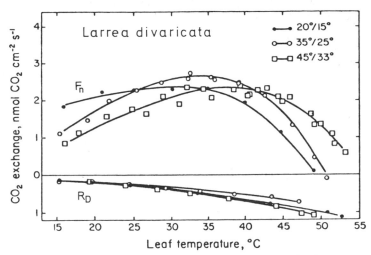

Fig. 3.40. Adaptation of photosynthesis and dark respiration of the desert shrub *Larrea divaricata* to the temperature in which it was grown (day/night: 20/15, 35/25 and 45/33° C). After Mooney et al. (1978). For data on wheat see Sawada and Miyachi (1974). The adaptive responses are primarily the result of changes at the cellular and subcellular level. Similar adaptive shifts in respiratory activity and in the temperature cardinal points of photosynthesis can be observed in the course of adaptation to the seasonal changes in temperature. Further examples are given by Larcher (1969b), Schulze et al. (1976), Strain et al. (1976), and Berry and Björkman (1980). See also the classical adaptation experiments of Harder (1925)

Modulative temperature adaptation occurs within a few days, or sometimes a few hours, by shifts in the substrate concentrations, by replacement of certain enzymes with isoenzymes having the same action but different temperature optima, and by incorporating into the biomembranes lipids resistant to cold or heat. The temperature dependence of the plant can thus be altered — particularly with respect to the optimum and the high-temperature limit for photosynthesis — by several degrees C (Fig. 3.40). *Respiratory activity* is enhanced during adaptation to cold and diminished during adaptation to heat. As a result, there is a better supply of energy at low temperatures, while the slower rise in respiratory rate at higher temperatures permits more economical consumption of metabolites. In the ideal case metabolism is so perfectly adjusted that respiratory intensity is kept at about the same level over a relatively broad range of temperatures (*homeostasis*). Figure 3.41 illustrates this process. Rapid modulative adaptation is particularly important to plants in locations where there are wide, abrupt fluctuations in temperature.

By evolutive temperature adaptations, photosynthesis and respiration become adjusted to the *average* temperature climate in the habitat of the plant. When *ecotypes* from different climatic zones are compared, particularly along an altitude gradient in the mountains, both the temperature optimum for net photosynthesis and respiratory activity prove to be matched to the habitat. The preferred temperature for photosynthesis — the genetically based ecotype-specific optimum range for CO_2 uptake — in evolutively adapted plants is clearly related to the average daytime temperature during

119

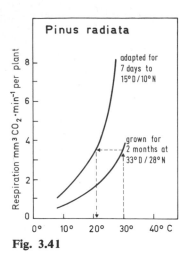

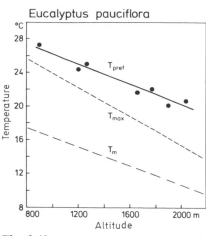

Fig. 3.41   **Fig. 3.42**

Fig. 3.41. Adaptation of the temperature-dependence of respiration of pine seedlings to two different temperatures. If warm-adapted seedlings (raised at 33° by day and 28° C by night) are transferred into a cool room (15° by day and 10° C by night), within a week activity is doubled. Furthermore, the temperature curve of respiration rises more steeply. As a result the plants now respire just as rapidly at 21° as formerly at 30° C. If the seedlings are subsequently returned to the warm room, the respiratory curve gradually resumes its original shape (not shown in the figure). After Rook (1969)

Fig. 3.42. Altitudinal gradients of the photosynthetic temperature optimum (preferred temperature T_{pref}) of *Eucalyptus pauciflora*, and the maximum (T_{max}) and mean (T_m) air temperature of the warmest month in the Snowy Mountains, Australia (Slatyer, 1978). See also the classical studies of Pisek and Winkler (1959) and Milner and Hiesey (1964)

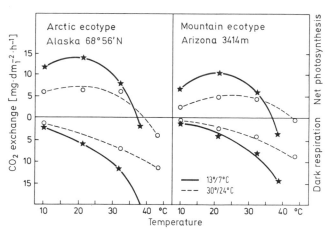

Fig. 3.43. Temperature dependence of net photosynthesis and dark respiration in arctic and mountain ecotypes of *Oxyria digyna* grown at low (day/night: 13°/7° C) and high (30°/24° C) temperatures (Billings et al., 1971). Photosynthesis and photorespiration of mountain ecotypes of *Trifolium repens* have been studied by Mächler and Nösberger (1977, 1978)

120

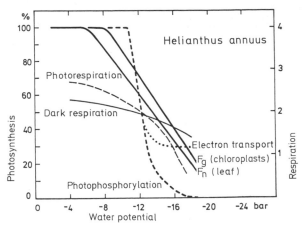

Fig. 3.44. Impairment of photosynthetic processes and respiration by water deficiency. The intensity of photosynthesis in sunflower leaves dried to different degrees is given as a percentage of that in leaves with adequate water supply. Respiratory intensity was determined for whole leaves and is given in mg $CO_2 \cdot dm^{-2} \cdot h^{-1}$. Photorespiration was determined by measurement of the amount of CO_2 released into CO_2-free air. After Boyer and Bowen (1970), Boyer (1971) and Keck and Boyer (1974)

the growing season (Fig. 3.42). Evolutive adaptation also permits modulative adaptation to *occur more readily*. For example, mountain ecotypes of *Oxyria digyna* exposed to heat exhibit a smaller reduction in photosynthetic capacity and a greater shift in the temperature optimum and upper limiting temperature for CO_2 uptake than do arctic ecotypes. Conversely, during exposure to cold the compensatory enhancement of respiration occurs more rapidly and effectively in arctic than in mountain ecotypes (Fig. 3.43).

3.2.3.3 CO_2 Exchange and Water Supply

Like carbon dioxide, water is used in the photosynthetic process, but it is not in this respect that water shortage can be a limiting factor; more important is the water necessary to maintain a high water potential in the protoplasm. The metabolic processes of the cell are critically dependent upon water in this sense. Loss of water has a *direct* inhibitory effect on the photosynthetic process — both on electron transport in the primary reactions and on the biochemical events and activity of the enzymes in the secondary reactions. The main result of loss of turgor is closure of the stomata and hence interruption of the CO_2 supply. In addition respiratory activity, photorespiration in particular, is reduced during water deficiency (Fig. 3.44).

CO_2 Exchange and Degree of Hydration of Thallophytes

In thallophytes the degree of hydration of the cells is matched to the humidity of the surroundings (cf. p. 206). These primitive plants rapidly soak up water when they are sprinkled, but they lose it again quickly through evaporation; thus their water content

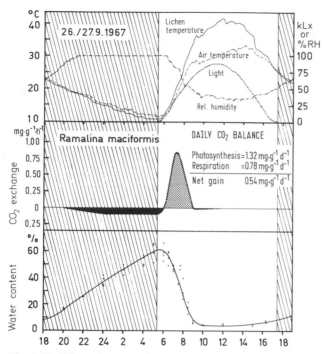

Fig. 3.45. CO_2 exchange and water content of the desert lichen *Ramalina maciformis* during the course of the day. The nighttime dew supplies the lichen with moisture, but in the morning it rapidly dries out again. *Stippled area,* CO_2 uptake; *Black areas,* CO_2 release; *Cross-hatched,* nighttime. After Lange et al. (1970)

fluctuates over short intervals with the meteorological conditions and stays in equilibrium with the water-vapor content of the air.

An ecologically important measure is the *humidity compensation point*. The minimum atmospheric humidity for net photosynthesis is about 70—90% relative humidity (*RH*) for aerial algae, 80—96% *RH* for lichens, and for those mosses that can extract water from the air, usually above 90% *RH*. As more water is taken in, the rate of photosynthesis rises rapidly, becoming maximal in the optimum turgor range — in lichens, between 50 and 80% of the water content at saturation. CO_2 uptake declines when the plants are full of water. During desiccation photosynthetic activity is gradually extinguished, and respiration is also suppressed.

The photosynthetic apparatus of the thallophytes is well suited to the frequent and pronounced fluctuations in the cellular water content. Completely dry thalli reactivate the photosynthetic process within minutes after they receive water again, even if they have been dried out for a long time. For lichens in dry habitats, this ability enables their very existence, e.g., desert lichens can utilize to the utmost the short period available between their imbibing water at night and drying out again in the morning. The course of CO_2 exchange during the day in a desert lichen, described in Fig. 3.45, illustrates this behavior under the conditions in the natural habitat. Throughout the night the thalli soak up moisture from the air, and in the early hours of the morning they obtain dew as

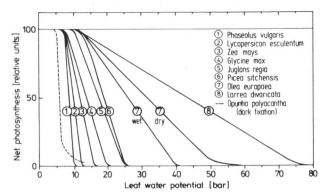

Fig. 3.46. Restriction of photosynthetic CO_2 uptake with increasing water stress in leaves. *1,* from Pospišilová et al. (1978); *2,* Brix (1962); *3* and *4,* Boyer (1970b); *5,* Davies and Kozlowski (1977); *6,* Beadle and Jarvis (1977); *7, wet,* well watered; *dry,* drought-adapted, Larcher et al. (1981); *8,* Odening et al. (1974); *Opuntia (stippled),* Gerwick and Williams (1978). Data on the dependence of net photosynthesis on water saturation deficit are given by Larcher (1969a)

well. After sunrise only three hours remain for the lichens to fix carbon; then they become dry and stiff again until nightfall.

Gas Exchange During Water Stress in Vascular Plants

The first effect of water deficiency upon vascular plants is on the stomata (see p. 92 and Table 3.3), the narrowing of which slows down CO_2 exchange. With increasing desiccation there is reduced hydration of the protoplasm in general, and thus reduced photosynthetic capacity. Normally CO_2 uptake is high only over a narrow range of the adequate water supply level; beyond this it begins to decline and eventually is entirely suspended (Fig. 3.46). There are therefore two critical points in the curve of gas exchange vs. water loss: the point of transition from full capacity to the *limited region* and the *null point for gas exchange.*

The first critical point comes at a level of water stress in which the stomata begin to close, causing the stomatal diffusion resistance to become greater than the residual (mesophyll) resistance. If water is supplied after this first critical point has been passed, recovery is rapid.

The second critical point is determined by marked or complete closing of the stomata as well as by the direct effect of water shortage on the protoplasm. Appreciable CO_2 uptake is no longer possible, though the CO_2 freed by respiration can be bound again. Once this state has been reached, a renewed water supply does not lead to an immediate recovery of photosynthesis. Recovery is delayed, and after severe desiccation the original photosynthetic capacity may, under certain conditions, never be achieved again (Fig. 3.47).

The sensitivity of CO_2 exchange to lack of water, and the positions of the foregoing two critical values, are to a large extent characteristic of a plant species, but they are also adaptable. The limiting values for a number of species representing different eco-physiologic types are shown in Fig. 3.46. Sciophytes are extremely sensitive to even

123

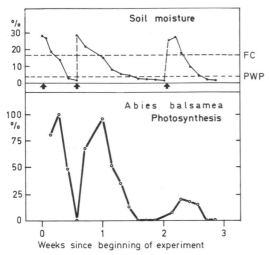

Fig. 3.47. Effect of drought on the net photosynthesis of year-old seedlings of the balsam fir. As a result of watering (*arrows*), the soil moisture was raised above field capacity (*FC*, see p. 215). Thereafter the soil dried out to below the permanent wilting percentage (*PWP*, cf. p. 216). After a brief period of drought, net photosynthesis recovered quickly and completely, but after the soil moisture had remained below the *PWP* for several days recovery was incomplete. After Clark (1961). For examples of drying-cycle experiments on various deciduous trees see Davies and Kozlowski (1977); on olive tree, Larcher et al. (1981); on eucalyptus, Collatz et al. (1976); on *Lolium,* Sheehy et al. (1975); on *Panicum,* Ludlow (1975); on soybean and sorghum, Rawson et al. (1978)

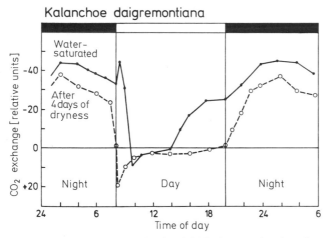

Fig. 3.48. CO_2 exchange in the CAM plant *Kalanchoe daigremontiana*, measured during experimental desiccation. *Negative values* indicate CO_2 uptake. Modified from Kluge (1976). For gas-exchange behavior during water deficiency in cacti see Szarek et al. (1973), Nobel (1977a), Hanscom and Ting (1978), and Gerwick and Williams (1978); in agave, Neales (1975) and Nobel (1976); in *Tillandsia*, Kluge et al. (1973); in *Sedum*, Kluge (1977)

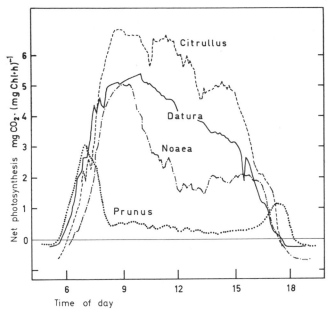

Fig. 3.49. Daily fluctuation in net photosynthesis of plants differing in sensitivity to drought and in the degree of drought stress to which they are exposed, at the end of the dry period in the Negev Desert. *Citrullus colocynthis* and *Datura metel* were watered, and the F_n values were reduced somewhat only during the hottest hours of the day and in the afternoon. The desert plant *Noaea mucronata*, after extensive CO_2 uptake in the morning, toward noon vigorously restricts its photosynthesis. *Prunus armeniaca* suffers considerably from lack of water toward the end of the dry season, and appreciable CO_2 uptake is possible only in the early morning and late evening. After E. D. Schulze et al. (1972). For examples of diurnal photosynthesis fluctuations typical of desert plants see Strain (1969), Stocker (1972, 1974), Szarek and Woodhouse (1976, 1978), and Voznesenskii (1977); see also the classical study by Harder et al. (1931)

slight losses of water. Heliophytes and herbaceous crops, on the other hand, can tolerate more or less severe water deficiency, depending on species and conditions of growth; in this regard there is little difference between C_4 and C_3 plants. As would be expected, plants of dry regions and desert plants in particular have especially high limiting values; many of these maintain gas exchange until the depression of water potential exceeds −40 bar. At the first sign of water deficiency CAM plants keep their stomata closed all day (Fig. 3.48). Under such conditions photosynthesis utilizes the internal CO_2 liberated by breakdown of the malate stored during the night. In fact, CO_2 uptake at night can at first be enhanced by dryness. But when the water potential falls below −10 to −15 bar, fixation of CO_2 is much reduced even in the dark.

The closing of the stomata when the water supply is impaired is primarily a water-conservation measure. The different forms this behavior can take are therefore most readily understandable from the point of view of water balance in the plant, which is also true of the daily rhythm of net photosynthesis in natural surroundings. In Fig. 3.49, typical *diurnal fluctuations in net photosynthesis with suboptimal water supply* are

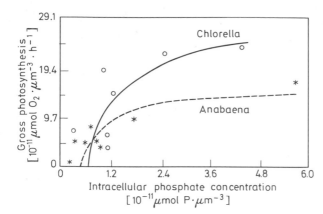

Fig. 3.50. Dependence of gross photosynthesis rates of *Anabaena* and *Chlorella* on intracellular concentrations of phosphorus, under conditions of light saturation (Senft, 1978)

summarized. The principle exemplified is the following: the more sensitive a species is to lack of water and the drier the conditions, the earlier in the day restrictions are imposed upon assimilative activity.

3.2.3.4 Gas Exchange and Mineral Nutrition

The influence of the *mineral nutrient supply* upon photosynthesis and respiration is extremely varied. In soils not seriously deficient in particular nutrients, the availability of minerals is less critical than the climatic factors. Nevertheless, it is almost always possible to enhance the yield of photosynthesis by the artificial provision of nutrients. In bodies of water, however, lack of minerals is very likely to be a limiting factor (Fig. 3.50). Conversely, minerals in excess are also harmful; at too high concentrations certain minerals (heavy-metal ions in particular, as well as air pollutants) impair photosynthesis. Mineral nutrients can influence carbon metabolism both directly and indirectly via the synthesis of new tissue and growth. Direct effects upon photosynthesis and respiration result from the fact that the minerals either are incorporated in metabolites, coenzymes and pigments or participate directly as activators in the process of photosynthesis (cf. Tables 4.1 and 4.2). Manganese, for example, acts as an activator of photolysis, and potassium is involved in the electron-transport system on the thylacoids. Nitrogen and magnesium are components of chlorophyll; various enzymes include iron, cobalt and copper, and phosphate is a component of nucleotides.

The lack of minerals, as well as alterations in relative amounts of the elements taken up, can affect the chlorophyll content and the number, size and ultrastructure of the chloroplasts; this applies even if the elements in question, e.g. iron, are not themselves incorporated into the chlorophyll molecule. In conditions of nitrogen and iron deficiency, chloroses are observed, which cause a diminution of CO_2 uptake to less than $^1/_3$ (Fig. 3.51). Lack of magnesium can have similar consequences. The chief result of insufficient chlorophyll is that the plants cannot make full use of intense light — they behave like sciophytes.

Mineral nutrients further affect gas exchange by influencing the behavior of the stomata, and by their effect on other properties of the leaves such as their anatomic structure, size, life span, and above all their number. Under nitrogen deficiency small

Fig. 3.51. Influence of nitrogen supply on net photosynthesis (F_n), respiration in the shoots (R_S), and respiration in the roots (R_R) of young spruce plants. Photosynthetic capacity changes in proportion to the chlorophyll content (*Chl*) of the needles. After Th. Keller (1971). For a review of the effects of mineral supply on CO_2 uptake see Nátr (1972, 1975)

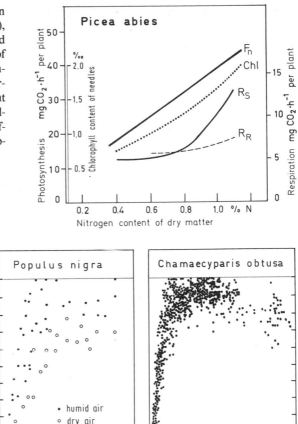

Fig. 3.52. Light-dependence of net photosynthesis in the open under different meteorological conditions. *Left graph*, The *circles* are data from mature poplars on clear days with low humidity; the *dots* are data taken on cloudy to slightly overcast days with high humidity. After Polster and Neuwirth (1958). *Right graph*, Data from pot-grown seedlings during summer (adjusted to a uniform 20° C to eliminate the influence of temperature). After Negisi (1966). For temperature-dependence under various field conditions see Helms (1965), for the full operational range under habitat conditions see Schulze and Hall (1982)

leaves develop, with stomata that are less movable, whereas too much nitrogen causes excessive respiration and thus reduces the photosynthetic yield.

3.2.3.5 The Interplay of External Factors Affecting CO_2 Exchange

Environmental factors do not, of course, act in isolation. The gas exchange rate of plants is an expression of the interplay of many internal and external environmental

Table 3.8. Reduction of CO_2 uptake (the average measured F_n as % of the maximum possible) in the natural habitat, by various inhibitory factors

A. Year-old seedlings of *Pinus densiflora* and *Cryptomeria japonica* in Japan (Negisi, 1966)

Factor	Average annual reduction		Reduction during main growing season (April to September)		Period of greatest effectiveness
	Pinus	*Crypto-meria*	*Pinus*	*Crypto-meria*	
Reduced photosynthetic capacity	−36%	−25%	− 8%	−10%	Winter
Lack of light at twilight and when the sun's elevation is low	−16%	−14%	− 5%	− 5%	Winter
Lack of light due to clouds	−11%	− 9%	−17%	−10%	Early summer
Temperature too low or too high	− 7%	−10%	−15%	−25%	Spring (cold) Midsummer (heat)
Total reduction	**−70%**	**−58%**	**−37%**	**−40%**	
Residual CO_2 uptake	+30%	+42%	+53%	+50%	

B. Mature beech in northwest Germany (E. D. Schulze, 1970)

Factor	Average reduction during growing season	Maximal reduction during periods of unfavorable weather
Lack of light at twilight	−22%	
Lack of light due to clouds and fog	−16%	−56%
Unfavorable temperature	− 3%	−15%
Dryness of air	− 2%	−13%
Total reduction	**−43%**	
Residual CO_2 uptake	−57%	

factors, the individual roles of which are not easy to unravel. Of these factors, one is usually *rate limiting* at any given time, though *the others continue to exert a sub-threshold influence.* For example, with increasing illumination the optimum and maximum temperatures for net photosynthesis are shifted toward higher values. This effect is advantageous for the plant, since intense irradiation is always associated with warming. If the strong radiation leads to overheating (or an intolerable loss of water),

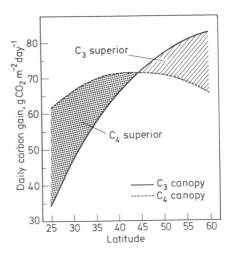

Fig. 3.53. Total daily carbon yield of identical C_3 and C_4 grass canopies at different latitudes in the northern hemisphere during July. Calculation based on environmental data (Ehleringer, 1978). Survey on the distribution of C_3 and C_4 grasses in relation to climate: Hattersley (1983)

then the light loses its role as the limiting factor and CO_2 uptake declines. As a result, in the field one does not obtain saturating curves like those measured in laboratory experiments (which attempt to treat one factor at a time), but rather curves with optimum (Fig. 3.52). In the range of optimum illumination in the field, light is sufficient to stimulate photosynthesis maximally but not so great that the disadvantageous side effects of strong radiation are noticeable. In this natural interplay the external factors are only rarely and briefly found to be optimal in themselves and yet so related to one another that maximum photosynthesis is achieved. On the *average*, the maximum daily values for CO_2 uptake reach only 70—80% of the actual photosynthetic capacity. This is true of herbaceous plants as well as trees.

Under the climatic conditions prevailing at intermediate latitudes, lack of *light* when the sun is low and obscured by clouds is the foremost factor limiting the yield of CO_2 assimilation by plants in the field (Table 3.8), especially for those in dense stands. When the *temperature is too low*, carbon incorporation by temperate-zone evergreens is considerably reduced in late autumn, winter and spring. In the high mountains, insufficient heat limits photosynthesis throughout the year. *Too-high temperatures* are of relatively little significance in the temperate zone, even in open habitats where it becomes quite hot. However, in the tropics and subtropics, heat-limitation of CO_2 uptake may play a role in natural selection; when the temperature is high and radiation intense, C_4 plants are capable of greater assimilation than C_3 plants, as long as the water supply is adequate to allow any carbon incorporation (Fig. 3.53).

From a global point of view, *lack of water* is the most important factor limiting assimilation. Study of diurnal fluctuations in CO_2 exchange by evergreen shrubs and subshrubs of maquis and bushland shows that the daily maxima of net photosynthesis are $\frac{1}{5}$ to $\frac{2}{3}$ lower during times of drought than during the rainy season. The amount of reduction varies with the sensitivity of the different species, with features of the habitat (especially the amount and accessibility of soil water), and with the time course of drought periods. An example of the different forms of behavior characterizing two desert shrubs is given in Fig. 3.54.

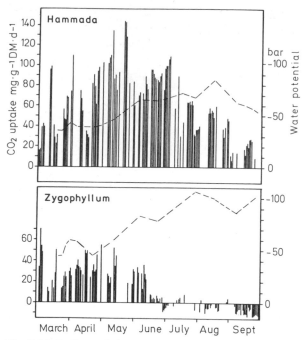

Fig. 3.54. Daily totals (*vertical lines*) of CO_2 uptake and release by desert plants exposed to different degrees of drought stress in their habitats. *Hammada scoparia* is a C_4 shrub that grows in loess-covered basins, whereas *Zygophyllum dumosum* is a C_3 dwarf shrub growing on south-facing slopes and stony flats. During the dry season *Zygophyllum* sheds its leaves; the water potential of the shoots (*dashed curve*) decreases sharply and the CO_2-exchange balance shifts to net loss of CO_2. *Hammada* is capable of withdrawing water from the soil even in dry periods, so that the water potential remains at a moderate level and the green shoots continue to fix CO_2, though to a lesser extent. After Lange et al. (1975). The annual variations in daily CO_2 balance of *Opuntia basilaris* (a CAM plant) are described by Szarek and Ting (1974)

3.2.4 The Gas Exchange Balance

Where the synthesis of organic dry matter by a plant in its habitat is concerned, the determining factor is not so much the brief peak values of photosynthesis, but rather the average total daily CO_2 uptake. The average values for net photosynthesis even under favorable climatic conditions, like those prevailing in the temperate zone, amount to *50—60% of the daily maxima* (i.e., 30—50% of the actual photosynthetic capacity).

3.2.4.1 The Time Factor

The decisive factor in carbon fixation is the time span over which a high rate of CO_2 acquisition is possible. This comprises the hours of daylight during the leaf-bearing time of year, insofar as assimilation is not blocked by frost, heat, or drought. From the sum total of daily carbon intake, the nightly CO_2 release from the leaves must first be subtracted. The resulting net consumption of CO_2 by photosynthesis in the course of

24 h is the daily balance and the sum of the daily balances gives the annual balance of CO_2 exchange.

The *daily balance* is positive if the intake during the day exceeds the loss at night, and it is greater, the more favorable the constellation of factors influencing photosynthesis during the day, and the shorter and cooler the night. When CO_2 uptake has been possible only briefly or inefficiently, there is little surplus to be assimilated. This is a routine occurrence during dry periods and in dry habitats (cf. Figs. 3.45 and 3.49), as well as after frost. In deep shade only slightly positive and — especially when the nights are warm — sometimes negative daily balances are common and in the extreme case can result in "starvation" of the plants. This is clearly one of the factors determining plant distribution.

The *annual balance* shows a greater gain, the longer the plants have been able to achieve large, or at least positive, daily balances. Considerable yields can be obtained despite moderate daily balances if the time span favorable to assimilation is long enough — as it is, for example, in warm-temperate, subtropical and tropical humid regions. Where assimilatory activity is possible only during a relatively short period — as in the high mountains, in the arctic, and in dry regions — the yields remain small even if the plants have a great photosynthetic capacity or can make particularly efficient use of the photosynthates.

On continental land masses, the time available for highly productive synthesis decreases in the following sequence, according to the growth form of the plants and the climatic conditions:

1. Evergreen plants of warm, humid regions, which carry on photosynthetic activity throughout the year (for example, in the tropical rain forest).

2. Evergreen plants in which there are seasonal variations in photosynthetic activity and the assimilation period is interrupted
a) by a cold season (for example, boreal conifers),
b) by a dry season (for example, shrubs of the Mediterranean maquis).

3. Seasonally green plants (deciduous woody plants and herbs)
a) which utilize fully the foliated season, in regions with high precipitation (for example, foliage trees in the temperate zone),
b) which utilize the foliated season only partially because of insufficient light (e.g., undergrowth in deciduous woodland),
c) which utilize the foliated season only partially because of dryness (e.g., steppe and dry woodlands).

4. Plants which take up carbon briefly between longer unfavorable periods; these include
a) tracheophytes in deserts with erratic precipitation (100—120 favorable days),
b) tracheophytes of the arctic and high mountains (60—90 favorable days),
c) mosses, lichens and aerial algae which take up carbon occasionally after they have been wet or when the humidity is high.

In the cold seas at high latitudes the production period is limited to the polar summer, which permits assimilative activity for several weeks, without interruption at night.

Table 3.9. The proportion of total mass (dry matter) of plants accounted for by assimilation organs, axial structures, and roots. Compiled from the original data of numerous authors

Plant	Green mass (photo-synthetically active organs)	Purely respiratory organs	
		Woody stems above ground	Roots and subterranean shoots
Evergreen trees of tropical and subtropical forests	ca. 2%	80—90%[a]	10—20%[a]
Deciduous trees of the temperate zone	1—2%	ca. 80%[a]	ca. 20%[a]
Evergreen conifers of the taiga and in mountain forests	4—5%	ca. 75%[a]	ca. 20%[a]
Alpine scrubwood	ca. 25%	ca. 30%[a]	ca. 45%[a]
Young conifers	50—60%	40—50%	ca. 10%[a]
Ericaceous dwarf shrubs	10—20%	ca. 20%[a]	60—70%[a]
Grasses	30—50%		50—70%
Steppe plants			
Wet years	ca. 30%		ca. 70%
Dry years	ca. 10%		ca. 90%
Desert plants	10—20%		80—90%
Arctic tundra			
Vascular plants	15—20%		
Cryptogams	>95%		
Plants of the high mountains	10—20%		80—90%

[a] The greater part of the mass is dead supporting structures.

3.2.4.2 Green and Non-Green Components of the Plant Mass

Plants consist not only of green, i.e., photosynthetically productive, tissues, but also of others that simply respire and must be nourished by the leaves. In the overall CO_2 budget, therefore, the respiration of all non-green tissues must also be taken into account. The situation is of course most apparent in cases where the axes of shoots, the roots, the flowers and the fruits of a plant make up a large proportion of its mass as compared with the foliage (cf. Table 3.9).

3.2.4.3 The Overall CO_2 Balance

The CO_2 balance of an entire plant involves the total gross photosynthesis ($W_L \Sigma_l F_g$, where l is the number of daylight hours in a year and W_L is the weight of the leaves) and the total annual respiration of leaves ($W_L \Sigma R_L$), shoot axes, flowers and fruits ($W_S \Sigma R_S$), and roots ($W_R \Sigma R_R$). That is,

$$CO_2 \text{ Balance} = W_L \Sigma_l F_g - W_L \Sigma R_L - W_S \Sigma R_S - W_R \Sigma R_R. \qquad (3.10)$$

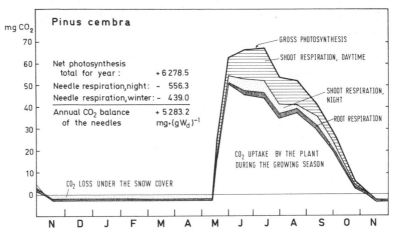

Fig. 3.55. Variation in the daily CO_2 balance of young stone-pines at the alpine tree line over a year. Part of the CO_2 gained by photosynthesis is lost the same day due to the respiration of the shoots and roots. In winter the daily balance is usually negative or at best zero: the CO_2 loss during the 6 winter months is subtracted from the CO_2 acquired during the growing season, to obtain the annual balance. After Tranquillini (1959). The CO_2 balance of tropical trees (the forest margin tree *Cecropia* and closed forest tree *Goethalsia*) is given by Allen and Lemon (1976)

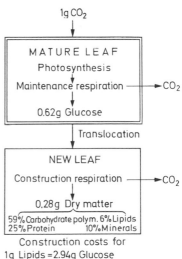

Fig. 3.56. The carbon costs of leaf construction. After Mooney (1972). For more detailed data see Penning de Vries (1972) and Schulze (1982)

To set up a complete gas-exchange balance for a plant is a laborious undertaking. It involves separation of the CO_2 exchange of the part of the plant above ground into divisions (in trees: top of the plant, base of the crown, stems) and measurement of respiration of underground parts by day and night over the whole year; in addition, the proportions by mass of the different organs are determined. It goes without saying that the external factors effective at the site (for example, light distribution, leaf temperatures, the humidity of the air near the leaves, temperature fluctuations in the vicinity of the stem and roots, availability of water, etc.) must also be noted if the carbon balance is to be meaningful. Fig. 3.55 shows an example of a yearly balance of CO_2 turnover, a typical one for young trees.

The units of gas-exchange balance are g or kg CO_2 per plant, per day or per year. Conversion factors can be used to express this quantity in terms of organic dry matter or carbon content (Fig. 3.56; see also p. XVII). Thus one can convert *from gas exchange to production of matter* by the plant. There is a close correlation between the amount by which the CO_2 balance is positive and the rate of increase of dry matter. The latter increases in proportion to the net photosynthetic capacity over a wide range. C_4 plants such as sugar cane, maize and millet achieve rates two or three times those of the C_3 crops sugar beet, alfalfa, and tobacco.

3.3 The Carbon Budget of the Plant

3.3.1 Dry-Matter Production

Assimilated carbon not lost by respiration (i.e., the surplus in the CO_2 budget) increases the dry matter of a plant and can be used for growth and for laying down reserves. The accumulation of carbon is discernible as an increase in weight which can be directly measured by weighing the harvested, dried plants. The *increase in mass of a plant due to the products of assimilation* is called dry-matter production.

The *production rate* (PR) or productivity expresses the increase in dry matter *per unit time* (day, week) during the period of production. The production rate of single plants can be given in terms of the "net assimilation rate" (*NAR*) or the "relative growth rate" (*RGR*).

The *net assimilation rate*, as formulated by Gregory (1926), is the increase in dry matter dW during a particular time interval dt, referred to the assimilation area A, the photosynthetic activity of which is responsible for the gain in matter.

$$\text{PR as } NAR = \frac{dW}{dt} \cdot \frac{1}{A}. \tag{3.11}$$

The *NAR* is given in g or mg dry matter or carbon per dm^2 leaf area of a plant and per day or week. The formula above considers instantaneous rates assuming that leaf area remains constant during the accumulation of matter. This is, however, not the case, for the total leaf area also continues to grow. Therefore the mean rate of increase is determined by a formula of Watson (1947):

$$NAR \quad \frac{W_2 - W_1}{A_2 - A_1} \cdot \frac{\ln(A_2/A_1)}{t_2 - t_1} \tag{3.12}$$

where W_1 and A_1 are the dry weight and leaf area, respectively, of the whole plant at time t_1. W_2 and A_2 are the dry weight and leaf area, respectively, at the later time t_2.

134

Table 3.10. Maximal and average net assimilation rates (mg dry matter per dm^2 leaf area per day) of vascular plants. Taken from original results and summaries of data by Blackmann and Black (1959), Jarvis and Jarvis (1964), Evans (1972), Bannister (1976), Alvim and Kozlowski (1977), Murata (1978), and Vong and Murata (1978)

Plant	Average for the growing season	Rate during the main growth phase
C_4 grasses	>200	400—800
Herbaceous C_3 plants		
Grasses	50—150 (180[a])	70—200 (270[a])
Dicotyledons	50—100	(60[b]) 100—600
Woody plants		
Tropical and subtropical cultivated plants	(5[c]) 10—20	30—50
Winter-deciduous trees of the temperate zone (young plants)	10—15	30—100
Conifers (young plants)	3—10	10—50
Ericaceous dwarf shrubs	5—10	ca. 15
CAM plants	2—4	ca. 10[d]

[a] Rice, [b] alpine plants, [c] cocoa (grown in shade), [d] pineapple.

With herbaceous plants a time interval (t_2-t_1) of 1—2 weeks is usually chosen. This equation results from calculating the average value of NAR over the time interval t_2-t_1, under the assumption that both A and W change linearly with time. The logarithmic function (ln, natural log) arises when the associated integral is evaluated.

The *relative growth rate* expresses the increase in dry matter per unit time with respect to the initial dry weight W of the plant:

$$PR \text{ as } RGR = \frac{dW}{dt} \cdot \frac{1}{W}. \tag{3.13}$$

The "dry weight" required here is that of the organic matter of the plant; it is therefore not accurate to use simply the value determined by weighing the dry plant, which includes not only carbon compounds but also minerals (these average 3—10% of the total dry weight). One must subtract from the dry weight of the whole sample the weight of the ashes after the sample is burnt.

The magnitudes of the highest possible and the average *net assimilation rates* depend primarily upon the morphological and physiological constitution of the plant. The mass of herbaceous plants increases most rapidly, that of woody plants about one-tenth as fast, and that of CAM plants most slowly (Table 3.10). By way of their effects on CO_2 exchange and the carbon balance, *environmental factors* also affect dry-matter production. Under increased radiation (higher intensity and/or longer exposure) production is greater; like photosynthesis, dry-matter production exhibits a temperature

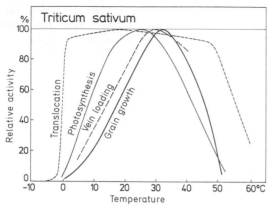

Fig. 3.57. The effect of temperature on various processes involved in the provision of photosynthates for growth, in wheat plants during grain development. The net photosynthesis of the flag leaf, the active export of photosynthates (vein loading) and the rate of carbon import by the grain proceed optimally within a narrow range of temperatures which is different in each case. Translocation in the phloem is largely independent of temperature, being interrupted only by frost or heat so great as to damage the plant. After Wardlaw (1974, 1976)

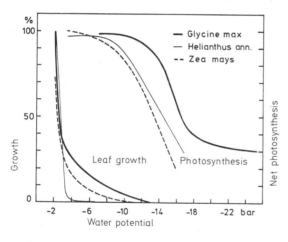

Fig. 3.58. Changes in leaf-area growth rate and net photosynthesis of soybean, sunflower, and maize as water deficiency increases. Leaf growth is rapidly restricted under far less severe drought stress than photosynthesis; this difference may be related to the decline in turgor that accompanies the onset of water deficiency. After Boyer (1970a)

optimum, and both water deficiency and inadequate or unbalanced provision of nutrients reduce the production of matter. Of course, it is not only the uptake of carbon that affects growth; transport processes during assimilation and the hormonally controlled growth activity of the individual organs are also crucial. These processes are affected by environmental factors in the same direction as CO_2 exchange, but often to a different extent (Figs. 3.57 and 3.58). Thus complete agreement between the variations in gas exchange and in production rate is not always to be expected.

3.3.2 Utilization of Photosynthates and the Rate of Growth

Plants consist largely of carbohydrates; carbohydrates provide the material for construction of the cell walls, and comprise 60% or more of the dry matter of higher plants.

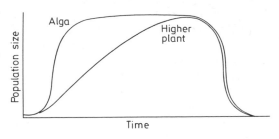

Fig. 3.59. Typical growth curve of populations of algae, as compared with that of higher plants (Vincent, 1971). For the growth dynamics of planktonic algae see Fogg (1975)

The carbohydrates produced in CO_2 assimilation must be distributed throughout the plant in a systematic way; distribution is controlled by demand (for energy, growth, or differentiation) and by coordinating mechanisms, some of which involve hormones. There are a number of characteristic *ways in which plants budget their photosynthates*, depending upon their level of organization and morphology. These differences are evident in the substances produced and the rate of growth.

3.3.2.1 Planktonic Algae

Planktonic algae exist as single cells or form colonies or simple groups of cells. The typical algal cell supplies only itself with carbon and need not produce any surplus for other cells. Within the cell there is a favorable ratio between the sites of production and cell components that consume photosynthates: in *Chlorella* the chromatophores take up about half of the volume of protoplasm. This being the case, it is not surprising that algal cells well supplied with nutrient elements and light accumulate large surpluses and grow rapidly (Fig. 3.59). They soon reach their terminal size and then proceed to divide. The autotrophic single-celled organism employs its yield of synthesized materials to increase the number of individuals, i.e., for reproductive processes. There is a direct relationship between photosynthetic yield and the number of divisions per day. The rate of growth in phytoplankton is thus usefully expressed as the increase in population density or the number of divisions per unit time.

3.3.2.2 Annual Plants

These are frequently rather small herbs that must make best use of a short period of time in which conditions are favorable for growth, flowering and the setting of fruit. They are found primarily in dry regions, where they complete their life cycles in a few weeks to months. These plants must employ their photosynthates in such a way that an abundance of tissue is formed in the shortest possible time. Annuals do this even when a rather long time is available for growth. Summer grain, sunflowers, and other annual crops thus yield particularly large harvests.

The operating principle of all annual plants consists in first using the greater proportion of the photosynthates for the formation of leaves, which then participate in production and increase the intake of the plant. While these photosynthetically active organs are being developed preferentially, the mass of those parts that only respire remains small, which improves the overall balance. In the *flowering phase*, the distribution system switches to favor the reproductive organs, which receive such a large share that all oth-

137

er parts of the plant are supplied with little more than needed to maintain themselves — the older leaves even shrivel up. Accordingly, in the course of the life cycle the proportions of leaves, axial structures, roots, and reproductive organs in a plant change considerably. The greatest change is in the fraction of the overall mass represented by leaves, which sinks from 30—60% during the elongation phase to 10—20% by the time the fruits are ripe. At this time, in the sunflower, 90% of the photosynthates produced daily moves into the fruits.

Under environmental *conditions conducive* to plant life, this way of investing assimilation products selectively guarantees both luxuriant growth and lavish fruiting. When local *conditions are less favorable*, on the other hand, particularly when there is a shortage of water or when the soil is poor in nutrients, the plant is forced to build up an extensive system of roots; this is done at the price of leaf-area development and leads to a smaller photosynthetic yield as well as deterioration of competitive ability (Table 3.9). Annual plants are primarily adapted to making use of advantageous — though short-lasting — situations, and are less able than other plants to endure prolonged unfavorable conditions.

3.3.2.3 Perennial Herbs

The herbaceous plants which live for several years usually at first undergo a development similar to that of the annuals. After their vegetative structures are formed, they lay down *reserve supplies* before proceeding to bloom. Toward the end of the first growing season the excess photosynthates are diverted to the stems and above all into the subterranean parts of the plant, which may develop into massive storage organs. Flowers are formed only after the plant has accumulated sufficient capital to draw

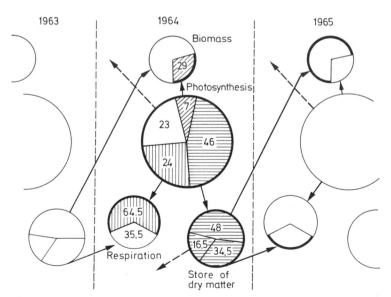

Fig. 3.60. Diagram of the utilization of photosynthates by the geophyte *Scilla sibirica*. After Goryshina (1969)

Table 3.11. Photosynthate budget of a specimen of *Scilla sibirica* (annual balance of dry matter in mg); Goryshina, (1969). For budgets of other geophytes see Iwaki and Midorikawa (1968) and Wassink (1972), for potatoes see Sale (1974)

	Intake	Consumption											
	Acquired by photosynthesis during the growing season	Respiration of parts above ground		Respiration of bulbs		Total used for respiration	Increase in biomass				Total consumption for growth and respiration	Reserves stored in the bulbs
		Leaves and shoots under snow	During the growing season	During dormancy	During the growing season		Root	Leaves and shoots under snow	Leaves and shoots in growing season	Total used for growth		
Photosynthesis during the current year	1610	–	178	–	201	379	–	–	118	118	497	740
Drawn from the stores in the bulbs	–	45	20	117	21	203	40	135	100	281	484	–
Total	1610	45	198	117	222	582	40	135	224	399	981	740

upon for that purpose. Photosynthates stored in one year are first used in the next to elaborate the shoot system (Table 3.11, Fig. 3.60). The synthesis of new substance builds up rapidly and thanks to the availability of stored material is largely independent of the factors affecting productivity in the spring. Once the plant is ready to flower, if the food supply is adequate, the flowers and fruits take precedence over the storage processes. Afterward, near the end of the season, photosynthates move preferentially to the subterranean parts of the plant, which increase correspondingly in weight.

Perennial plants have the advantage wherever the period of time favorable to production is not long enough to permit sufficient assimilation for flowers and fruits to be formed, as well as in cases where the plants bloom so early that the necessary materials cannot be provided by the available mass of leaves. This applies, for example, to *spring geophytes*, many of which open their flowers before the leaves have unfolded. *Alpine plants* must accomplish flowering and ripening of seeds during the short mountain summer; their accumulation of photosynthates is subject to many uncertainties. Finally, these considerations also apply to *steppe plants*, which must utilize the times between winter cold and summer drought for their life cycle. A trait all these species have in common is the presence of storage organs such as rhizomes, tubers, root thickenings, and bulbs. Moreover, these species frequently develop an extensive root system, so that the subterranean dry mass amounts not uncommonly to twice and sometimes even to four times the mass of the parts above ground (Table 3.9).

3.3.2.4 Trees

The tree — the most differentiated and largest life form among plants — manages its carbon supplies in a way suited to its long lifetime. Even in youth a large fraction of the photosynthates is used for growth of the *trunk*. In the first years of life the leaf mass can make up half of the overall dry substance of the plant, but with increasing size the ratio of leaf mass to stems is altered, the leaf mass growing only slightly while trunk and branches become steadily thicker and heavier. Foliage comprises only 1%—5% of the total mass of mature trees (Table 3.9), and these leaves must thus supply the materials for maintenance and growth to parts of the tree amounting to many times their own weight. The consequence is a modest acquisition of carbon and increase in mass as compared with herbs, but that is no disadvantage in view of the long lifespan. Even after maturity the tree increases its mass of wood from year to year; from the standpoint of the photosynthate budget, this represents inaccessible capital, since it is permanently withdrawn from the metabolism of the tree. Organic matter tied up as wood can be used in metabolism only by other components of the ecosystem; the tree stores such matter not for itself but as a member of a food chain. On the other hand, their growth from necessitates this great expenditure on supporting tissue by woody plants. It procures decisive competitive advantages for the trees over herbaceous plants in areas with long production periods; the herbs are slowly but surely overshadowed by the ever taller woody plants.

In correspondence with their size and differentiated infrastructure, the *allocation of photosynthates* in trees takes place according to a complicated scheme.

In deciduous trees, the carbohydrate stores are emptied shortly before the leaves begin to unfold, the substances being sent to the buds and later to the new shoots (Fig. 3.61).

140

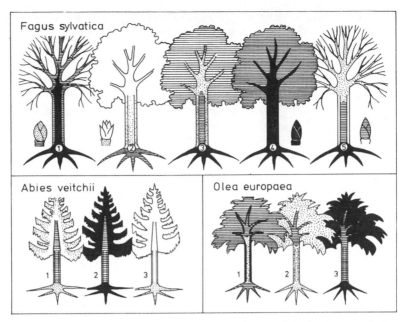

Fig. 3.61. Variations in starch deposition by trees throughout the year. Maximal accumulation of starch is indicated by *black*, large amounts by *cross-hatching*, and small amounts by *stippling*; in the parts left *white*, starch is present in traces or not at all. *Fagus sylvatica* (Central Europe): *1*, just before leaf emergence in the spring; *2*, during leaf unfolding; *3*, midsummer; *4*, just before abscission in the autumn; *5*, conversion of starch to soluble carbohydrates at low temperatures during winter. After Fischer (1891), Gäumann (1935), and K. Kober (unpubl.). *Abies veitchii* (Japan): *1*, during growth of new shoots in spring; *2*, late summer; *3*, during winter frost. After Kimura (1969). *Olea europaea* (Northern Italy): *1*, during shooting and flowering in spring; *2*, during a dry period in midsummer; *3*, in winter after the end of the rainy season. After Thomaser (1975). For the storage dynamics of Atlantic dwarf shrubs see Stewart and Bannister (1973), and Grace and Woolhouse (1973); of chaparral species, Mooney and Hays (1973); of mountain plants, Larcher (1977) and Zachhuber and Larcher (1978)

About a third of the reserve materials serves for the building up of assimilation surfaces, which very soon operate with a positive balance and in their turn contribute to the further formation of the leaves and shoots in the new growth. After the foliage is completely formed, it supplies the tree with photosynthates. As a rule flowers and developing fruit are supplied preferentially, next in order is the cambium, and last the newly forming buds and the depots of starch in roots and bark. Floral primordia are formed in numbers depending on the amount of material left. At the end of the growing season the surplus photosynthates are moved into the woody tissue and the bark of branches, trunk and roots and stored. In the tropics and in dry regions the trees pass through several seasonal storage periods — four in the case of fig trees, for example.

Evergreen woody plants of the temperate zones do not produce new shoots as soon as the winter dormant period is past, for they still have the assimilation organs of the previous year. When buds do begin to open, the carbon taken up by these old organs in the

141

spring can meet a large part of the demand, and the rest comes from reserves in axis and roots. Because of this "head start" the new leaves mature relatively rapidly — even though evergreen leaves as a rule incorporate three times as much dry matter for a given area as does the delicate foliage of deciduous trees. Moreover, sufficient photosynthates remain for cambial growth. Evergreen leaves — and the associated extension of the productive period — become particularly advantageous wherever a long winter, or a summer dry season, restricts the growing season. In the mountains, in the northern forest belt, and in regions where aridity limits tree growth, evergreen woody plants generally gain dominance over deciduous plants. Only in regions where the unfavorable season is extremely harsh and prolonged (subarctic, Eastern Siberia, semideserts) do seasonally green trees and shrubs again predominate.

As a consequence of the principle by which photosynthates are distributed in woody plants, there is competition between the abundant *setting of fruit* and vegetative growth that strengthens the plant as a whole; when photosynthates are only sparsely produced it is likely that vegetative, but not reproductive, buds will be formed for the following year. Thus nutrition and the bearing of fruit regulate growth of the wood and the frequency with which a tree flowers. The amount of photosynthates used for reproduction is considerable: 5—15% in pines, 20% or more in beeches, and in apple trees as much as 35% of the net annual yield of photosynthesis. For this reason many temperate-zone trees can bear large quantities of fruit only at intervals of several years. Broad-leaved trees as a rule do so every 2—3 (5) years, and conifers every 2—6 (10) years. Near the polar limits of a tree's distribution, and in mountain locations, the fructification intervals are a great deal longer. In the tropics all transitions between the extremes are represented; some trees (e.g., the coconut palm) bear fruit continually, others one, two, or several times a year (coffee, cocoa, *Artocarpus*), and still others at intervals of two or more years.

3.3.3 Translocation of Photosynthates

The products of CO_2 assimilation are constantly being translocated within a plant — from the leaves and other photosynthetically active tissues (green bark, and parts of flowers such as the awn) to sites where they are consumed or stored, and from the storage depots to growth zones and into seeds and fruits.

The products are conducted through the lumens of sieve tubes, narrow passages that also present high filtration resistance. Nevertheless, the rate of movement of this stream of photosynthates is considerable, since the sieve-tube sap is usually very concentrated.

Translocation follows the concentration gradient that becomes established between sink and source — i.e., sites where the requirement is great (*attraction centers*) and those where the *synthesis or mobilization of photosynthates* occurs (A. L. Kursanov). Fully developed leaves preferentially supply the consumer with the greatest attraction; the products of the lowest leaves are translocated to the root system, and those higher in the plant to the growth zones of the shoot and especially to flowers and ripening fruit (Fig. 3.62). In herbaceous dicotyledons the photosynthates flowing out of a *single* leaf are distributed among *several* consumers (flowers, fruit), and conversely each consumer

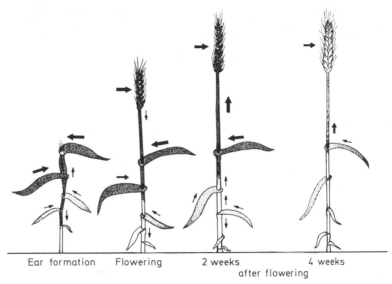

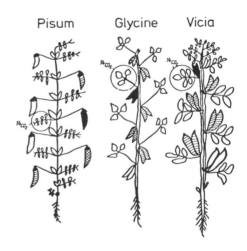

Fig. 3.62. Formation and allocation of photosynthates in wheat plants. The darker stippling indicates regions of particularly productive assimilation, and the thickness of the arrows shows the relative rate of transport of the products. After Stoy (1966)

Pisum   Glycine   Vicia

Fig. 3.63. Redistribution of photosynthates, from the leaves to the ripening fruits, in pea, soybean, and broadbean plants. The black pods contain 35%–60% of the photosynthates formed in the leaf primed with $^{14}CO_2$, the pods with lattice shading, 10%–35%, the cross-hatched pods 5%–10%, and the remaining pods less than 5%. After Bartkov and Zvereva (1974); see also Pate et al. (1977)

draws on a combined stream from several leaves (Fig. 3.63). This balances the supply of materials in the plant and ensures that the storage tissues are equally filled. Detailed knowledge of the physiological mechanisms regulating the distribution of photosynthates in the plant — in particular the provision of photosynthates to seeds, fruit and storage organs — is important for ecological analysis of its *reproductive capacity*, and critical in planning ways to improve the harvest of agricultural crops. In grain the proportion of the dry matter in the whole shoot that is found in the harvested kernel (the *harvest index*) varies from about 25% (older varieties of maize, rye) to 50%

143

(rice, barley). In Fabales the harvest index ranges from about 30% (soybean) to 60% (*Phaseolus* bean). By breeding to emphasize a distribution that favors provision of photosynthates to the seeds, it has been possible to increase the weight of the kernel in maize from 24% to 47% of the total shoot mass, while that of rice has been raised from 43% to 57%.

3.4 The Carbon Budget of Plant Communities

3.4.1 The Productivity of Stands of Plants

The quantity of dry matter formed by the vegetation covering a given area is called *community production* or *primary production* (*PP*). It is expressed with reference to the *area of ground* covered, in tons of organic dry matter per hectare (t · ha^{-1}) or g · m^{-2}. Production in this case is greater the higher the assimilation rates of the species of which the stand is composed, the more completely the light passing through an extensive system of *assimilating surfaces* is absorbed (the leaf-area index, *LAI*), and the longer the time in which the plants can maintain a positive gas-exchange balance (duration of the *assimilation period*). The rate of production — the growth per unit time — can be determined for communities as well as for single plants. When a homogeneous stand is concerned, this is called the *crop growth rate* (*CGR*). Stands of C_4 grasses in the tropics and subtropics achieve maximal *CGR*s of 50—70 g DM · m^{-2} · d^{-1}, whereas in the less intensely irradiated temperate zone the maximum is about 20—30 g DM · m^{-1} · d^{-1}. Cultivated stands of C_3 plants produce at most 15—30 g DM · m^{-2} · d^{-1}, depending on the species, though the average over the entire growing season is a third to half of this. The maximal *CGR* of pineapple plantations (CAM plants) is in the range 10—15 g DM · m^{-2} · d^{-1}. (Data surveys are given by Bartholomew and Kadzimin, 1977, Sale, 1977, and Monteith, 1978).

The community production rate (*PPR*) is calculated from the growth rate of the single plants (*NAR*, *RGR*) by multiplying this by parameters reflecting the density of the stand. For example,

$$PPR \text{ as } CGR = NAR \cdot LAI . \tag{3.14}$$

The *leaf-area index* is optimal for production when the PhAR is absorbed as completely as possible during its passage through the canopy of leaves. In stands of cultivated plants this is frequently the case with a *LAI* of about 4—8 (Fig. 3.64). If the density of foliage were less, the light available to individual plants, and thus their *NAR*, would be greater, but with respect to the yield per unit ground area an open stand of plants is less productive than a closed stand. If the plants are too closely spaced and the foliage overlaps too extensively, the light in the most shadowy places is no longer sufficient to keep the CO_2 balance positive at all times; thus the yield per unit area will be reduced.

The density of the foliage of individual plants and the closeness of the plants (that is, the degree of cover) are more than just important factors affecting production — in fact,

Fig. 3.64. Relationship between the net assimilation rate of maize plants and the production rate of stands of maize (crop growth rate) as a function of leaf-area index. After W. A. Williams et al. (1965)

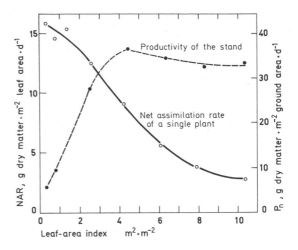

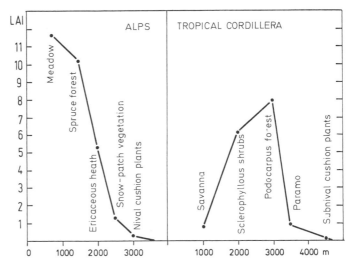

Fig. 3.65. Leaf-area index of various types of vegetation at a sequence of altitudes in the Alps and in the coastal Cordillera and the Andes of Venezuela. Where closed communities give way to open communities that cover only a small fraction of the ground area, the LAI decreases sharply. After Vareschi (1951, 1953)

each is itself *affected by production.* With an unfavorable food supply and a scarcity of water, the plants lack the raw materials for the synthesis of an extensive leaf system, and the *LAI* remains insufficient. In dry regions, on poor stony soil, and in areas with a very short growing season there arise open plant communities, the *LAI* of which falls to minimal values as the amount of cover decreases (see Table 3.13). This can be seen especially clearly if a series of communities at different altitudes is considered (Fig. 3.65). When the closed plant cover gives way to an open one, the *LAI* decreases rapidly and irregularly from place to place; in such cases the *LAI* depends primarily upon the amount of cover and reflects to a smaller extent the degree to which the leaves overlap. In very dense plant communities (for example, certain coniferous forests), an increase of the LAI above 12—14 is prevented by lack of light.

A relationship similar to that between LAI and stand production exists between the latter and the amount of chlorophyll per m^2 of ground (see Table 3.13). A related measure, used chiefly to characterize the degree of overlapping of photosynthetically active layers in bodies of water, is the chlorophyll content in the plankton-filled column of water below 1 m^2 of water surface.

3.4.2 Carbon Balance in Plant Communities

3.4.2.1 The Production Equation

The *carbon balance* of a plant community is determined from the difference between intake and output. *Intake* is measured as the overall quantity of carbon fixed by photosynthesis in the course of a year. This gross primary productivity PP_g cannot be measured for land plants in the field; therefore (as for gross photosynthesis) a rough estimate is computed from the net community production and the respiration of the stand.

$$PP_g = PP_n + R .\tag{3.15}$$

The yield of net primary production is used for building up organic matter, part of which becomes detritus in the course of the year and is lost (L) or is grazed by consumers (G). These *deductions* from the net yield include the shedding of leaves, flowers, fruits and dead branches, the decay of dead roots, consumption by animals and parasites, and the release of photosynthates in fluid form via the excretions of roots and consumption by symbionts (mycorrhiza). The remaining *net yield* goes to increase the plant mass per unit area of ground (the biomass B); it represents the annual change in the dry matter comprising the stand (ΔB).

$$PP_n = \Delta B + L + G .\tag{3.16}$$

This production equation is ascribable to P. Boysen-Jensen, who as early as 1932 clearly distinguished these relationships and thereby initiated the analysis of causative factors in the field of production ecology.

All the quantities in the production equation can be determined directly; their sum is the measure ordinarily used to express net primary productivity (for an example of such a calculation see Table 3.12). The measurements are not simple to make in natural stands of plants; data for production by forests and other perennial, many-layered plant communities should be considered as guidelines only, unless they are confirmed by different procedures, applied at the same time.

3.4.2.2 The Proportion of Losses Due to Respiration

A considerable fraction of the amount of carbon obtained by photosynthesis is respired and thus is unavailable for incorporation into the tissue of the plant. The "operating

146

Table 3.12. Productivity and loss of organic dry matter in forests (annual balance); all data in metric tons dry matter per hectare per year. Further data for tropical forests are given by Müller and Nielsen (1965) and Odum and Pigeon (1970); for Mediterranean sclerophyllous woodland see Lossaint and Rapp (1971) and Rapp (1971), and for temperate-zone forests Whittaker and Woodwell (1968, 1971), Duvigneaud and Denaeyer-de Smet (1970), and Froment et al. (1971)

Stand	Beech wood, Denmark, 60 years old		Tropical rain forest, Thailand	
Authors	Mar-Möller et al. (1954)		Kira et al. (1964) Yoda (1967)	
LAI	5.6		11.4	
Increase in biomass, ΔB		in % P_g		in % P_g
Foliage	0		0.03	
Stems	5.3		2.9	
Roots	1.6		0.2	
Total	6.9	35%	3.13	2%
Loss, L				
Foliage	2.7		12.0	
Stems	1.0		13.3	
Roots	0.2		0.2	
Total	3.9	20%	25.5	20%
$PP_n = \Delta B + L$	10.8	55%	28.6	22%
Consumption in respiration				
Foliage	4.6		60.1	
Stems	3.5		32.9	
Roots	0.7		5.9	
Total	8.8	45%	98.9	78%
$PP_g = PP_n + R$	19.6	100%	127.5	100%
k_{PP}	2.23		1.29	

cost" of respiration, which must be subtracted from the gross primary production, can be expressed by the *productivity coefficient* k_{PP}. Unlike the photosynthetic coefficient [k_F; cf. Eq. (3.8)], k_{PP} is an average over daytime and nighttime and over phases of activity and rest, and above all it takes into account the CO_2 balance of the entire plant, including both green and nongreen parts.

$$k_{pp} = \frac{PP_g}{R} \div \frac{PP_n + R}{R} \qquad (3.17)$$

k_{PP} has the value 2 if the same amount of photosynthate is metabolized in respiration as is retained as dry matter. As a dimensionless quantity, k_{PP} may be expressed in the form of a percentage or a fraction (as in Table 3.12). A coefficient of 2 corresponds to use of 50% (or 0.5) of the photosynthates for respiration. In plankton populations k_{PP} is about

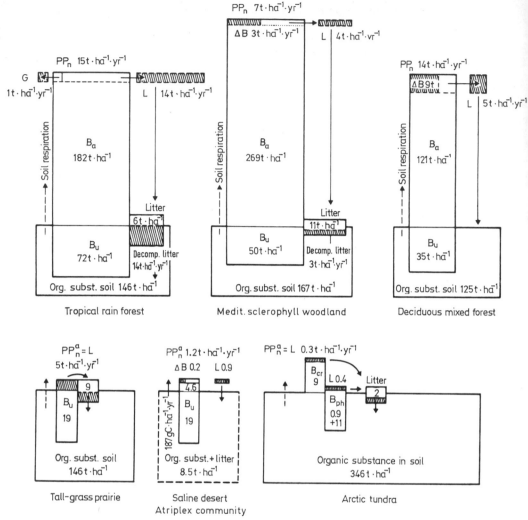

Fig. 3.66. Content and turnover of organic dry matter in various stands of plants. *Areas in heavy outline*, content of organic dry matter in t · ha⁻¹; *cross-hatched areas*, annual turnover of dry matter in t · ha⁻¹ · yr⁻¹. B_a, biomass above ground; B_u, underground biomass; B_{cr}, biomass of cryptogams; B_{ph}, biomass of phanerogams; PP_n, net production; *PP_n^a, net production above ground;* ΔB, annual increase in biomass; L, annual loss of dry matter as detritus; G, annual loss of dry matter by grazing. In all these stands of plants, by far the largest fraction of the total net production enters the soil as detritus from the plants. Depending on the rate at which this is decomposed, varying amounts of litter (forming mats in the prairie) and of organic soil components accumulate. In deserts the amount of litter fluctuates considerably from year to year; in especially dry years it can be larger (in part due to drought damage to the plants) than the production of dry matter. *Upper row*, virgin tropical rainforest in Puerto Rico (Odum and Pigeon, 1970), evergreen oak wood in southern France (Rapp 1971) and winter-deciduous mixed forest in Belgium (Duvigneaud and Denaeyer-de Smet, 1970). *Lower row* tall-grass prairie in Missouri (Kucera et al., 1967), saline desert with *Atriplex confertifolia* (C₄ plant) in Utah (Caldwell et al.,

10, and in stands of herbaceous plants it is 2—5. Forests and dwarf-shrub heaths with a relatively large proportion of photosynthetically unproductive mass in the temperate zone use 40—60% of the gross primary production for respiration (K_{PP} 1.5—2.5), while the corresponding value for the moist-warm tropics is about 75% (k_{PP} 1.3).

3.4.2.3 Loss as Detritus and by Grazing, and its Effect on the Carbon Budget of Plant Communities

The fraction of the annual net primary production represented by the losses L and G is a critical factor in the carbon balance of a plant community. Depending on the yield of net primary production and the amount lost, the organic mass composing the community may either increase (ΔB positive), stay the same ($\Delta B = 0$), or decrease (ΔB negative). These relationships are illustrated in Fig. 3.66. Which of these possibilities is realized depends chiefly upon the species composition of the community, its dynamic state (age and stage of succession), and the degree of stress imposed by natural influences and those of civilization.

1. Woodlands and Dwarf-Shrub Heaths. When woody plants colonize new areas (e.g., in reforestation and in heath with burning cycles) the initial phase is one of development. Because most of the plants are young, the mass of foliage must feed a relatively small mass of axes and roots; k_{PP} is favorable, and the net primary productivity is therefore large. There is a considerable surplus of organic matter, which visibly increases the mass of the stand from year to year (Fig. 3.67).

The productive phase of growth gives way, with increasing age of the stand, to the mature phase, in which ΔB at first stays positive and later fluctuates about zero. This reduced rate of growth is brought about not by increased losses, but by the decline in net production as development of the community proceeds. The larger the trees grow, the smaller is the ratio of green to non-green tissues. As a result, the yield of photosynthesis suffices only for renewing the leaves and for the respiration of the enormously enlarged mass of shoot and root systems. In deciduous forests the increase in wood comes to a halt when the mass of leaves falls to less than 1% of the total mass. In stands of woody plants that naturally reseed themselves, particularly in virgin forest, individuals of all age classes are present simultaneously and the above sequence does not occur. Here a steady state is reached in which PP_g and losses $(R+L+G)$ are in balance, so that ΔB always tends toward zero.

2. Grasslands. In the course of the production period the phytomass grows rapidly, and at the same time parts of the shoot system and the roots die off or are eaten. In

◁——

1977) and arctic tundra in Canada (Bliss, 1975). For further examples see the following publications: Atlantic ericaceous heath, Gimingham (1972), Chapman et al. (1975), and Wielgolaski (1975); alpine dwarf-shrub heath, Larcher (1977), Schmidt (1977), tundra, Rosswall and Heal (1975), Bliss (1977) and Tieszen (1979); halophyte communities, Ketner (1972) and Eckhardt et al. (1977); stands of swamp plants, Dykyjová and Květ (1978); Mediterranean subshrub steppe *(Phrygana)*, Margaris (1976); desert vegetation, Rustanov (1972), Evenari et al. (1975b), Szarek (1979), and Sen (1982)

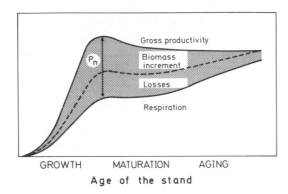

Fig. 3.67. Diagram of the variation, with advancing age of a uniform stand of trees, in gross primary production, net primary production (*dotted area*), rate of increase in biomass (*dotted area above the dashed line*), rate of loss (*dotted area below the dashed line*) and respiration. After Kira and Shidei (1967)

GROWTH   MATURATION   AGING

Age of the stand

herbaceous plant communities left in their natural state, at the end of the growing season the leaves turn yellow and dry up, and parts of both the shoot and the root systems are withdrawn. In the steppes, this loss accounts for more than half of the phytomass formed during the year, and in desert plant communities consisting primarily of ephemers, it may be 60—100%. In fields that are regularly mown or grazed, the biomass is continually removed during the growing season, so that G exceeds L. The difference between losses due to grazing and those due to shedding of foliage consists primarily in the fact that when the plants are eaten, assimilation organs are removed while still capable of full photosynthetic activity, and excessive grazing (G more than half of PP_n) endangers the existence of the stand. On the other hand, once the growing season is concluded, the entire biomass above ground can die off without harm to the ecosystem.

In herbaceous stands, whether left undisturbed or under cultivation, ΔB fluctuates with steadily decreasing amplitude around zero. In dry regions with variable precipitation, ΔB fluctuates in successive years between large positive and negative values, but when averaged over many years it is also approximately zero.

3. Plankton Populations. In populations of plankton, the first effect of net primary production is an increase in the numbers of the (usually short-lived) algae in the euphotic zone. This supply of matter in the community serves to feed a horde of consumers; G is high, on the average $^2/_3$ of the net primary production The loss L is less, but is hard to estimate quantitatively. In the case of phytoplankton, L consists of those cells that sink below the compensation depth (cf. Fig. 3.30), either drawn by gravity or carried by water currents. The velocity of sinking depends upon the size, shape and density of the organisms, as well as upon water movements and temperature. In cool waters at 6° C, most algae sink at an average rate of 3 m per day; at 20° C they sink twice as fast. A characteristic of aquatic ecosystems is the sequential appearance of a productive growth phase and a protective equilibrium phase (as is found in woodlands), as well as a high proportion of loss due to grazing (as is found in mown or grazed meadows). Thus in an aquatic ecosystem $G \geqq L$ and the sum of the two exceeds ΔB.

3.4.3 The Net Primary Production of the Earth's Plant Cover

Although detailed studies have been made (particularly within the International Biological Program), and despite the fact that we can now use photographs from airplanes

Table 3.13. Net primary production and related characteristics of the biosphere (Whittaker and Likens, 1975)

Ecosystem type	Area (10^6 km^2)	Net primary production (dry matter)			Biomass (dry matter)			Chlorophyll		Leaf-surface area	
		Normal range (g/m^2/year)	Mean (g/m^2/year)	Total (10^9 t/year)	Normal range (kg/m^2)	Mean (kg/m^2)	Total (10^9 t)	Mean (g/m^2)	Total (10^6 t)	Mean (m^2/m^2)	Total (10^6 km^2)
Tropical rain forest	17.0	1000–3500	2200	37.4	6–80	45	765	3.0	51.0	8	136
Tropical seasonal forest	7.5	1000–2500	1600	12.0	6–60	35	260	2.5	18.8	5	38
Temperate forest:											
Evergreen	5.0	600–2500	1300	6.5	6–200	35	175	3.5	17.5	12	60
Deciduous	7.0	600–2500	1200	8.4	6–60	30	210	2.0	14.0	5	35
Boreal forest	12.0	400–2000	800	9.6	6–40	20	240	3.0	36.0	12	144
Woodland and shrubland	8.5	250–1200	700	6.0	2–20	6	50	1.6	13.6	4	34
Savanna	15.0	200–2000	900	13.5	0.2–15	4	60	1.5	22.5	4	60
Temperate grassland	9.0	200–1500	600	5.4	0.2–5	1.6	14	1.3	11.7	3.6	32
Tundra and alpine	8.0	10–400	140	1.1	0.1–3	0.6	5	0.5	4.0	2	16
Desert and semidesert scrub	18.0	10–250	90	1.6	0.1–4	0.7	13	0.5	9.0	1	18
Extreme desert—rock, sand, ice	24.0	0–10	3	0.07	0–0.2	0.02	0.5	0.02	0.5	0.05	1.2
Cultivated land	14.0	100–4000	650	9.1	0.4–12	1	14	1.5	21.0	4	56
Swamp and marsh	2.0	800–6000	3000	6.0	3–50	15	30	3.0	6.0	7	14
Lake and stream	2.0	100–1500	400	0.8	0–0.1	0.02	0.05	0.2	0.5	–	–
Total continental:	149	–	782	117.5	–	12.2	1837	1.5	226	4.3	644
Open ocean	332.0	2–400	125	41.5	0–0.005	0.003	1.0	0.03	10.0		
Upwelling zones	0.4	400–1000	500	0.2	0.005–0.1	0.02	0.008	0.3	0.1		
Continental shelf	26.6	200–600	360	9.6	0.001–0.04	0.001	0.27	0.2	5.3		
Algal beds and reefs	0.6	500–4000	2500	1.6	0.04–4	2	1.2	2.0	1.2		
Estuaries (excluding marsh)	1.4	200–4000	1500	2.1	0.01–4	1	1.4	1.0	1.4		
Total marine:	361	–	155	55.0	–	0.01	3.9	0.05	18.0		
Full total:	510	–	336	172.5	–	3.6	1841	0.48	243		

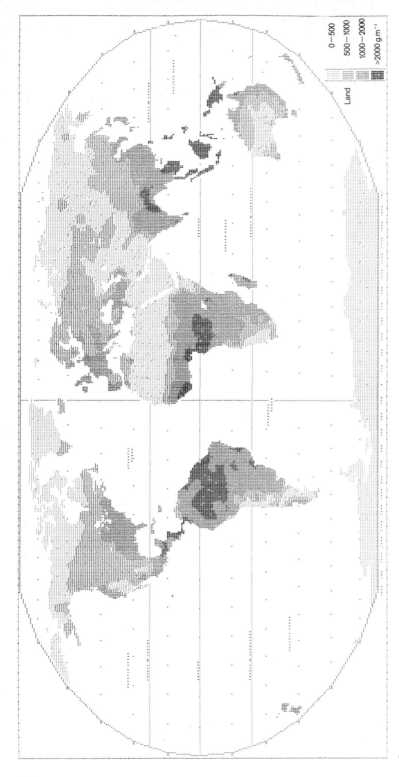

a

3.68a and b. Annual net primary production on land (Lieth, 1972) and in the oceans (map generated by computer from the data of Hsiao, Van Wyk and Lieth). The advantage of computer maps is that new data can rapidly be incorporated as it becomes available

Land

0—500
500—1000
1000—2000
>2000 g.m⁻¹

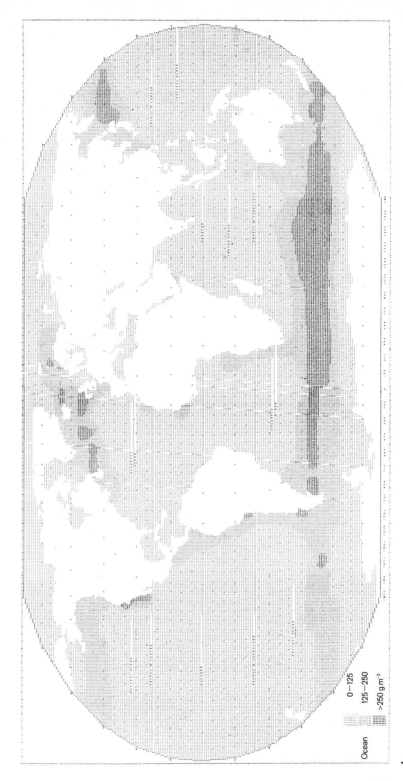

Ocean

0—125

125—250

>250 g·m⁻²

b

153

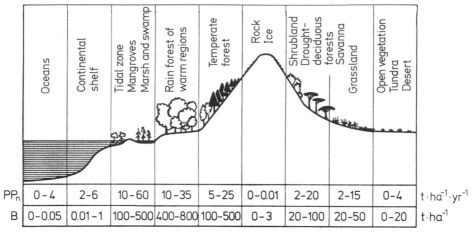

PP$_n$	0-4	2-6	10-60	10-35	5-25	0-0.01	2-20	2-15	0-4	t·ha^{-1}·yr^{-1}
B	0-0.05	0.01-1	100-500	400-800	100-500	0-3	20-100	20-50	0-20	t·ha^{-1}

Fig. 3.69. Differences in annual net primary production (PP_n) and in phytomass (B) in different biomes over the earth. All numbers refer to t dry matter. Based in part on E. P. Odum (1971) with figures estimated from calculations by Bazilevich and Rodin (1971) and Whittaker and Likens (1975)

and satellites for analysis of production, estimation of the net primary production of the plant cover remains difficult and imprecise. Thus there is considerable variation among the estimates made by different authors. According to the calculations of Whittaker and Likens (1975) about $170 \cdot 10^9$ tons of dry matter are fixed on earth per year, roughly two-thirds on land and the remaining third in the oceans.

High net primary production is limited to those regions of the continents and the oceans that offer the vegetation an optimal combination of water, warmth and nutrient salts. On land this is found in the tropics and in water, in the zone between 40° and 60° north and south latitudes (Fig. 3.68). The most abundant production, however, is found in the transitional zones where land and water meet — in shallow water near the coasts and on coral reefs, in rain forests, water meadows and swamps in warm countries (Table 3.13 and Fig. 3.69). The greater part of the earth's area, both land and water, permits only moderate production. On the *continents* it is almost always the water supply that limits yield, though other factors are nutrient deficiency and, at latitudes north or south of 50° and in the mountains, a shortening of the production period due to cold. In tropical *seas* lack of nutrients, and in seas near the poles lack of light, limits productivity. On a smaller scale, too, there are differences in productivity, sometimes quite pronounced, between adjacent regions; these depend on local variations in food supply, type and composition of the plant communities, and the degree of intervention by man. *Intensive cultivation* can achieve yields in a given region that far exceed those of the primary production of the plant communities that would have been found there naturally (Table 3.14). The highest yields per unit area are obtained from stands of the tropical C$_4$ fodder grass *Pennisetum purpureum* and intensively cultivated sugar cane, with an annual production of 80—85 t dry matter per hectare under the most favorable conditions. The highest known dry-matter production has been achieved experimentally by algae in culture tanks (equivalent to about $100 \, t \cdot ha^{-1} \cdot yr^{-1}$). With land plants, too, production

Table 3.14. Maximal annual dry-matter production by stands of crop plants. From lists of data by Lieth (1962), Gifford (1974), Loomis and Gerakis (1975), Bassham (1977), Cooper (1977), Wolverton and McDonald (1979), and Eagles and Wilson (1982). Data for high-yield plantations are from Siren and Sivertsson (1976), and Moraes (1977)

Plant	Greatest yield per unit area (total plant mass in t $W_d \cdot ha^{-1} \cdot yr^{-1}$
C_4 grasses	
Sugar cane	70—80
Maize (subtropics and tropics)	ca. 30
Maize (temperate zone)	ca. 15
Pennisetum purpureum	ca. 85
Sorghum bicolor	ca. 45
Cynodon dactylon	ca. 30
C_3 grasses	
Wheat	18—30
Rice	22
Barley	15
Rye	10
Festuca rubra	15
Dactylis glomerata	10
Fabales	
Medicago species	30
Soy bean	10
Root crops	
Manioc	ca. 40
Sugar beet	ca. 30
Potato	ca. 20
Helianthus tuberosus	20
Trees in high-yield plantations (energy farming)	
Eucalyptus	ca. 55
Salix smithiana	52[a]
Poplar hybrids	35—40[b]
Hevea brasiliensis	25—35
Water plants	
Water hyacinth	154
Seaweeds	30—55
Sewage algae	35—70

[a] Wood accounts for 61% of total production ($32 \ t \cdot ha^{-1} \cdot yr^{-1}$).
[b] Wood accounts for 51% of total production ($17-21 \ t \cdot ha^{-1} \cdot yr^{-1}$).

can be considerably increased — in some cases doubled — on a small scale by culture in tanks of nutrient solution (*hydroponics*). On the average, though, throughout the world, the agricultural yields are far less than could be achieved, chiefly because of inadequate techniques of cultivation, extensive rather than intensive use of the land, incomplete utilization of the local production period, and failure to use the best varieties.

Table 3.15. Energy content of the dry matter of plants. From the original publications of many authors and unpublished measurements

Plant material	kcal·g⁻¹	kJ·g⁻¹
Planktonic algae	4.6—4.9	19.3—20.5
Seaweeds	4.4—4.5	18.4—18.9
Lichens, mosses	3.4—4.6	14.2—19.3
Most herbaceous vascular plants		
Shoots	3.8—4.3	15.9—18.0
Roots	3.2—4.7	13.4—19.7
Seeds	4.4—5.0	18.4—21.0
Fabales	4.4—4.9	18.4—20.5
Halophytes	3.7—4.4	15.5—18.4
Epiphytic vascular plants	3.8—4.0	15.9—16.8
Trees in seasonally green forests		
Leaves	3.9—4.8	16.3—20.1
Trunk wood	4.2—4.6	17.6—19.3
Roots	4.0—4.7	16.8—19.7
Trees in evergreen tropical forests		
Leaves	3.6—4.1	15.9—17.2
Trunk wood	3.9—4.2	16.3—17.6
Roots	3.9—4.2	16.3—17.6
Mangrove trees		
Leaves	4.2—4.3	17.6—18.0
Trunk wood	4.2—4.3	17.6—18.0
Roots	3.9—4.1	16.3—17.2
Sclerophyllous woody plants		
Leaves	4.8—5.2	20.1—21.8
Trunk wood	4.5—4.7	18.9—19.7
Roots	4.2—4.7	17.6—19.7
Evergreen conifers		
Needles	4.9—5.0	20.5—21.0
Trunk wood	4.7—4.8	19.7—20.1
Ericaceous dwarf shrubs		
Leaves	5.0—5.6	21.0—23.5
Shoot axes	5.1—5.8	21.4—24.3
Garrigue shrubs	4.5—5.4	18.9—22.6
Desert shrubs		
Leaves	4.9—5.4	20.5—22.6
Shoot axes	4.5—5.0	18.9—21.0
Cacti		
Bark	3.7—4.0	15.5—16.8
Core	4.2—4.7	17.6—19.7
Plant constituents		
Oxalic acid	0.7	2.9
Glucose	3.7	15.5
Starch, cellulose	4.2	17.6
Proteins	5.5	23.0
Lipids	9.3	38.9
Lignin	6.3	26.4
Terpenes	11.2	46.9

3.4.4 Energy Conversion by Vegetation

The efficiency (ε) of the conversion of radiant energy to chemical energy by photosynthesis is given by

$$\varepsilon = \frac{\text{stored chemical energy} \cdot 100}{\text{absorbed radiant energy}}. \tag{3.18}$$

The *efficiency coefficient of photosynthetic energy utilization* indicates the percent of the absorbed radiant energy fixed in the form of chemical bonds by the conversion of carbon dioxide to carbohydrates. The photosynthetic process uses up *15.9 kJ per gram of carbohydrate assimilated* ($= 3.8$ kcal). The product of this conversion factor and the gross rate of photosynthesis gives the amount of energy bound. The utilization of radiation by photosynthesis in single leaves under particularly favorable circumstances attains efficiencies up to 15% (in C_4 grasses, up to 24%), but leaves usually operate at efficiencies of 5—10% or even less. With cultures of algae efficiencies of 12—15% have been achieved.

The *efficiency of productivity in stands of plants* is computed from the energy content of gross primary production and the PhAR absorbed per unit area of ground in the same period of time. If the absorbed PhAR has not been measured, it can be estimated as 47% of the total incident short-wave radiation between 0.3 and 3 μm, for which figures are available (cf. Table 2.1).

The *energy bound* in the plant mass is determined from representative samples; the energy content is high if the carbon content of the organic dry matter is high, and it differs according to species and organ of the plant (Table 3.15) and the time of year. One may take an energy content of roughly 18.5 kJ (4.5 kcal) per gram organic dry matter as a mean.

On the average, with all the varying temporal and spatial conditions affecting assimilation throughout the production period, plant communities operate with efficiencies below 2%—3%, and at most, under intensive agriculture, up to 9% (maize in subtropical regions). Relatively unproductive plant communities achieve efficiencies below 1%, and the same is true for a large part of the oceans. In comparison with the possible values of the photosynthetic efficiency coefficient achievable, the yield under ordinary conditions is extremely meager. The vegetation cannot make full use of all the daylight hours; it is often in those regions exposed to considerable incident radiation (near the Tropics of Cancer and Capricorn) that there is little precipitation and hence low productivity. A world map of the utilization of radiation by assimilation would show a narrow strip near the equator and another encompassing the temperate latitudes, representing the zones in which photosynthesis is most efficient.

4 The Utilization and Cycling of Mineral Elements

Plants require a large number of inorganic elements derived from minerals or mineralized by decay of organic matter. The mineral nutrients are taken up in the form of ions and incorporated into the plant mass or stored in the cell sap. After the combustion of the organic dry matter in the laboratory, the inorganic compounds remain as ash. In the ashes of plants, one finds all of the chemical elements occurring in the lithosphere. Some of these are essential for life; such substances include the *macronutrients* N, P, S, K, Ca, Mg and Fe, large quantities of which are required, as well as the *trace elements* or *micronutrients* Mn, Zn, Cu, Mo, B and Cl. In addition there are elements that are essential only for certain plant groups: Na for the Chenopodiaceae, Co for the Fabales with symbionts, Al for the ferns, Si for the diatoms, and Se for some planktonic algae.

As has been known since the time of J. Liebig, a nutrient available in inadequate concentration is a yield-limiting factor (the *law of the factor in minimum*). However, a nutrient "in minimum" is not the only factor that determines yield. If metabolism is to be well regulated, production of new tissue rapid, and development unimpaired, both macronutrients and trace elements must be taken up by the plant not only in sufficient quantities but in suitable proportions (rules describing this relationship have been developed by scientists in agriculture and forestry). The various species of plants differ considerably in their nutrient requirements. The requirements of agricultural plants

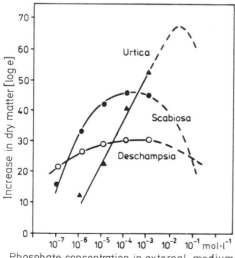

Fig. 4.1. The effect of phosphate supply on dry-matter production by a ruderal plant (*Urtica dioica*), a meadow plant (*Scabiosa columbaria*) and a plant adapted to nutrient-poor acid soils (*Deschampsia flexuosa*). *Urtica* requires particularly large amounts of phosphate for growth; *Deschampsia* utilizes the available phosphate only to a moderate extent, but fairly uniformly over a wide range of concentrations. After Rorison (1969)

have been studied in considerable detail, but little experimental research has been done on the specific needs of wild plants — even though it is precisely this sort of investigation that can provide important insights into the causes of characteristic patterns of distribution within the earth's plant cover. An example of the differing phosphate requirements of species with contrasting characteristics in respect to mineral nutrition is given in Fig. 4.1.

4.1 The Soil as a Nutrient Source for Plants

4.1.1 The Mineral Nutrients in the Soil

Plant nutrients occur in the soil in both dissolved and bound form. Only a tiny fraction (less than 0.2%) of the nutrient supply is dissolved in the soil water. Most of the remainder, almost 98%, is bound in organic detritus, humus, and relatively insoluble inorganic compounds or incorporated in minerals. These constitute a nutrient reserve which becomes available very slowly as a result of weathering and mineralization of humus. The remaining 2% is absorbed on soil colloids.

4.1.1.1 Adsorptive Ion Binding and Ion Exchange in the Soil

Colloidal clay particles and humic substances, because of their surface electrical charges, attract ions and molecular dipoles and bind them reversibly. Soil colloids thus act as ion exchangers. Their *exchange capacity* depends upon the active surface area of the micelles; in the clay mineral montmorillonite this area is 600—800 m^2 per gram, and in humic substances it can be 700 m^2 per gram or more. Both clay minerals and humic colloids have a net negative charge, so that they retain primarily cations. There are also certain positively charged sites where anions can accumulate, but there are always more cations adsorbed than anions. As a rule, the more highly charged ions are attracted more strongly — for example, Ca^{2+} more strongly than K^+ — and among ions with the same charge those with little water of hydration are retained more firmly than strongly hydrated ions. By this kind of adsorptive binding, masses of ions accumulate on the surface of the much swollen micelles of clay and humus.

This coating of ions amounts to an intermediate stage between the fixed soil phase and solutions in the soil. If ions are added to or withdrawn from the soil solution, *exchange takes place*. Those ions that adhere more firmly are attracted more strongly by the colloid and displace other ions from its surface. The tendency for adsorption decreases in the order Al^{3+}, Ca^{2+}, Mg^{2+}, NH_4^+, K^+, and Na^+ for cations, and for anions it decreases from PO_4^{3-} through SO_4^{3-} and NO_3^- to Cl^-. Heavy metal ions can also be adsorbed, though only in trace amounts.

The *adsorptive binding of nutrient ions* offers a number of advantages. Nutrients freed by weathering and the decomposition of humus are captured and protected from leaching. Moreover, the concentration of the soil solution remains low and relatively constant, so that the plant roots and soil organisms are not exposed to extreme osmotic conditions; yet when they are needed the adsorbed nutrient ions are readily available to the plants.

An equilibrium exists with respect to the soil solution, the soil colloids, and the reserves of mineral substances in the soil; it is complex and capable of adaptation. This system controls the exchange of mineral substances and ensures a continual supply of nutrient elements. The concentration of hydrogen ions in the soil solution exerts a great influence upon this ion-exchange equilibrium.

4.1.2 The pH of the Soil ("Soil Reaction")

4.1.2.1 Soil pH and the Availability of Nutrients

Most soils in humid regions are weakly acid to neutral, though bog soils are markedly acidic (pH about 3) and the saline and alkali soils of arid regions are basic. Acidification takes place in a number of ways: by removal of the bases through leaching, by the withdrawal from solution of exchangeable cations, by the release of organic acids by the plant roots and microorganisms, and above all by dissociation of carbonic acid, which accumulates in the soil as a product of respiration and fermentation. Depending on the parent rock and the degree of saturation of the adsorption complexes with cations, the soil is buffered to within a certain pH range. Calcareous soils are buffered primarily by the system $CaCO_3/Ca(HCO_3)_2$, — that is, by the salt of a strong base and a weak acid; they are therefore weakly alkaline. The pH of the soil changes in the course of the year (especially in association with the distribution of precipitation), and there are also local differences, particularly between the different horizons in the soil. Therefore to characterize a habitat the pH must be measured over the entire year and if possible for the entire depth profile, or at least in the zone most densely penetrated by roots.
The pH of the soil has an effect upon its structure, on the processes of weathering and humification, and above all on the mobilization of nutrients and the exchange of ions. The most important of these relationships are summarized in Fig. 4.2. In very acid soils too many Al, Fe, and Mn ions are liberated, and Ca^{2+}, Mg^{2+}, K^+, PO_4^{3-} and MoO^{2-} are depleted or occur in a form difficult for plants to take up. In more alkaline soils, on the other hand, Fe and Mn ions, PO_4^{3-}, and certain trace elements are fixed in relatively insoluble compounds, so that the plants are more poorly supplied with these nutrients.

4.1.2.2 Soil pH and the Plant

Soil pH has a direct effect on the viability of plants, in addition to its effect on the nutrient supply. Below pH 3 and above pH 9, the protoplasm of the root cells of most vascular plants is severely damaged. Moreover, the increased concentrations of Al^{3+} in very acid soils, and of borates in alkaline soils, act to poison the roots.
Different species display characteristic tolerance limits and requirements in their physiological behavior with respect to soil pH (Fig. 4.3). Some *Sphagnum* species prefer a strongly acid milieu; they are very sensitive to OH^- ions and succumb even in the neutral range. According to a classification system set up by H. Ellenberg, they are to

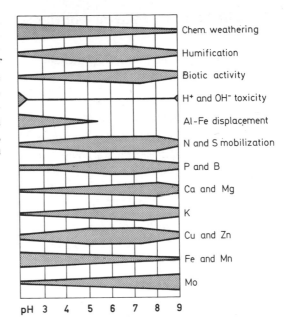

Fig. 4.2. Influence of the soil pH on soil formation, mobilization and availability of mineral nutrients, and the conditions for life in the soil. *The width of the bands* indicates the intensity of the process or the availability of the nutrients. After Truog (1947). For a detailed treatment of the effect of pH on soil properties and plant functions see, e.g., Rorison (1969) and Etherington (1975)

be regarded as strongly acidophilic, with a narrow tolerance span. The hair grass *Deschampsia flexuosa*, the presence of which is an indicator of acid soil, develops optimally at a pH between 4 and 5, but can also grow in the neutral range and will tolerate weakly alkaline soils. This species is "acidophilic-basitolerant". Similar behavior is shown by *Calluna vulgaris* and *Sarothamnus scoparius*. Converse behavior is found in coltsfoot (*Tussilago farfara*), which has an optimum in the neutral-to-alkaline range, but good tolerance down to pH 4; coltsfoot is classified as basiphilic-acidotolerant. *Most vascular plants* are amphitolerant, having a broad optimum in the range between weak acidity and weak alkalinity, and are able to exist between pH 3.5 and 8.5.

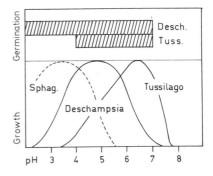

Fig. 4.3. Influence of the pH on the growth of a bog moss (*Sphagnum rubellum*) and on germination and growth of plants of wood hair grass (*Deschampsia flexuosa*) and coltsfoot (*Tussilago farfara*) cultivated in nutrient solution. After Olsen as cited in Ellenberg (1958)

161

4.2 The Role of Mineral Nutrients in Plant Metabolism

4.2.1 The Uptake of Mineral Nutrients

4.2.1.1 The Withdrawal of Nutrient Ions from the Soil

A root takes nutrients from the soil (Fig. 4.4):

1. by absorption of nutrient ions from the soil solution. These ions are available directly, but their concentrations in the soil solution are very low: NO_3^-, SO_4^{2-}, Ca^{2+} and Mg^{2+} are present in concentrations below $1000\ mg \cdot l^{-1}$, while there is less than $100\ mg \cdot l^{-1}$ of K^+ and less than $1\ mg \cdot l^{-1}$ of phosphate ions. The soil solution replenishes its ionic content by drawing ions from the solid phase of the soil;

2. by exchange absorption of adsorbed nutrient ions. By releasing H^+ and HCO_3^- as dissociation products of the CO_2 resulting from respiration, the root promotes ion exchange at the surface of the clay and humic particles and obtains in return the ions of nutrient salts;

3. by freeing bound nutrient stores via excreted H^+ ions and organic acids. Nutrient elements fixed in chemical compounds, chiefly heavy metals, are liberated and form chelated complexes. Metal chelates are protected from being bound again but are readily taken up by plant roots. The excretion of H^+ and acids depends upon the intensity of respiration and thus upon the availability to the roots of oxygen and carbohydrates, and upon the temperature. Furthermore, some ions are freed more readily than others, and different plant species vary in their ability to make them available.

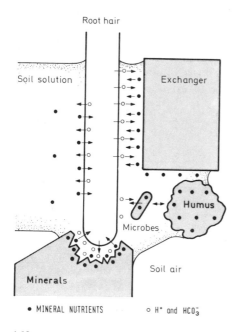

Root hair

Soil solution

Exchanger

Humus

Microbes

Soil air

Minerals

• MINERAL NUTRIENTS   o H^+ and HCO_3^-

Fig. 4.4. Mobilization of mineral nutrients in the soil and the uptake of mineral substances by the root. After Finck (1969)

4.2.1.2 The Uptake of Ions by Roots

The rate at which nutrients are supplied to a plant depends on the concentration of diffusable minerals in the rooted soil strata and on the ion-specific rates of diffusion and mass transport. Nitrate, as a rule, rapidly reaches the root surface; phosphate and potassium ions, with lower diffusion coefficients, move more slowly. The ions of nutrient salts from the soil solution first move, with the inflowing water, into the interconnected system of cell walls and intercellular spaces in the parenchyma of the root cortex (*apoplasmatic transport*); there they are adsorbed, owing to the charges on the surfaces of the cell walls and at the outer boundaries of the protoplasts. This is a purely passive process, following the concentration and charge gradients between the soil solution and the interior of the root.

4.2.1.3 Ion Uptake into the Cell

Because of their hydration, ions would be hindered in entering the living cell if it were not for special properties of cell membrane systems that assist ion transport. Usually such transport depends on large protein molecules or complexes of such molecules ("*carrier systems*"), which can either move within the membrane or occupy fixed positions. Because the ions are electrically charged, passage through the membrane is possible only if two ions of opposite charge are transported at the same time, if an ion within the cell is exchanged for another with the same charge, or if there is already an electrical potential difference across the membrane (e.g., at the membranes of mitochondria). The binding sites of the carrier systems in many cases are *specific* for certain ions or for groups of closely related ions.

Ion transport does not only occur passively, down a concentration gradient. It is very often necessary for a plant to accumulate ions by transporting them against a concentration gradient (Fig. 4.4). Such transport requires a supply of energy. This *active transport* (performed by "*ion pumps*") therefore depends on processes – respiration and photosynthesis – that provide energy either directly by way of the *membrane potentials* generated by the flow of electrons, or indirectly by *hydrolysis of ATP*. In certain cases there is a measurable increase in respiration during ion uptake ("*salt respiration*"; H. Lundegardh and H. Burström). Ion pumps are active over the entire outer boundary of the protoplasm (in the plasmalemma and tonoplast), as well as within the protoplasm wherever ions are transported between compartments.

The following characteristic properties of the uptake of nutrient salts can be explained in terms of both passive and active ion transport:

The Ability to Concentrate Ions. Plant cells are capable of taking up ions against a concentration gradient and accumulating them, particularly in the vacuoles, at concentrations much greater than those in the external solution. This ability is particularly important for aquatic plants, which must draw their nutrient elements from extraordinarily dilute solutions.

Preference. Plant cells are adapted to take up preferentially certain nutrient ions that they require. Thus cations are preferred to anions in the uptake process, and among the cations some are accumulated in higher concentrations than others. When necessary, electrical neutrality can be maintained by ion exchange (H^+, HCO_3^-). This selectivity is

163

a characteristic of the physiological constitution of a plant, and its nature varies with the species.

Limits to Selectivity. Plant cells cannot entirely exclude salts that are not required or are injurious, even if the plant is damaged as a result; the biomembranes are not very permeable to ions, but the permeability is never zero. Consequently, with pronounced differences in concentration on the two sides of a membrane, some ions do leak through. Especially when the outside concentration is high — for example, in saline soils — the cells are flooded with ions (e.g., Na^+ and Cl^-) in quantities greater than are desirable.

4.2.2 The Translocation of Minerals in the Plant

Nutrient ions begin their movement through the plant in the rhizodermis and root cortex. There they enter the cytoplasm and move through it into the cell sap. The vacuoles serve exclusively as depots, playing no role in intercellular ion transport. Nutrient salts that infiltrate the vacuole passively, or are actively excreted into it, remain stored there until they are actively pumped back into the cytoplasm. *Ion transport* from cell to cell bypasses the vacuoles, proceeding along a continuous chain of living protoplasts that contact each other directly by way of plasmodesmes. E. Münch (1930) has called this continuum of living protoplasts the *symplast*. In contrast to apoplasmatic transport along cell walls, which is blocked by hydrophobic elements embedded in the walls of the root endodermis (Casparian strip), the symplast route leads all the way to the conductile system of the central cylinder (Fig. 4.5). In the central cylinder sieve tubes and companion cells take up ions directly by way of contact with the symplast. The ions

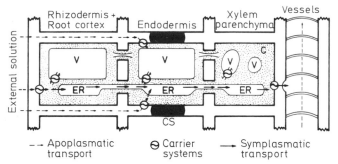

Fig. 4.5. Diagram of ion transport from the *external solution* to the long-distance conducting system in the central cylinder of the root (*vessels*). From the rhizodermis to the endodermis (the Casparian strip, *CS*) ions, together with water, are transported apoplasmatically — that is, in the cell walls and in water-filled intercellular spaces. After they have been taken into the living protoplasts (cytoplasm, *C*) ions are transported symplasmatically, over the endomembrane system (ER) and through plasmodesmas. Vacuoles (*V*) are spaces for the excretion and accumulation of substances; they are not part of the symplast. The pathways for transport in the opposite direction, which is mainly passive, are not shown. Modified from Lüttge (1973) and Läuchli (1976). A schematic diagram of ion transport through the whole plant is given by Weatherley (1969). For active transport processes see Bowling (1976)

Table 4.1. Occurence, uptake, distribution, incorporation and fuction of macronutrients. From Wallace (1951), Mengel (1968), Epstein (1971), Etherington (1975), Bonner and Varner (1976), Bowling (1976), Baumeister and Ernst (1978)

Bio-element	Bound form in soil	Accessible form in soil	Taken up as	Incorporation in plant	Function in plant	Sites of accumulation	Transport-ability	Deficiency symptoms
N	Organically bound, nitrate	supplied by microbial decomposition; NH_4^+ adsorbed on clay minerals and humus; NO_3^- in solution	NO_3^-, NH_4^+, (urea)	Free as NO_3^- ion (vacuoles) in organic compounds, in protein, nucleic acids sec. plant substances	Essential component of protoplasm and enzymes	Young shoots, leaves, buds, seeds, storage organs	Good, primarily in organically bound form	Stunting or dwarfism; spindly appearance; shoot: root ratio shifted toward roots; premature yellowing of old leaves, sometimes reddening
P	Organically bound, phosphates of Ca, Fe, Al	As PO_4^{3-}, HPO_4^{2-}, rel. insoluble, adsorbed and in chelated complexes. Microbial release slight	HPO_4^{2-}/HPO_4^-	Free as ion, in esteric compounds, nucleotides, phosphatides, phytin	Basal metabolism and synthesis (phosphorylation)	More in reproductive organs than in vegetative (pollen granules)	Good, in organically bound form	Disturbance of reproductive processes (delayed flowering), stunting, dark green or bronze-violet discoloration of leaves and needle-tip drying in conifers
S	Organically bound, sulfur-containing minerals, sulfates of Ca, Mg and Na	SO_4^{2-} readily soluble, little adsorbed	SO_4^{2-} from soil (SO_2 from air)	Free as ion, bound as SH- or SS-group and as ester, in protein, coenzymes, sec. plant substances	Component of protoplasm and enzymes	Leaves, seeds	Good in organic form, poor as ion	Similar to N-deficiency, intercostal chlorosis of young leaves

Table 4.1 (continued)

Bio-element	Bound form in soil	Accessible form in soil	Taken up as	Incorporation in plant	Function in plant	Sites of accumulation	Transport-ability	Deficiency symptoms
K	Feldspar, mica, clay minerals	Adsorbed > > dissolved	K^+	Dissolved as ion (primarily in cell sap) and adsorbed	Colloidal effect (promotes hydration). Synergists: NH_4^+, NA^+. Antagonist: Ca^{2+}. Enzyme activation (photosynthesis, nitrate reductase), osmoregulation (stomata)	Division zones, young tissue, bark parenchyma, sites of intense metabolism	Good	Disturbed water balance (tip drying), curling of edges of older leaves, old conifer needles dropped prematurely, root rot
Mg	Carbonate (dolomite), silicate (augite, hornblende, olivine), sulfate chloride	Dissolved > > adsorbed; deficient in acid soils	Mg^{2+}	As ion dissolved and adsorbed, bound in complexes, organically bound in chlorophyll and pectates, component of enzymes and ribosomes	Regulation of hydration (antagonist to Ca^{2+}); basal metabolism (photosynthesis, phosphate transfer); synergists: Mn, Zn	Leaves	Good in part	Stunted growth, interveinal chloroses of old leaves

Ca	Carbonate, gypsum, phosphate, silicate (feldspar, augite)	Adsorbed > dissolved; deficient in very acid soils	Ca^{2+}	As ion; as salt dissolved; crystallized and encrusted as chelate; organically bound in pectates	Regulation of hydration (antagonists: K^+, Mg^{2+}); enzyme activator (amylase, ATPase), regulator of growth in length	Leaves, tree bark	Very poor	Disturbance in growth by division (small cells), tip drying, leaf deformation, impaired root growth
Fe	Sulfides, oxides, phosphates, silicates (augite, hornblende, biotite)	Adsorbed > mobilized; fixed in chalk soils	Fe^{2+}, Fe (III)-chelate	In metal-organic compounds; component of enzymes (heme, cytochrome, ferredoxin)	Basal metabolism (redox reactions), nitrogen metabolism, chlorophyll synthesis	Leaves	Poor	Straw-yellow interveinal chloroses, in extreme case white coloration of young leaves (veins green) apical-bud formation suppressed

Table 4.2. Occurrence, uptake, distribution, incorporation and function of trace elements

Bio-element	Bound form in soil	Accessible form in soil	Taken up as	Incorporation in plant	Function in the plant	Site of accumulation	Trans-port-ability	Symptoms of deficiency
Mn	Amorphous oxide (MnO_2), carbonates, in silicates	Adsorbed > dissolved; better available in acid soils; accumulates under reducing conditions	Mn^{2+}, Mn-chelate	In metal-organic compounds and complexes; component of enzymes	Basal metabolism (oxidases, photosythesis, phosphate transfer), stabilizes chloro-plast structure, nitrogen metabolism; nucleic-acid synthesis synergists: Mg, Zn	Leaves	Poor in part	Inhibition of growth, chloroses and necroses on young leaves, leaf abscission
Zn	Phosphates, carbonates, sulfides, oxides, in silicates	Adsorbed > soluble; mobilization acid > basic	Zn^{2+} Zn-chelates	Bound in complexes	Chlorophyll formation, enzyme activator, basal metabolism (dehydrogenases), protein breakdown, biosynthesis of growth regulators (IAA)	Roots, shoots	Poor	Stunted growth, white-green discoloration of older leaves, disturbances in fructification
Cu	Sulfides sulfates, carbonates	Adsorbed, mobilization acid > basic, strong fixation of humus	Cu^{2+} and Cu-chelates	Bound as complexes, component of enzymes (plasto-cyanin, phenol oxidases)	Basal metabolism (photosynthesis, oxidases); nitrogen metabolism; sec. metabolism	Woody axes of shoots	Poor	Tip drying, leaf curl, spotty chloroses of young leaves

Mo	Molybdates, in silicates	Adsorbed, mobilization basic > acid	MoO_4^{2-}	In metal-organic compounds; component of enzymes	Nitrogen fixation (reductases), phosphorus metabolism, iron absorption and translocation		Poor	Disturbance of growth and deformation of shoots, browning of leaf edges
B	Tourmaline, borates	Adsorbed > soluble, availability acid > basic	HBO_3^{2-} $H_2BO_3^-$	Bound to carbohydrates as complexes; esteric binding	Carbohydrate transport and metabolism; phenol metabolism, activation of growth regulators (growth of pollen tubes)	Leaves, tips of shoots	Poor	Disturbance of growth (meristem necroses), diminished branching of roots, phloem necroses, disturbances of fructification, excessive cork formation
Cl	Salt, silicates	Soluble > adsorbed	Cl^-	Free as ion, mostly stored in cell sap	Colloidal effect (increases hydration); enzyme activation (photosynthesis)	Leaves	Good	Leaf curl, root thickening

flow passively into the water-filled dead tracheae and tracheids, following the concentration gradient; in addition, they are actively excreted into the vessels by parenchyma cells.

Moving through the *xylem*, the nutrient salts are distributed by the transpiration stream at higher levels in the plant. At the end points of the vascular network they diffuse through the cell walls to the surface of the protoplasts of the bundle parenchyma and are actively transported into the parenchyma. Cellular nutrient transport is again effected by the symplast, and in the process some of the salts are again stored in vacuoles. In the nutrient translocation chain, the rate-limiting stage is the conduction of ions through the symplast in the roots; the transpiration stream is usually capable of carrying much greater quantities of salts.

In addition to the xylem, another important route in nutrient translocation involves the *phloem*. The two long-distance translocation systems are linked at many sites, particularly in the roots and in the nodes of the stems. Along with the stream of metabolites, inorganic materials are shifted to sites where the need is greatest. Translocation via the sieve tubes is involved especially in the redistribution of mineral substances already incorporated into the plant. The various substances differ in the ease with which they can be redistributed (Tables 4.1 and 4.2); nutrients bound in organic compounds, such as those of N, P and S, can be readily shifted, as can the alkali ions and Cl$^-$. More difficulty is encountered in the case of the heavy metals and the ions of alkaline earths, especially calcium. The latter thus accumulates steadily in the leaves, which mark the end of the xylem translocation route.

4.2.3 Utilization and Deposition of Minerals in the Plant

The inorganic bioelements may be incorporated into the plant tissues, become components or activators of enzymes, or (through effects associated with colloid chemistry) regulate the degree of hydration of the protoplasm. A survey of the specific ways the elements are incorporated and operate, and of the sites in the plant where they are concentrated, is given in Tables 4.1 and 4.2. Details of the biochemistry of mineral incorporation can be found in textbooks of plant physiology and applied plant sciences.

4.2.3.1 The Incorporation of Nutrients

During the growing season, much of the total uptake and incorporation of minerals has been completed before the rapid increase in mass begins (Fig. 4.6). The most important nutrient elements must be made available at an early stage, and it is clear that an inadequate supply of minerals restricts the production of organic matter from the very start. The unfolding leaves of trees accumulate the chief nutritive elements — N, P, K and others — for later use. Eventually the rate of uptake of organic matter exceeds that of minerals; the ratio of organic dry matter to inorganic components begins to shift in favor of the organic matter, although the absolute quantities of mineral substances in individual leaves are not reduced. There is a decrease in absolute mineral content only if nutrients are transported out of the leaves. With increasing age, the elements Ca, S, and

Fig. 4.6. Mineral uptake and production of dry matter by growing and ripening oat plants. Mineral uptake is shown as the cumulative % of the total amount taken up in the season, and the production of dry matter as the cumulative % of the total increase in dry weight for the season. N and K are taken up at an especially high rate by young, rapidly growing plants, whereas Ca-uptake essentially parallels the formation of new tissue. After Scharrer and Mengel as cited by Mengel (1968)

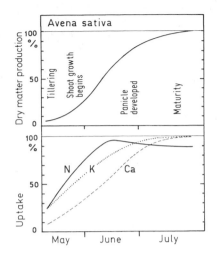

the other less easily moved elements Fe, Mn, and B accumulate in the leaves. On the other hand, the more mobile elements N, P, and above all K are most concentrated in young leaves, their concentrations declining as the leaves mature and age. As a result, during the course of a year there is a characteristic increase of the Ca : K ratio in the leaves.

4.2.3.2 The Ash Content of Dry Matter and the Composition of Plant Ash

A survey of the ash content of various plant groups is given in Table 4.3; the *average composition of plant ash* is detailed in Table 4.4. There are rather large quantities (1—5% of the dry matter) of the elements N, K, Ca, and in some plants Si. Mg, P, and S are found in amounts between 0.1 and 1%, and the trace-element content lies between 0.02% (Fe) and a few ppm. The proportions of the various bioelements can be characteristic of certain plant species and families, as well as of specific organs and ages of a plant. Trees and shrubs, as a rule, as well as some mountain plants contain more N than K, whereas in herbaceous plants the opposite tends to be true. The *ratio Ca : K* is particularly characteristic; in Caryophyllaceae, Primulaceae and Solanaceae potassium predominates, and in Fabaceae, Crassulaceae, and Brassicaceae there is more calcium. Grasses, sedges, palms, ferns and horsetails build up higher concentrations of Si than of Ca, so that this element can amount to as much as $^2/_3$ of the total ash; in diatoms, the skeletons of which are composed of silicates, Si can comprise more than 90% of the ash. Most plants contain somewhat more P than S, but the inverse ratio is also found: Brassicaceae always contain considerably more S than P. Finally, there are a number of plants that store large amounts of Na — an element normally found at the bottom of the list, just ahead of the trace elements. Chief among these plants are those growing on saline soils; these include many Chenopodiaceae as well as Brassicaceae and Apiaceae.

Within a given plant the leaves and cortical tissues produce the most ash and the woody organs, the least. In the foliage the elements stored preferentially are N, P, Ca, Mg, S, and in the case of grasses and palms Si as well. Flowers and fruits store mainly K, P and

Table 4.3. Average ash content of the dry matter in various groups of plants. From measurements by numerous authors

Bacteria	8—10%
Fungi	7— 8%
Planktonic algae without skeletal material	ca. 5%
Diatoms	up to 50%
Seaweed	10—20%
Mosses	2— 4%
Ferns	6—10%
Grasses	6—10%
Herbaceous dicotyledons	6—18%
Halophytes	10—55%
Cacti	10—16%
Ericaceous dwarf shrubs	
Leaves	3— 6%
Shoots	1— 2%
Broad-leaved trees	
Leaves	3— 4%
Wood	ca. 0.5%
Bark	3— 8%
Conifers	
Needles	ca. 4%
Wood	ca. 0.4%
Bark	3— 4%

S, the cortex of tree trunks contains relatively large amounts of Ca and Mn, and the wood of some (especially tropical) species stores Si and Al.

From the content and composition of the ash one can infer — in addition to the properties of the plant species — something about the nutrient supply in the place where the plants grew. Plants on particularly nutrient-poor soil — and to an even greater extent, those on acid soils — are low in ash (1—3% of the dry matter), as are epiphytes. Conversely, plants on saline soils have a high ash content (up to 55% of the dry matter), and their ash contains above-average amounts of Na, Mg, Cl, and S. Ruderal plants, growing on nutrient-rich soils, also contain relatively large amounts of Na, as well as a good deal of nitrate stored in the cell sap. Since the plants, though able to absorb nutrient salts in the soil preferentially, cannot entirely exclude any salt, the composition of the ash reflects the geochemical peculiarities of the habitat. A high nitrogen content is found particularly in plants of cold regions (tundra, taiga, cold-winter deserts, and the high mountains), calcium accumulates in plants on calcareous soils and in the vegetation of dry subtropical regions, and there are elevated amounts of Al, Fe, and Mn in plants on acid soils, of Si in tropical rain forests and savannas, and of Cl and S in halophytes. High concentrations of heavy metals characterize plants growing near ore deposits. Analysis of ash to measure habitat-dependent mineral accumulation can help in determining the presence of nutrient deficiencies and improper fertilization of crops; moreover, knowledge of their mineral content allows wild plants to be used as indicators of nutrient availability and ore deposits.

Table 4.4. Average content of mineral bioelements in plants, in the soil and in sea water; all data in parts per thousand. From compilations by Kalle (1958), Finck (1968), Fortescue and Marten (1970). Lists of data for various plant groups are given by, e. g., Höhne (1963), Duvigneaud and Denaeyer-de Smet (1973), Abd El Rahman et al. (1975), Golly et al. (1975), Klinge (1976), Baumeister and Ernst (1978)

Element	Land plants $(g \cdot kg^{-1}$ dry matter) Range	Mean	Stored in soil $(g \cdot kg^{-1}$ DM) Mean	Marine organisms $(g \cdot kg^{-1}$ DM) Mean	Sea water $(g \cdot l^{-1})$
N	10–50	20	1	50	0.0003
P	1–8	2	0.7	6	0.00003
S	0.5–8	1	0.7	10	0.9
K	5–50	10	14	10	0.4
Ca	5–50	10	14	5	0.4
Mg	1–10	2	5	4	1.3
Fe	0.05–1	0.1	38	0.4	0.00005
Mn	0.02–0.3	0.05	0.9	0.02	0.000005
Zn	0.01–0.1	0.02	0.05	0.2	0.000005
Cu	0.002–0.02	0.006	0.02	0.05	0.00001
Mo	0.0001–0.001	0.0002	0.002		
B	0.005–0.1	0.02	0.01	0.02	0.005
Cl	0.2–10	0.1	0.1	40	19.3

4.2.3.3 Nutrient Requirements and Excess Minerals

With respect to the quantities of inorganic elements available to plants, three basic nutritional states can be distinguished: deficiency, adequate supply, and injurious excess (Fig. 4.7).

When suffering from *nutrient deficiency* the plants are stunted, and in some cases they flower, bear fruits, and age prematurely. If the deficiency involves only some of the vital elements, or if the plant species requires extraordinary amounts of certain elements, *specific* deficiency symptoms can appear. These are best known for cultivated plants, but they can also be found in wild plants. The most important symptoms of specific nutrient deficiency in cultivated plants are given in Tables 4.1 and 4.2.

Fig. 4.7. Schematic representation of the relationship between inorganic nutrition (concentration of minerals in the plant) and the production of dry matter. In the range marked "dilution effect" the mineral concentration falls as dry matter is rapidly produced, though the amount of minerals is not reduced. After Drosdoff and after Prevot and Ollagnier as cited by Smith (1962)

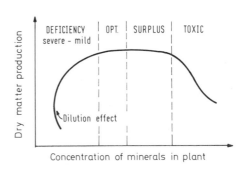

With an *adequate nutrient supply*, the actual amounts of the nutrients available can vary over wide ranges without noticeable effects on growth and development. Once the plant's requirements have been met, a moderate excess of certain nutrients seems to offer no further advantage for growth. But it cannot be excluded that other ecologically important properties conferring competitive advantages — such as resistance to parasites or extreme climatic situations — are in fact promoted.

In the range of *excessive concentrations*, inorganic nutrients can act as poisons, particularly if only one is present in excess. In nature such situations arise in saline and alkali soils, in ruderal habitats, in serpentine soils, and especially in soils rich in heavy metals and the rubble heaps near mines. Aquatic plants can be harmed by phosphate-rich drainage water.

4.2.4 The Elimination of Minerals

The ascending sap carries minerals into the shoot, where they accumulate in the cell walls or in the vacuoles. Small quantities of minerals are removed as components of the various materials eliminated by the plant. Most of the minerals incorporated are released, however, only when *parts of the plant die and are shed*. Loss and replacement of leaves and bark are thus a necessary and regular process of elimination in perennial plants.

Frey-Wyssling (1949) has proposed a scheme in which three processes of direct elimination are distinguished (Fig. 4.8). *Recretion* is the elimination of salts in the same form in which they were taken up. Recretion takes place over the entire surface of a plant, from which salts are washed away by rain. K^+, Na^+, Mg^{2+}, and Mn ions are leached out relatively easily. Many species growing in saline habitats have glands specialized for the elimination of salt, and species of saxifrage eliminate calcium through the hydathodes. *Secretion* is the release of assimilation products such as the sugar in nectar and the carbonic and amino acids given off by the roots. Planktonic algae secrete into the water quite large amounts of soluble carbohydrates, C_4 acids, and amino acids. *Excretion* is the elimination of products of intermediate metabolism and end products of catabolism.

The ecological significance of secretions and excretions lies not so much in regulating the composition of the plant as in their role in biotic interference; for example, they act

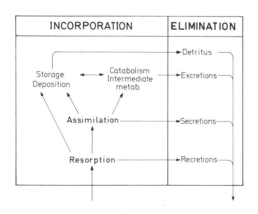

Fig. 4.8. Model of the turnover of inorganic matter in plants. Based in part on Frey-Wyssling (1949)

as attractants of pollinators, as deterrents to herbivores, and as inhibitors in intraspecific and interspecific competition (allelopathy and antibiosis). Substances with *allelopathic* effects are organic compounds given off by plants which injure other plants or prevent them from becoming established in the vicinity. In most cases these are ethylene, etheric oils, phenol compounds, alkaloids, glycosides, and cumarin derivatives released into the air, given off by the roots, or washed out of the shoot and into the ground by rain.

4.3 Nitrogen Utilization and Metabolism

Among the macronutrients, nitrogen is especially significant. In terms of quantity, it is fourth among the bioelements (see Fig. 1.1). Herbaceous plants contain 2—4% on the average, the leaves of deciduous trees 1—4%, evergreen needles and sclerophyll foliage 1—2%, and shoots and roots 0.5—1% nitrogen. Not uncommonly, then, increase in mass of a plant is limited by the supply of nitrogen. When insufficient nitrogen is available, greater amounts of carbohydrate are converted to storable forms (starch and fat) and utilized in secondary metabolism (e.g., increased lignin synthesis). If the nitrogen deficit becomes severe the plants are stunted, the individual cells are small and their walls thickened (nitrogen-deficiency sclerosis or peinomorphosis), and as a rule reproductive events and senescence set in before the normal time.

4.3.1 The Nitrogen Metabolism of Higher Plants

4.3.1.1 Nitrogen Uptake

Green plants utilize inorganically bound nitrogen; that is, they are autotrophic with respect to nitrogen as well as to carbon. Nitrogen is taken up from the soil as nitrate or ammonium ion. Most plants can meet their nitrogen requirements with either NO_3^- or NH_4^+, as long as the pH in the space occupied by the roots is suitable. Like all ion absorption (see Chap. 4.2.1), that of nitrogen requires energy and is thus dependent on respiration; plants growing on cold, poorly aerated soils often suffer from nitrogen deficiency.

4.3.1.2 Nitrogen Assimilation

The nitrogen taken up is incorporated into carbon compounds in amino groups, forming *amino acids* (Fig. 4.9). Amino acids are the basic compounds from which proteins, nucleic acids and other nitrogen compounds are synthesized.

The first step is the reduction of nitrate to nitrite, catalyzed by a chain of enzymes and cofactors; among these, *nitrate reductase* plays the decisive role. The activity of nitrate reductase is organ-specific (in trees it is greatest in the roots, and in many herbaceous plants — especially ruderal plants and nitrophytes — in the leaves; see Fig. 4.11) and age-dependent (activity is maximal during the young stages and in growing organs). It

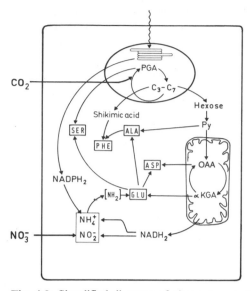

Fig. 4.9. Simplified diagram of nitrogen assimilation and its association with the basal metabolism of the cell. The reducing agents involved in the transfer of nitrate to the amino group are $NADH_2$ (provided via respiration) and $NADPH_2$ (via photosynthesis). The different amino acids arise by amination of intermediate and end products of the respiratory citric acid cycle and the dark reactions of photosynthesis. *PGA*, 3-phosphoglyceric acid; *Py*, pyruvate; *OAA*, oxaloacetate; *αKGA*, α-ketoglutaric acid; *GLU*, glutamic acid and related amino acids; *ASP*, asparagine and related acids; *ALA*, alanine and related acids; *PHE*, phenylalanine and related acids; *SER*, serine and related acids. Shikimic acid is derived from erythrose-4-phosphate (C_4 intermediate stage in the Calvin cycle). For a more detailed description of the pathways for synthesis see Magalhaes et al. (1974), Magalhaes (1975), and Beevers (1976); C_4 plants are treated by Raghavendra and Das (1978) and Neyra and Hageman (1978)

is induced by nitrate (substrate induction) and regulated according to the daily alternation of light and darkness (with peak activity at about the middle of the light period). *Nitrite reductase* is involved in reducing nitrite to NH_4. The energy and the "reducing power" for assimilative nitrate and nitrite reduction are provided by respiration ($NADH_2$) and — in chloroplast-containing cells — by photosynthesis ($NADPH_2$).

The actual process of assimilation is the *reductive amination of α-keto acids*; in higher plants the first of these is α-keto-glutaric acid, an intermediate product in the respiratory citric-acid cycle. Glutamic acid is formed, and this can transfer its NH_2 group to other α-keto acids produced in glycolysis and the citric-acid cycle (transamination). From these *primary amino acids*, others are derived; their carbon chains are obtained from intermediate products of carbohydrate metabolism, including the Calvin cycle and the oxidative pentose-phosphate cycle. Amino acids are also formed during photosynthesis (glycine, serine, and alanine in C_3 plants, asparaginic acid in C_4 plants) and during photorespiration (glycine, serine).

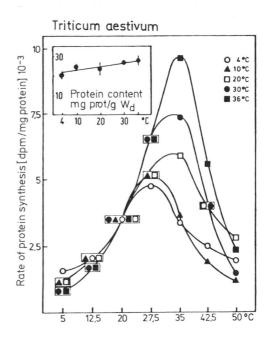

Fig. 4.10. The variation in temperature-dependence of protein synthesis by wheat plants acclimated to different temperatures (indicated by the *symbols*). As a result of adaptation to warm conditions, the temperature optimum for protein synthesis is shifted to higher temperatures and the protein content of the shoot (*inset*) increases. After Weidner and Ziemens (1975)

Protein metabolism is also organ-specific and age-dependent; organs and tissues that are growing and storing materials typically synthesize protein at an especially high rate, whereas in aging leaves and parts of flowers protein breakdown predominates. Among the environmental influences, effects on protein metabolism are exerted chiefly by temperature and stress factors such as drought and excess salinity.

The *temperature optimum* for protein synthesis, as a rule, is limited to a narrow band, because all the metabolic events that precede or participate in the production of protein (active nitrogen uptake and translocation, basal metabolic processes that provide metabolites, amino-acid synthesis, transcription and translation) are themselves temperature-dependent, with different temperature coefficients — despite which their turnover rates must be mutually compatible. As a result, protein synthesis is characterized by the capacity for especially flexible and prompt *adaptation* (Fig. 4.10). This is a central prerequisite for temperature adaptation of the plant at all levels: molecular, functional, and morphological.

When a plant is under *stress* protein synthesis is inhibited and protein breakdown accelerated, so that there is a considerable increase in free amino acids and amides. A distinguishing characteristic of stress-disturbed protein metabolism is alteration of the proportions of the amino acids, in particular a significant increase in proline concentration (cf. Chap. 4.4.2.2).

4.3.1.3 Nitrogen Distribution in the Plant

Some of the inorganic nitrogen compounds taken up are assimilated in the roots, while the rest are carried into the shoot by the sap flow and there incorporated into organic compounds. In addition, nitrate is stored in the cell sap of root and shoot. The amino

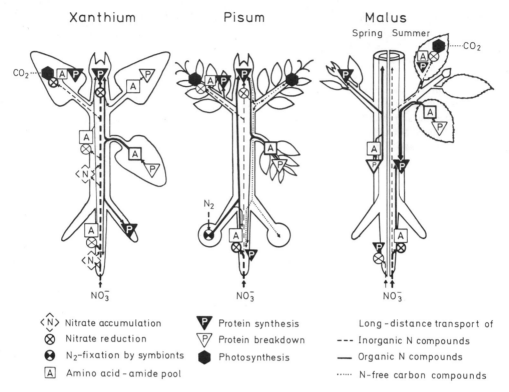

Xanthium   Pisum   Malus

⟨N̂⟩ Nitrate accumulation

⟨N̂⟩ Nitrate accumulation   ▼ Protein synthesis   Long-distance transport of

⊗ Nitrate reduction   ▽ Protein breakdown   --- Inorganic N compounds

⊗ N₂-fixation by symbionts   ⬣ Photosynthesis   —— Organic N compounds

Ⓐ Amino acid-amide pool   ······ N-free carbon compounds

Fig. 4.11. Diagram of nitrogen distribution in a one-year-old ruderal plant (*Xanthium pennsylvanicum*), a member of the Fabales with N_2 fixation in the root nodules (e.g., *Pisum sativum* or *Lupinus albus*) and a temperate-zone deciduous tree with intermittent shooting in spring (e.g., *Malus domestica* or *Fagus sylvatica*). For the herbaceous plants, the situation during the main growth phase is illustrated; during flowering and ripening of fruit the reproductive organs receive most of the mobile nitrogen compounds. For the apple tree (*Malus domestica*) two phases are shown: on the *left* the movement of amino acids, amides and soluble proteins from storage tissues in trunk and branches into the tips of the new shoots, and on the *right* the replenishment of the storage tissues in summer and autumn. The *thickness of the arrows* indicates the relative intensity of nitrate reduction and transport processes. Long-distance transport of photosynthates, which supplies the nitrogen-assimilating tissues with carbon skeletons, is shown only for *Pisum*. Simplified diagram based on data of Thomas (1927), Gäumann (1935), Wallace and Pate (1967), Pate (1976), and Beevers (1976)

acids may be used where they are formed, for the biosynthesis of macromolecules (protein, nucleic acids), or may be transferred to other organs. The most important *translocatable forms* are the amino acids glutamic acid and asparaginic acid, as well as their amides glutamine and asparagine; in certain species and families citrullin and allantoic acid are also found. When carbohydrate is abundant or nitrogen limited, amino acids predominate; when there is a good supply of nitrogen, amides are the predominant form. Translocatable organic nitrogen compounds are also produced in large amounts when proteins are broken down in aging or stress-damaged parts of a plant and — in woody plants — during the mobilization of storage proteins. Long-distance transport of these substances occurs by way of the phloem. In leaves, growing parts of shoots and ripening fruits the organic transport forms of nitrogen provide amino groups for the synthesis of amino acids and for transamination, and serve as building blocks for protein synthesis and cell growth.

Vascular plants exhibit various *types of nitrogen distribution* (Fig. 4.11), depending on the site of most active nitrate reduction, the nature of the chief nitrogen compounds transported and stored, and the intensity and direction of protein metabolism at different times.

Although *annual herbs* on nitrogen-rich soils (for example, *Xanthium pennsylvanicum*, species of *Chenopodium*, and various Apiaceae) accumulate large amounts of nitrate in the cell sap of the roots and older parts of the shoot, little is assimilated. When nitrogen is most actively taken up, just before flowering, the nitrate-nitrogen concentration in the stems can amount to 20—40%, and in the roots to 15—20%, of the total nitrogen content of the respective organ. Most of the nitrogen taken in by the roots is translocated in inorganic form, as nitrate, along the conducting elements of the xylem to the leaves and the tips of the shoots, where nitrate reductase activity is greatest in this type of plant. The growing parts of the plant also receive, by way of the phloem, the amino acids and amides that are formed in excess by leaves especially productive photosynthetically or are liberated by protein breakdown in the lowest leaves that have already begun to turn yellow. The phloem is also the pathway by which the root is supplied with organic nitrogen compounds when excess amounts of these are produced in the shoot.

In *Fabales* with root nodules containing symbiotic nitrogen-binding bacteria (cf. p. 181), inorganic nitrogen is assimilated primarily in the root, and is conducted to the rest of the plant in the *organic nitrogen-transport form*. The main products of the nodules are amines; these rise into the shoot through the xylem. Again, a third source of nitrogen for the growing tips of the shoot and later for the flowers and fruits is proteolytically derived amino acids.

In *trees* — to the extent that they have been studied — nitrate reduction and amino-acid synthesis also occurs mainly in the root; only a little nitrate reaches the parts above ground, where it is reduced and assimilated in the leaves. In midsummer and autumn the amino acids and amides formed in the root are gradually shifted to the trunk and branches; there they accumulate chiefly in the bark. Prior to abscission the products of protein breakdown migrate out of the leaves and are stored in the axial parts of the tree. Thus a considerable supply of amino acids and proteins is available for the growth of new leaves and shoots the following spring; these reserves are sufficient to supply all the nitrogen-containing precursors required.

4.3.1.4 Nitrogen Excretion

In the carbon metabolism of green plants the inorganic carbon source, CO_2, is taken from the air and the end product —again CO_2 — is returned to the atmosphere. By contrast, higher plants take inorganic nitrogen compounds from their surroundings but break down the products of assimilation to the inorganic level (NH_3) themselves only in exceptional cases and in negligibly small quantities. Rather, the nitrogen is eliminated primarily in an organically bound form: roots release amino acids and other organic nitrogen compounds, and a greater amount of nitrogen is lost by the falling of leaves and fruits.

The return of the organic nitrogen compounds to the original inorganic form is accomplished by nitrogen-heterotrophic organisms (animals, many fungi and bacteria). Only through their activity is the nitrogen cycle of the higher plants completed; it is thus dependent on the nitrogen metabolism of microorganisms.

4.3.2 Nitrogen Fixation by Microorganisms

There are diazotroph microorganisms that can make use of the exceedingly inert atmospheric nitrogen. These N_2-fixing organisms represent the most ecologically significant level of N-autotrophy. All of them are prokaryotic — cyanophytes and bacteria — some living free in the soil and others as symbionts.

4.3.2.1 Nonsymbiotic Nitrogen Fixation

In the free-living microorganisms, nitrogen fixation was first demonstrated by S. Winogradsky for the soil bacteria *Clostridium pasteurianum* and *Azotobacter chroococcum*. But there are many other species of bacteria that incorporate molecular nitrogen, chief among them the photoautotrophic bacteria and certain hydrogen-oxidizing bacteria living in water, as well as some cyanophytes of genera which form heterocysts — for example, *Nostoc, Anabaena, Calothrix*, and *Mastigocladus*. These blue-green algae are self-sufficient with respect to the essential elements, being autotrophic for carbon as well as for nitrogen. Nitrogen-fixing blue-green algae are found in bodies of water and are among the first to colonize raw soils, particularly in mountains and the arctic, in thermal springs and in other extreme habitats even on the leaves of trees.

Binding of atmospheric nitrogen is begun by the reductive splitting of the N_2 molecule. This strongly endergonic reaction is catalyzed by the *nitrogenase* system, a complex of two proteins, one with iron and one with molybdenum and iron as activators. The energy and the electrons necessary for reduction are supplied by respiration (Fig. 4.12). The consumption of easily accessible organic food is correspondingly high. Under laboratory conditions *Azotobacter* requires 50—100 g glucose per gram of nitrogen fixed, *Clostridium* 100—200 g, and *Klebsiella* 200 g. Under *natural conditions*, therefore, it is to be assumed that nitrogen fixation by free-living microorganisms is limited chiefly by the supply of suitable food substrates. On raw-humus soils containing organic substances difficult to decompose, nitrogen-fixing bacteria do not thrive. Because nitrogenase activity is optimal at 20°—30° C and comes to a halt near 0° C, the rate of nitrogen fixation in cold regions is low; in the subarctic and arctic, depending on location, it reaches 0.1—2 kg $N_2 \cdot ha^{-1} \cdot yr^{-1}$. The performance of free-living nitrogen-fixing organisms is best in warm, permanently damp habitats; blue-green algae in rice paddies bind 50—70 kg $\cdot ha^{-1} \cdot yr^{-1}$. Algae living on the Australian Barrier Reef process up to 30 kg $N_2 \cdot ha^{-1} \cdot yr^{-1}$. After green-manuring of fields in the temperate zone, one can ex-

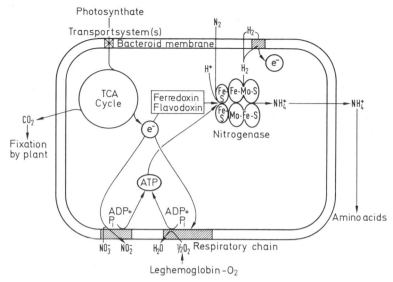

Fig. 4.12. Diagram of aerial nitrogen fixation by symbiotic bacteria. *TCA-Cycle*: respiratory citric acid cycle. After Shanmugam et al. (1978)

pect about $2 \, kg \, N_2 \cdot ha^{-1} \cdot yr^{-1}$, whereas the yield in the subtropics can amount to 50 kg $N_2 \cdot ha^{-1} \cdot yr^{-1}$. Nitrogen-fixing microorganisms also live as epiphytes; on the leaves of tropical trees they are able to bind $1.5-8 \, kg \, N_2 \cdot ha^{-1} \cdot yr^{-1}$.

4.3.2.2 Symbiotic Nitrogen Fixation

Symbionts which fix nitrogen solve the problem of carbohydrate supply by living in the cells of autotrophic plants. In this way they manage to turn over a considerable amount of nitrogen, which is also to the advantage of the plant in which they live. J. S. Pate and coworkers measured the carbon and nitrogen balances of certain annual *Fabales* (pea, lupin, cowpea). They found that during the developmental phase, up to the onset of flowering, $^1/_3$ to $^2/_3$ of the carbohydrates formed in the shoot are translocated to the roots and their nodules; of this amount, about $^1/_3$ is respired, $^1/_5$ used for nodule growth, and the remaining 40—50% returned to the shoot together with the fixed nitrogen. For every gram of nitrogen, in the form of amino acids and amides, 4 g of carbon in the form of carbohydrates are required at the time when the host plant and nodule bacteria show the greatest synthetic activity. For this reason the amount of N_2 fixed by the symbionts depends very much on the supply of photosynthates. When photosynthesis by the host plant is impaired, bacterial nitrogen fixation falls off, whereas an increase in photosynthetic production (for example, by exposure to elevated CO_2 concentrations) can triple the yield of organically bound nitrogen.

Host and symbiont form an ecological unit, just as much an independent entity with respect to acquisition of the carbon and nitrogen compounds needed for protoplasmic development as are the blue-green algae. This mutual adjustment also finds expression in the temperature-dependence of nitrogenase activity by the symbionts, which is

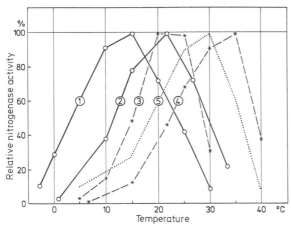

Fig. 4.13. Temperature-dependence of nitrogenase activity in root nodules with N_2-fixing symbionts and in blue-green algae. *1, Astragalus alpinus*, Norway; *2, Medicago sativa*, greenhouse culture at 16° C; *3, Alnus glutinosa*, England; *4, Casuarina equisetifolia*, Malaysia; *5, Anabaena cylindrica*, Cambridge collection. After Granhall and Lid-Torsvik, 1975 (*1*) and Waughman, 1977 (*2—5*). The temperature-dependence curve for nitrogenase activity of subarctic lichens with blue-green algae resembles Curve *2* (cf. Kallio, 1975)

clearly related to the prevailing temperatures within the range of distribution of the host plant (Fig. 4.13). The symbiotic N_2-fixing organisms bind nitrogen at a higher rate than do the free-living microorganisms; on the average one can expect fixation of 200 kg N per hectare per season, and under optimal conditions the yield is twice as great. The most important symbiotic N_2-fixing organisms are bacteria of the genus *Rhizobium*, which live in nodules of the roots of legumes. There are a few species and many races of *Rhizobium*, all specialized for certain species of Fabales. Sometimes free-living nitrogen-fixing bacteria maintain a loose symbiotic relationship (*association symbiosis*), colonizing the root network of various plants and the mycorrhiza of trees. *Azotobacter paspali*, in association symbiosis with the C_4 grass *Paspalum notatum*, binds 90 kg $N_2 \cdot ha^{-1} \cdot yr^{-1}$, twice as much as free-living N_2-fixers under comparable conditions. Another group of nitrogen-fixing symbionts comprises the *Actinomycetes*, primarily those of the genus *Frankia*, which form root nodules in *Alnus, Myrica, Hippophae, Elaeagnus, Casuarina, Ceanothus*, and certain other woody plants. All these shrubs and trees grow in nitrogen-poor soils. Blue-green algae (for example, *Nostoc* and *Anabaena*) enter reciprocal trophic relationships with mosses, lichens, ferns and higher plants.

4.4 Habitat-Related Aspects of Mineral Metabolism

4.4.1 Calcicole and Calcifuge Plants

Some species of plants can be found only on chalk soils, and others only on silicaceous and sandy soils poor in calcium. The attempt to discover the causes of this striking habi-

Table 4.5. Plants with susceptibility to lime-chlorosis and to aluminium toxicity (Grime and Hodgson, 1969)

Calcifuge plants susceptible to lime-chlorosis	Calcicole plants susceptible to aluminium toxicity
Deschampsia flexuosa	*Hordeum vulgare*
Holcus mollis	*Agrostis stolonifera*
Paspalum dilatum	*Festuca pratensis*
Lathyrus montanus	*Beta vulgaris*
Galium saxatile	*Medicago sativa*
Eucalyptus dalrympliana	*Asperula cynanchica*
Eucalyptus gomphocephala	*Scabiosa columbaria*
Eucalyptus gunnii	*Lactuca sativa*

tat-dependence of plant distribution was one of the first tasks undertaken by analytical ecologists. As early as the beginning of the last century F. Unger, on the basis of observations of the Tyrolean mountain flora, tried to explain the relationships involved. The problem is extraordinarily complex, and even today it cannot be regarded as solved.

Calcareous soils differ from others in the following major ways: they are usually more permeable to water and therefore dryer and warmer than silicaceous soils, but they are primarily distinguished by the fact that they contain much greater amounts of Ca^{2+} and HCO_3^-. Because of this, calcareous soils are buffered toward a higher pH than other soils, and show a neutral to weakly alkaline reaction. Nitrogen is more rapidly mineralized in calcareous soils; P, Fe, Mn and the heavy metals are less accessible than in acid soils. Silicon does not appear to be involved in the question. It seems reasonable to assume that all these edaphic factors affect the inorganic metabolism and the vigor of plants, and that interspecific differences in resistance and ability to compete under a given constellation of such influences affect the composition of the plant community.

Plants with a marked *preference for calcareous soils* may well be damaged by iron, manganese, and above all by aluminium ions, which are liberated preferentially in acid soils. Calcifugous plants, by contrast, are capable of binding the heavy metal ions in certain complexes, and a surplus of Al^{3+} does them no harm. On the other hand, signs of phosphorus and iron deficiency ("lime-chloroses"; Table 4.5) appear if these species are transplanted to calcareous soils. Strict calcifuges are also hypersensitive to HCO_3^- and Ca^{2+}. Peat moss and calcifugous grasses such as *Deschampsia flexuosa* produce large quantities of malate in their roots if the concentration of HCO_3^- is too great; this acts to inhibit growth and can lead to root damage. Plants differ widely in their management of calcium (Fig. 4.14). Many Brassicaceae and some Fabales take up large amounts of Ca^{2+} and store it in the cell sap. Polygonaceae, Chenopodiaceae, most Caryophyllaceae and the representatives of various other plant families, on the other hand, cannot tolerate very large concentrations of dissolved calcium. They bind the Ca^{2+} in the vacuoles by way of oxalate; the more Ca^{2+} they are forced to take up, the

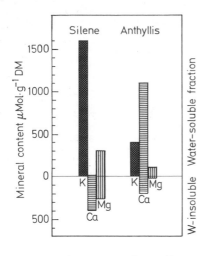

Fig. 4.14. Content of inorganic matter in the leaves of *Silene vulgaris* and *Anthyllis vulneraria* from chalky habitats. *Silene* binds the Ca taken up in water-insoluble form, as the oxalate, and contains a great deal of K. In contrast, *Anthyllis* can tolerate a high Ca level, and there is more Ca than K in the water-soluble fraction. After Kinzel (1969), Horak and Kinzel (1971), see also Kinzel (1980)

more oxalate accumulates. Representatives of these families can maintain themselves on calcareous soils only if their acid metabolisms permit sufficient amounts of oxalate to be removed from circulation.

4.4.2 Plants of Saline Habitats

4.4.2.1 Habitat Characteristics

All saline habitats have in common a higher than normal content of readily soluble salts. The ocean, salt lakes and saline ponds are *aquatic* saline habitats; on land there are *saline soils* under both humid and arid climatic conditions. In regions of heavy precipitation, it is possible for the soil to become salty in the spray region of the tidal zone, on dunes, and in marshes; saline soils can also be found in the vicinity of flowing waters that have been in contact with salt deposits. Moreover, the salt content of the soil can be increased if streets are salted to keep them free of ice. The salinity of the soil is greatly increased in dry areas, if the evaporation from the soil is greater in the course of a year than the amount of precipitation that infiltrates the soil (cf. Fig. 5.26). Especially large amounts of salt accumulate near the ocean (mangrove belt), as well as in places where the water table is high, in low regions with no drainage (salt flats), and in intensively irrigated areas where drainage is insufficient.

Only in the open ocean does the salt content remain constant. Even in the tidal zone the salinity varies widely, and in terrestrial saline habitats the easily transported salts continually shift up and down as the water moves in the soil. During the growing season salts steadily accumulate within the plant cover as residues of evaporation; after the plant parts have died off these are washed out and returned to the soil. A survey of salt turnover and movement in halophyte habitats is given in Fig. 4.15.

4.4.2.2 The Effects of High Salt Concentrations on Plants

The vigor of plants in saline habitats depends upon the concentration and the chemical composition of the soil solution. Saline soils in humid regions contain predominantly

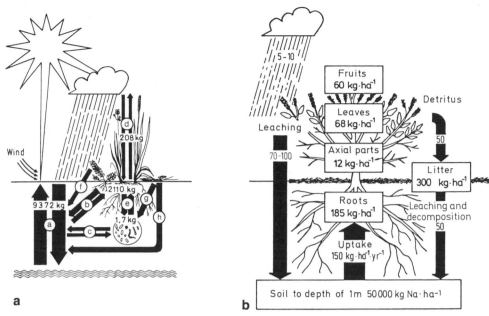

Fig. 4.15. a Circulation of sodium chloride in a halophyte habitat under humid climatic conditions (Germany). Just under the surface of the soil is a saline ground-water horizon; during the growing season the soil solution contains 1.9% NaCl. *a*, NaCl transport in the soil by capillary ascent during evaporation and by percolating water after precipitation; *b*, NaCl uptake by vascular plants (*Triglochin maritima, Juncus gerardi, Glaux maritima*) and NaCl release after the roots die in fall and winter; *c*, NaCl uptake by bacteria in the soil and release by dying bacteria; *d*, distribution and accumulation in the above-ground parts of the plants in spring and summer; *e*, salt uptake and release by bacteria in association with halophytes; *f*, leaching from living plants; *g*, NaCl uptake by bacteria during decomposition of dead parts of plants; *h*, leaching from plant detritus and litter. All quantities in kg NaCl per ha in the top 10 cm of soil. After Steubing and Dapper (1964). **b** Sodium turnover and content of Na^+ in a halophyte habitat under arid climatic conditions (*Atriplex confertifolia* community, saline desert in Utah). Turnover is given in kg $Na^+ \cdot ha^{-1} \cdot yr^{-1}$, content in kg $Na^+ \cdot ha^{-1}$. After Breckle (1976, and pers. comm.)

NaCl. Neutral salty soils of this sort also occur in dry regions, but more often the saline soils of steppe and desert contain sulfates and carbonates of Na, Mg, and Ca, which tend to make them more alkaline (solonez soils with pH 8—10). These salts affect the plants through the osmotic retention of water and the specific ionic effect on the protoplasm.

Osmotic Effect of the Salts. Salt solutions retain water, so that as the salt concentration increases water becomes less and less accessible to the plants. A 0.5% NaCl solution has an osmotic potential of −4.2 bar, that of a 1% solution is −8.3 bar, and that of a 3% solution (the concentration of sea water) is −20 bar. Plants can draw water from a saline substrate only if they can produce an osmotic potential lower than that of the soil solution. Plants adapted to life in saline habitats (*halophytes*) achieve this by *accumulation of salt in the cell sap*. By this means they compensate for the osmotic

185

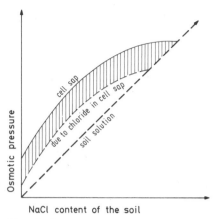

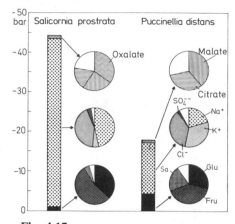

Fig. 4.16   **Fig. 4.17**

Fig. 4.16. Diagram of the relationship between the osmotic pressure of the soil solution (as the salt content increases) and the ability of halophytes to compensate. With increasing NaCl concentration in the soil, the osmotic pressure in the soil solution increases linearly. Halophytes overcome the correspondingly greater difficulty in withdrawing water from the soil, by taking up salt in excess if necessary and accumulating it in the cell sap (up to a certain limiting value given by the intersection of the curve showing the fraction due to chloride in the cell sap with the straight line). Added to the osmotic effect of this salt storage is that of other cell-sap components, primarily sugars (the osmotic effectiveness of nonchlorides in the cell sap is indicated by the *cross-hatched region of the curve*). Only when the salt concentration in the soil is very high are the halophytes no longer able to compensate for the raised osmotic pressure of the soil. After Walter (1960)

Fig. 4.17. Total osmotic potential of the cell sap of *Salicornia prostrata* (Na-Cl halophyte) and of *Puccinellia distans* (K-Cl glycohalophyte), with component potentials as indicated. *Blocks*, components due to inorganic ions (*stippling*), organic ions (*cross-hatched*) and soluble carbohydrates (*black*). *Circles*, the proportions of various ions and sugars involved in each component potential. *Glu*, glucose; *Fru*, fructose; *Sa*, saccharose. *White sectors*, the remaining ions and sugars. After Albert (pers. comm.), based on data of Albert and Popp (1977, 1978)

potential in the saline soil (Fig. 4.16). The concentrations of salt accumulated, as well as the *ion ratios* Na^+/K^+ and Cl^-/SO_4^{2-} in the cell sap of halophytes are very often specific to the family and species of the plant (Fig. 4.17). Many dicotyledonous halophytes accumulate more Na^+ than K^+, whereas in most monocotyledonous halophytes the amount of K^+ equals or exceeds that of Na^+. The anion most commonly accumulated is Cl^- (for example, in *Salicornia*, species of *Suaeda*, *Atriplex confertifolia* and many grasses and rushes). Sulfate is the predominant ion accumulated by Plumbaginaceae, *Plantago maritima*, *Lepidum crassifolium*, species of *Tamarix* and various other dicotyledons. Different species of a single genus can be of different types: in *Salsola kali* $K^+ > Na^+$ and $Cl^- > SO_4^{2-}$ and in *Salsola turcmanica* $Na^+ > K^+$ and $Cl^- > SO_4^{2-}$, while *Salsola rigida* combines $Na^+ > K^+$ with $SO_4^{2-} > Cl^-$. Among halophytes the monocots, grasses in particular, contain less salt than the dicots; in the former ("glycohalophytes") the very low osmotic potential in the cell sap is maintained by storage of soluble carbohydrates along with the salts.

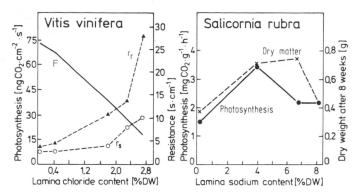

Fig. 4.18. Changes in photosynthesis, CO_2 diffusion resistances, and dry-matter accumulation in response to increasing sodium and chloride levels and to salt stress. The photosynthesis of the salt-sensitive grapevine is reduced primarily because of impairment of the nonstomatal processes. Obligate halophytes like *Salicornia*, by contrast, require NaCl for normal metabolic function. After Downton (1977) and Tiku (1976). The effect of salt concentration on germination and dry-matter production in *Salicornia* is also described by Kreeb (1965); for salinity effects on photosynthesis in littoral seaweeds see Zavodnik (1975)

Specific Ionic Effects and Salt Stress. Excess of Na^+ and, to an even greater extent, of Cl^- has a marked tendency to cause protoplasmic swelling; moreover, it affects the *activity of enzymes* (for a survey see Flowers et al., 1977), so that quantitative and qualitative changes occur in both basal and constructive metabolism (Fig. 4.18). These changes result in inadequate energy production by photophosphorylation and phosphorylation in the respiratory chain, in disturbances of nitrogen assimilation, in alterations in the amino acid pattern (relative increase in proline), and in abnormalities of protein metabolism that cause the formation of toxic intermediate and end products. When the NaCl content of the soil is high the *uptake of mineral nutrients* (K^+ and Ca^{2+} in particular) is reduced. Dry-matter production and growth rate decline (Fig. 4.19), with root growth especially impaired. The buds of *salt-damaged* woody plants appear late and are stunted, and the leaves are small; cells die and cause necroses in roots, buds, leaf margins, and shoot tips. The leaves become yellow and dry before the growing season has ended, and finally whole portions of the shoot dry out — often in sectors, correlated with the drying of the associated root sectors. Moreover, hardening to freezing temperatures may be less successful.

4.4.2.3 Salt Resistance

Salt resistance is the ability of a plant to withstand the presence of excess salts (especially Na^+, Cl^- and SO_4^{2-}) without serious impairment of vital function. Salt *tolerance* is a property of the protoplasm that enables it to tolerate more or less well, depending on species, tissue type and level of vitality, the changed ionic ratios associated with salt stress and the toxic and osmotic effects associated with increased ion concentration (see Fig. 4.21). Resistant protoplasts can survive 4—8% NaCl, whereas salt-sensitive protoplasts are destroyed in solutions of as little as 1—1.5% NaCl. One

187

Table 4.6. Sensitivity of nonhalophytic trees and shrubs to NaCl under winter conditions. ● = sensitive to soil salinity; × = sensitive to salt spray. Toxic limits for chloride content of leaves in early summer: for salt-sensitive deciduous trees and shrubs, 0.3—0.5% Cl in dry matter; for sensitive conifers, 0.2—0.4%; for salt-resistant deciduous trees and shrubs, 0.8—1.6%; for resistant conifers about 0.6%. The table includes the measurements of many authors, taken from the lists of species in Sucoff (1975) and Meyer (1978)

Relatively salt-sensitive	Relatively salt-tolerant
Deciduous trees and shrubs	
● × *Acer negundo*	*Acer platanoides*
× *Berberis vulgaris*	*Betula papyrifera, pendula*
× *Cornus mas* and *sanguinea*	*Caragana arborescens*
● *Evonymus europaeus* and *alatus*	*Elaeagnus angustifolia*
● × *Fagus sylvatica*	*Fraxinus excelsior* and *americana*
● *Juglans nigra*	*Gleditsia triacanthos*
● *Ligustrum vulgare*	*Hippophae rhamnoides*
× *Lonicera caerulea*	*Lonicera periclymenum*
× *Lonicera tatarica*	*Lonicera xylosteum*
● *Platanus acerifolia*	*Lycium halimifolium*
● *Populus nigra* cv. *italica*	*Populus canadensis, alba, tremuloides*
× *Prunus serotina*	*Potentilla fruticosa*
× *Rosa canina*	*Ribes aureum* and *nigrum*
× *Rosa rugosa*, certain varieties	*Rosa rugosa*, certain varieties
× *Sambucus nigra*	*Robinia pseudacacia*
● *Salix purpurea*	*Salix alba, fragilis, viminalis*
× *Syringa vulgaris*	*Symphoricarpus racemosus*
● *Tilia americana*	*Tilia platyphylla*
× *Viburnum lantana* and *opulus*	*Ulmus americana* and *pumila*
Conifers	
× *Abies balsamea*	*Juniperus chinensis* cv. *pfitzneriana*
× *Chamaecyparis pisifera*	*Larix decidua* and *leptolepis*
● × *Picea abies* and *glauca*	*Pinus mugo*
● *Pinus strobus*	*Pinus nigra*
× *Pinus sylvestris*	*Pinus ponderosa*
● × *Pseudotsuga menziesii*	
× *Taxus baccata*	

important factor in protoplasmic salt resistance seems to be that the salt ions taken into the cell are *nonuniformly compartmented*. Most are stored in the vacuole, leaving the cytoplasm relatively low in salt. As a result, the salt stress to which the cytoplasmic enzyme systems are directly exposed is less. Salt compartmentation is brought about and maintained by ion pumps in the boundary layers of the cytoplasm. The osmotic disequilibrium between cytoplasm and vacuole thus produced is compensated by the accumulation of organic compounds in the cytoplasm.

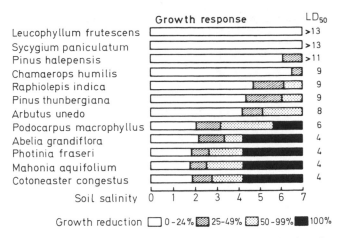

Fig. 4.19. Salt resistance of ornamental shrubs and trees (indicated on the *right* by the dose for 50% lethality) and growth reduction on saline soils. Soil salinity and LD_{50} are given in terms of the electrical conductivity (siemens $\cdot$ cm^{-1}) of the saturation extract (Francois and Clark, 1978)

There are some extraordinarily *salt-resistant organisms*. The plant flagellate *Dunaliella salina* lives in concentrated salt pools; the enzymes of the halophilic bacterium *Pseudomonas salinarum* and the yeast *Debaryomyces hansenii* remain functional even at concentrations of 20—24% NaCl. Halophytic vascular plants can thrive on soils with salt concentrations of 2—6% or more (in the extreme case, up to 20%), and can hold in their cell sap NaCl corresponding to a 10% solution. Of course, if plants are to colonize saline habitats the *process of germination* and the salt resistance of the seedlings must be adjusted accordingly. Only a few halophytes — *Tamarix, Suaeda depressa*, and *Halocnemum strobilaceum,* for example — germinate with a soil-solution salinity of 3% or more; the limit for germination and growth is usually 2% or less. The seedlings and young plants of halophytic species are thus more salt-sensitive than the full-grown plants. Moreover, young plants are exposed to the greatest danger, for in the upper layers of soil occupied by their roots, as a rule, the salt content is highest.

For *nonhalophytes*, salt-resistance is a factor in the natural selection determining *colonization along the shore* (the zonation of vegetation perpendicular to the beach). Agronomists, gardeners and foresters are concerned with the *salt-resistance of crop plants* in their efforts to utilize potentially arable saline soils in dry regions, especially in the subtropics. Protection and proper management of the environment requires an understanding of the resistance of trees, shrubs, and grasses to the de-icing salts (NaCl, $MgCl_2$, $CaCl_2$) spread in winter on adjacent streets and highways. The salt resistance of certain crops grown as food for animals and humans is given in Fig. 4.20. Table 4.6 summarizes the ability of trees and shrubs commonly planted along roads to endure salt stress. The resistant species are not only less vulnerable to direct salt damage; growth and yield of these plants are also less affected than are those of more sensitive species.

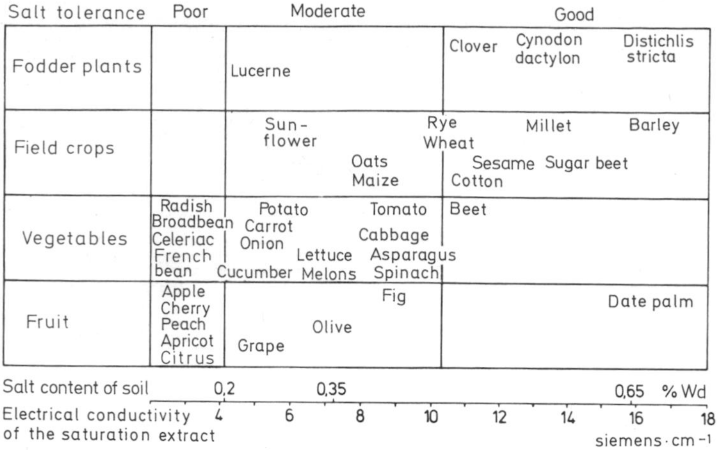

Salt tolerance   Poor   Moderate   Good

	Poor	Moderate		Good		
Fodder plants		Lucerne		Clover	Cynodon dactylon	Distichlis stricta
Field crops		Sun-flower	Rye Wheat	Millet	Barley	
			Oats Maize	Sesame Sugar beet		
			Cotton			
Vegetables	Radish Broadbean Celeriac French bean	Potato Carrot Onion Cucumber	Tomato Cabbage Lettuce Melons Asparagus Spinach	Beet		
Fruit	Apple Cherry Peach Apricot Citrus	Grape	Olive	Fig	Date palm	

Salt content of soil   0,2   0,35   0,65   % Wd

Electrical conductivity   4   6   8   10   12   14   16   18
of the saturation extract   siemens·cm^{-1}

Fig. 4.20. Salt resistance of various crop plants. The different species are placed in the table above the soil salt content at which a 50% reduction in yield is to be expected. In grain the increase in green mass is restricted more strongly than the kernel yield. Salt tolerance depends on state of development; barley becomes more resistant as it grows, and maize becomes more sensitive. After Kreeb, Peterson, Strogonov and Thorne as cited by Kreeb (1965)

4.4.2.4 Regulation of Salt Content in Halophytes

Halophytes must be able to take up and accumulate salts, in order to withdraw osmotically bound water from the soil; on the other hand, if these salts accumulated progressively throughout the plant's life, the inevitable result would be a reduction in yield and, eventually, toxic effects. Even a high degree of tolerance fails when stress is prolonged and continually increasing. Under such conditions *protective and compensatory mechanisms* are vital — mechanisms that shield the protoplasm from the effects of the stressor, or at least weaken or delay them. There are various means by which halophytes may regulate their salt turnover (Fig. 4.21).

Salt Filtration: Some mangrove trees (e.g., *Rhizophora*) greatly reduce the salinity of the water in their conducting systems by ultrafiltration at the plasmalemma of the root parenchyma cells.

Salt-Transport Prevention: In the mimosacean *Prosopis farcta*, the transport of salt into the leaves is prevented. The salt ions, Na^+ in particular, are taken up by the roots, but they are retained there or in the trunk. A similar principle of regulation can be observed in various crop plants, Fabales in particular.

Salt Elimination: There are various means by which a plant can rid itself of excess salt: exudation and recretion at the surface of the shoot, and shedding of plant parts heavily loaded with salt. Salt-recreting glands actively eliminate salts, thus keeping their accumulation in the leaves within certain limits. Such glands are found in various mangrove plants (for example, *Avicennia*), species of *Tamarix*, *Glaux maritima*, various Plumbaginaecea, and halophilic grasses such as *Spartina*,

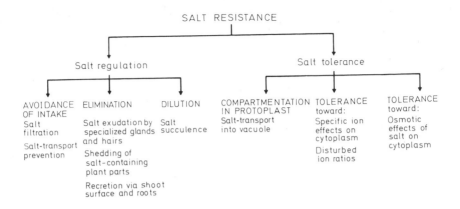

SALT RESISTANCE

Salt regulation | Salt tolerance

AVOIDANCE OF INTAKE	ELIMINATION	DILUTION	COMPARTMENTATION IN PROTOPLAST	TOLERANCE toward:	TOLERANCE toward:
Salt filtration	Salt exudation by specialized glands and hairs	Salt succulence	Salt-transport into vacuole	Specific ion effects on cytoplasm	Osmotic effects of salt on cytoplasm
Salt-transport prevention	Shedding of salt-containing plant parts			Disturbed ion ratios	
	Recretion via shoot surface and roots				

Fig. 4.21. Components of the salt resistance of vascular plants. After Steiner (1934), Henkel and Shakhov (1945), Levitt (1972), Waisel (1972), Kreeb (1974a), and Flowers et al. (1977)

Distichlis, etc. The vesicular hairs of some *Atriplex* species in arid regions and of *Halimione* collect chlorides in the cell sap, then die off and are replaced by new hairs. Desalination can also be achieved by the *abscission of older leaves* that have accumulated considerable amounts of salt. Meanwhile, young leaves capable of accumulating salt while performing their normal function grow to replace those that fall. This feature is characteristic of halophytic rosette plants such as *Plantago maritima*, *Triglochin maritimum* and *Aster tripolium*.

Succulence: Since the essential factor in the action of salts is not the absolute quantity but rather the concentration, a progressive accumulation of salts during the growing season can be compensated if the cells steadily draw in water; in the process, they become considerably distended. The salt concentration in the cell sap then remains fairly constant. Chloride ions are responsible for the development of this type of succulence. There is as yet no satisfactory explanation of the physiological processes that bring about expansion of the cells. Succulence is widespread among halophytes; it is found in those of wet saline environments (*Salicornia* and other coastal plants of the family Chenopodiaceae, and the mangrove plant *Laguncularia*) as well as in xerohalophytes from dry regions. In the latter case salt succulence includes features typical of drought-adapted succulents.

4.4.3 Plants on Soils Rich in Heavy Metals

Soils covering ore-bearing rock or slag heaps contain heavy-metal ions (especially Zn, Pb, Ni, Co, Cr and Cu) and Mn, Mg, Cd and Se in amounts toxic to most plants. Heavy-metal contamination also occurs in industrial zones (due to the emission of metal dust and the presence of Cd, Hg and other metal salts in bodies of water), along roads (Pb) and where fungicides are used (Cu, Zn, As).

191

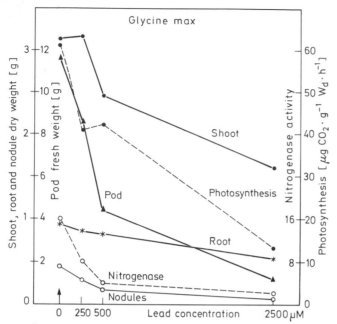

Fig. 4.22. The influence of lead on photosynthesis, nitrogen fixation, and dry-matter production by soybean plants. The lead was added to the nutrient solution in dissolved form. Nitrogenase activity was determined via C_2H_4 reduction, and is given as μmol $C_2H_4 \cdot$ plant$^{-1} \cdot$ h^{-1}. Data from Huang et al. (1974). For further examples of the effects of lead on enzyme activities and metabolic processes see Bazzaz et al. (1974) and Mathys (1975); on extension growth, Lane et al. (1978)

Most plants are sensitive to heavy metals; opening of the stomata is impaired, photosynthesis depressed, respiration disturbed and growth slowed (Fig. 4.22). But habitats with high concentrations of heavy metals in the soil can be colonized by plants capable of depositing surplus heavy metals in their cell walls or rendering them *harmless*, either by binding to SH groups in the boundary layers of the cytoplasm or by compartmentation and the formation of complexes with organic acids, phenol derivatives, mustard oils and other organic compounds in the cell vacuole (Table 4.7). Because *toxicity* of heavy-metal ions depends chiefly on the *inactivation of vital enzymes*, these avoidance mechanisms offer protection which can be very effective. The heavy-metal sensitivity of enzymes is element-specific. In *Silene vulgaris*, for example, nitrate reductase is 50% inhibited by Cu at a concentration of 0.001 mM, and by Cd and Zn at concentrations of about 0.01 mM; phenol oxidase is inhibited to the same extent by about 0.1 mM Cu and Pb but hardly affected by Mn, and enolase is inhibited by 0.05—0.1 mM Cd, Zn and Mn. In spinach chloroplasts, $PbCl_2$ in concentrations above 20 mM disturbs porphyrin metabolism and thus the biosynthesis of chlorophyll.

The ability to develop *heavy-metal resistance* is genetically based but can be modified by adaptation. *Chemo-ecotypes* display characteristic patterns of isoenzymes; they show element-specific increases in the ability of the protoplasm to resist the raised concentrations of heavy metals in their tissues that occur when they grow on soils rich in

Table 4.7. Distribution of zinc in organs, tissues and cell compartments of metallophytes growing in zinc-rich soil (Ernst, 1976). For the compartmentation of lead, see Malone et al. (1974)

Plant, organ	Vacuole, cytoplasm	Cell organelles	Cell wall
Cardaminopsis halleri			
Leaves	82%	6%	12%
Roots	38%	5%	57%
Silene vulgaris			
Leaves	64%	10%	26%
Roots	18%	10%	72%
Agrostis tenuis			
Leaves	48%	11%	41%
Roots	38%	10%	52%
Minuartia verna			
Leaves	46%	8%	46%
Roots	20%	8%	72%

Table 4.8. Protoplasmic heavy-metal resistance of toxicophytes and plants with normal heavy-metal sensitivity. The measure of resistance is the concentration of $ZnSO_4$ and $CuSO_4$ solutions (in mM) at which epidermis cells die after 48 h (Biebl, 1947; Url, 1956; Ernst, 1972b). For heavy-metal toxicity see also Antonovics et al. (1971), Bremner (1974) and Fox et al. (1978)

Plant	Zn tolerance	Cu tolerance
Flowering plants		
Silene vulgaris		
Zn-Cu ecotypes	40	0.04
Zn ecotypes	40—200	0.004
Cu ecotypes	0.4	0.08
Nonadapted ecotypes	0.4	0.004
Normally sensitive flowering plants	0.05—0.5	ca. 0.005
Mosses		
Bazzania trilobata	180	
Mielichhoferia elongata		100
Normally sensitive mosses	ca. 0.5	ca. 0.005

these elements. The greater the exposure to a certain element, the more tolerance is developed toward that element (Table 4.8).

Some of the heavy-metal plants (*metallophytes*, in the terminology of Duvigneaud and Denaeyer-de Smet, 1963) that can be used as indicator plants for ore deposits near the surface, and are also suitable for planting in mining and industrial areas otherwise diffi-

Table 4.9. Examples of extremely high mineral content in metallophytes and toxicophytes (As, Se). Concentrations are given in mg · kg⁻¹ dry matter (Cannon, 1960; Duvigneaud and Denaeyer-de Smet, 1973; Ernst, 1972a, 1975, 1976; Gaudet, 1975; Jaffré et al., 1976; Porter and Peterson, 1975; Kelly et al., 1975; Wild, 1974)

Plant	Location	Element	Concen-tration	Degree of accumu-lation[a]
Eichhornia crassipes	Tropical waters	Fe	14,400	10
Minuartia verna	Germany	Cu[1]		
Leaves			1,030	147
Roots			1,850	109
Thlaspi alpestre ssp *calaminare*	England	Zn[2]		
Leaves			25,000	208
Roots			11,300	140
Minuartia verna	Yugoslavia	Pb[3]		
Leaves			11,400	950
Roots			26,300	970
Minuartia verna	Germany	Cd[4]		
Leaves			348	3,480
Roots			382	3,820
Jasione montana, leaves	England	As[5]	31,000	
Mechovia grandiflora, leaves	Katanga	Mn	7,000	7
Acrocephalus robertii, leaves	Katanga	Co	1,490	50
Psychotria douarrei	New Caledonia	Ni[6]		
Leaves			45,000	
Roots			92,000	
Pearsonia metallifera	Rhodesia	Cr[7]		
Leaves			490	98
Roots			1,620	162
Astragalus preussi, leaves and roots	Utah (USA)	U	70	116
Astragalus racemosus, leaves	Colorado (USA)	Se[8]	15,000	

[a] *Degree of accumulation*, according to Duvigneaud, is defined as

$$\frac{\text{Mineral concentration in plants on contaminated soils}}{\text{Mineral concentration in plants on normal soils}} \cdot 100 .$$

1. Other cuprophytes: *Haumaniastrum robertii* in Katanga, *Becium homblei* and *Indigofera dyeri* in Rhodesia, *Polycarpea spirostylis* in Australia, *Gypsophila patrini* in Central Asia, ecotypes of *Silene vulgaris*, *Gladiolus* species in Africa, and some mosses (e. g., *Mielichhoferia* species).
2. Other calamine plants in Europe: *Minuartia verna*, *Viola calaminaria* and ecotypes of *Silene vulgaris* and *Armeria maritima*.
3. Other lead-accumulating species: *Agrostis tenuis*, *Festuca ovina*, *Erianthus giganteus*, *Cerastium holosteoides*, also *Calluna vulgaris*.
4. Other cadmium-accumulating species: *Thlaspi alpestre* ssp. *calaminare*.
5. Other arsenic-accumulating species: *Calluna vulgaris*, *Agrostis tenuis*.

cult to landscape, are shown in Table 4.9. There are also species with such great genetic variability that they have evolved a number of specialized ecotypes which can be resistant to *several* heavy metals and other elements — for example, *Agrostis tenuis, A. canina, Festuca ovina, Plantago lanceolata* and *Silene vulgaris* are resistant to Zn, Cu, Cd, Ni and in some cases also to As, Al, F, and NaCl. Metallophytes take up large amounts of heavy-metal ions and store them at concentrations of $0.5-8$ g $\cdot$ kg^{-1} (in the extreme case, up to 25 g $\cdot$ kg^{-1}). This is a hundred to a thousand times the normal concentration of trace elements in the plant; in fact, it is on the order of magnitude of the concentrations of the chief nutrients phosphorus and sulfur. The great capacity of the tropical floating plant *Eichhornia crassipes* to accumulate metal ions is exploited for the detoxification of lakes and streams.

4.5 The Toxic Effects of Environmental Pollutants

4.5.1 Toxic Substances in the Environment

By many of their activities — industry, traffic, the agricultural and domestic use of chemicals, and above all the extravagant consumption of fossil fuels — humans release into their environment phytotoxic substances that are taken up by the plants from air, water, or soil.

The pollutants appearing in the environment at a particular place and time are called *immissions*; the concentration of a pollutant in the environment is the *immission concentration*. *Atmospheric pollutants* particularly dangerous to plants are SO_2, hydrogen halides (HF, HCl), ozone and peroxi-acetyl-nitrate (*PAN*, produced from automotive and industrial fumes under strong radiation). There are other harmful substances that reach plants through the air — nitrogen oxides, ammonia, hydrocarbons, tar fumes, soot, and dust. Plants growing in *water* are endangered by poisonous chemicals in sewage (cyanide, chlorine, and hypochlorite, phenol and benzol derivatives, heavy-metal compounds, etc.), by detergent additives (sulfonates, phosphates) and by the seepage from sewage plants, garbage dumps and cultivated fields, which can cause eutrophication or poisoning.

4.5.2 Pollution Injury

The effects of many forms of pollution can be determined by the product of their concentration and the period of exposure (Fig. 4.23); but the relationship is linear only

---

◁———

6. Other nickel-accumulating species: *Hybanthus austrocaledonicus, H. floribundus, H. caledonicus, Sebertia acuminata* in Australia and Caledonia.
7. Other chromium-accumulating species: *Sutera fondina, Dicoma niccolifera, Convolvulus ocellatus* in Rhodesia.
8. Other selenophytes: *Aster xylorrhiza, Stanleya* species, various species of *Astragalus* in North America, *Neptunia amplexicaule, Acacia cana.*

195

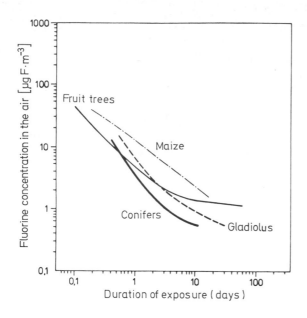

Fig. 4.23. Relationship between concentration of and duration of exposure to noxious fluoride gases, in the production of acute and chronic damage to various plants. After McCune (1968) and National Research Council USA (1971) as cited by Keller (1975). For adaptation to air pollution and breeding for resistance to intoxication see Bradshaw and McNeilly (1981)

over a certain range. The lower limit is set by the concentration threshold below which there are no observable changes even after prolonged exposure. At the other end of the range, when a certain high concentration is exceeded, even very brief exposure causes severe damage.

Under exposure to *high concentrations* (for example, more than 1 cm³ SO$_2$ · m⁻³ air or 0.1 cm³ HF · m⁻³) plants suffer *acute injury*, with externally visible symptoms such as chlorophyll bleaching, leaf discoloration, necrosis of areas of tissue and organs, or the death of the entire plant. Damage of this sort is in general apparent only in the immediate vicinity of the pollution source. With *low pollutant concentrations* (e.g., 0.05—0.2 cm³ SO$_2$ · m⁻³ or 0.001—0.002 cm³ HF · m⁻³) there is at first no externally visible poisoning. Nevertheless, it can be shown that chemical, biochemical, fine-structural, and functional changes have occurred (Fig. 4.24).

Early recognition of latent or already established pollution damage is facilitated by the following symptoms: accumulation of the toxic material in the plant, changes in pH at the surface of the shoot (e.g., the bark) and in the tissues, diminished or increased activity of certain enzymes (e.g., enolase inhibition and enhanced peroxidase activity under HF stress; Fig. 4.25), increase in compounds with SH groups (especially glutathione) and in phenols, lowered ascorbic-acid level in the leaves, depression of photosynthesis, stimulation of respiration, low dry-matter production, changes in permeability, disturbances in water balance, impaired hardening to freezing, and defects in fertility (e.g., pollen sterility).

Under *prolonged exposure* the transient disturbances of metabolism develop into *chronic injury* with irreversible consequences (Fig. 4.26). Plants cultivated in farm and garden show reduced productivity and hence diminished yield, and the quality of the harvest is lower because of biochemical responses to poisoning. In trees growth, including cambial growth, is less vigorous. The analysis of annual rings (Fig. 4.27)

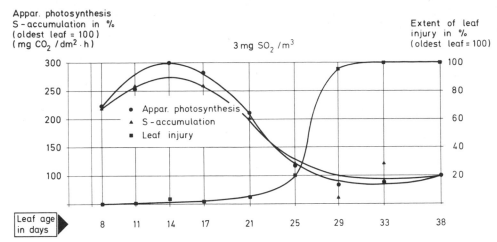

Fig. 4.24. Accumulation of sulfur, depression of net photosynthesis and extent of necrosis in the leaves of *Alnus glutinosa*, as a result of SO_2 pollution. After Guderian (1977). Inhibition of RuBP carboxylase by sulfite ions is described by Ziegler (1972). For SO_2 effects on stomatal diffusion resistance see Majernik and Mansfield (1970). Biscoe et al. (1973), Farrar et al. (1977), Klein et al. (1978), Noland and Kozlowski (1979), Unsworth (1981), on plant water potential Halbwachs (1971)

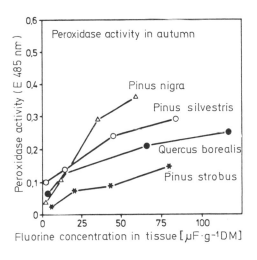

Fig. 4.25. Peroxidase activity in needles and leaves as a function of fluorine content. The peroxidase activity was determined colorimetrically at 485 nm, 2 min after addition of p-phenylenediamine and hydrogen peroxide to the tissue extract. From Keller (1974). For a review of the effects of air pollutants on enzymes see Hoarsman and Wellburn (1976), Rabe and Kreeb (1979)

gives a clear record of progressive chronic pollution injury; not only the amount, but also the structure of the wood is changed. Over the years the foliage becomes more sparse and it is more difficult to supply the shoots at the top of the crown with water. Branches dry out, and gradually the tree dies.

The *symptomatology of pollution damage* is varied and quite unspecific; a given substance affects different plants in very different ways, and a particular symptom can be produced by a variety of substances. Often only a few of the possible effects appear.

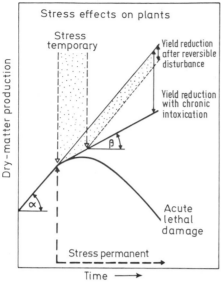

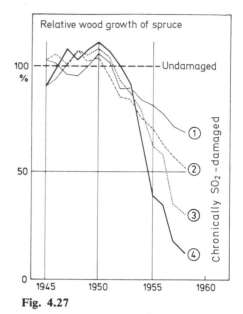

Fig. 4.26

Fig. 4.27

Fig. 4.26. Diagram of the effects of environmental stress on dry-matter production by plants. With brief exposure to low-intensity stress (low pollutant concentrations) within the tolerable limits, physiological processes are transiently altered (reversible disturbance), whereas after longer exposure the vital functions are permanently affected (chronic intoxication). After recovery from a reversible disturbance dry-matter production occurs at the same rate as before the stress (lines sloping at angle α), but with chronic poisoning the rate of growth is permanently reduced (angle $\beta < \alpha$) and the yield correspondingly less. When the tolerable limit is exceeded lethal damage is soon apparent and tissues die (loss of biomass). Modified from Härtel (1976). Connections between visual symptoms, biochemical deviations, and yield reduction are indicated by Jäger and Klein (1977). Surveys of pollution injury to vital functions are given by Mudd and Kozlowski (1975), Mansfield (1976), Guderian (1977), Hällgren (1978), Linzon (1978), and Taylor (1978)

Fig. 4.27. Reduced wood production in stands of firs exposed to SO_2. The degree of reduction is expressed by a relative index (average annual-ring thickness in pollution-damaged stands as a percentage of the thickness in healthy stands). The relative growth index is computed by analysis of bore samples by the methods of tree-ring chronology. Levels of damage: *1*, beginning of needle loss; *2*, advancing needle loss, considerably damaged stands; *3*, severe damage, with dying twigs and branches; *4*, dying trees dried at the tops. After Vinš (1962). Further data on effect of pollutants on annual-ring growth are given by Fritts (1976) on xylem structure by Liese et al. (1975), and Grill et al. (1979)

4.5.3 Pollution Resistance and Bioindicators of Pollution Stress

Knowledge of specific resistance to pollutants is of practical significance when plants are to be established in areas of concentrated population and industry (Table 4.10). Various species, and even the varieties of a species, differ in sensitivity. Moreover, we have learned that plants in general respond more sensitively than humans to envi-

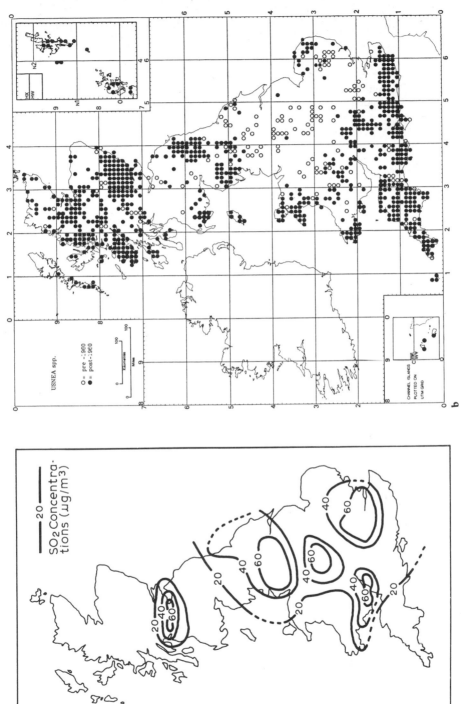

Fig. 4.28a and b. SO$_2$ exposure and damage to lichens in the British Isles. **a** Annual mean sulfur dioxide concentration at sites remote from local pollution. From Saunders and Wood (1973). **b** Pre- and post-1960 distribution of *Usnea* species (Hawksworth et al., 1973)

199

Table 4.10. Sensitivity of agricultural, ornamental and forest plants to chronic exposure to noxious gases. After Guderian (1977), expanded by data from Taylor et al. (1975). The list represents only a general survey; sensitivity can be quite different in various species of the same family and even among individual ecotypes or varieties. See species lists given by Garber (1967), Berge (1970), Treshow (1970), Naegele (1973), Mudd and Kozlowski (1975), Dässler (1976), Steubing (1976), and the bibliographies of Garber (1973) and Thomas (1973). For color illustrations of pollution damage see Jacobson and Hill (1970)

Plant	SO_2	HF	HCl	NO_x
Field crops				
Grain	×	×	×	●
Brassicaceae	O	O	×	O
Beet	O	×	O	
Potato	O	O	×	×
Sunflower	O	O	×	
Fodder plants				
Grasses and fodder grain	×	●	×	
Clover, alfalfa	×	●	×	●
Cabbage	O	●	O	
Vegetables				
Brassicaceae	O	×	O	O
Fabales	×	●	×	●
Apiaceae	O	O	O	●
Chenopodiaceae	●	×	×	
Cucurbitaceae	O	O		
Solanaceae	O	O	×	
Asteraceae and Cichoriaceae	×	×		●
Allium	O	●	×	
Ornamental plants				
Liliales	O	●	×	O
Ranunculaceae	×	×	×	
Rosaceae	×	×	×	●
Fabales	×	×	×	
Geraniaceae	O	O		●
Araliaceae	O	O		
Caryophyllaceae	×	×		
Ericaceae	O	O	O	×
Solanaceae				●
Asteraceae and Cichoriaceae	O	O	O	●
Fruit trees				
Pome fruits	●	●	●	●
Stone fruits	×	●	×	
Juglans	●	●	●	
Corylus	×	●	●	
Vitis	×	●	●	
Citrus	O			

Table 4.10 (continued)

Plant	SO$_2$	HF	HCl	NO$_x$
Forest trees				
Abies, Picea, Pinus sylvestris	●	●	●	×
Pseudotsuga, Cedrus	●	●	●	
Pinus nigra, Thuja, Juniperus, Taxus	○	○	○	○
Tilia, Fraxinus, Fagus, Sorbus acuparia	×	×	×	
Carpinus	×	×	●	○
Quercus	○	○	×	○
Betula				●

● Very sensitive, × slight sensitivity, ○ relatively resistant.

Table 4.11. Sensitivity of lichens and mosses to SO$_2$. Selected examples from data of Dässler and Ranft (1969), Hawksworth and Rose (1970), Ferry et al. (1973), Türk et al. (1974), and Winkler (1977)

Sensitivity	Lichens	Mosses
High sensitivity	*Evernia prunastri*	*Atrichum undulatum*
	Lobaria pulmonaria	*Brachythecium albicans*
	Parmelia species	*Dicranum rugosum*
	Pseudevernia furfuracea	*Mnium undulatum*
	Ramalina farinacea	*Plagiothecium curvifolium*
	Teloschistes flavicans	*Plagiothecium succulentum*
	Usnea florida	*Rhytidiadelphus squarrosus*
Slight sensitivity	*Buellia punctata*	*Ctenidium molluscum*
	Lecanora conizaeoides	*Dicranella heteromalla*
	Lecanora varia	*Marchantia polymorpha*
	Lepraria incana	*Pohlia nutans*
	Xanthoria parietina	*Rhytidiadelphus triquetrus*

ronmental poisons, so that plants can be used for the *bioindication* of environmental danger. Particularly sensitive species can serve as indicators (Table 4.11) and those particularly resistant to pollution, as accumulators — plants that collect large amounts of pollutant without harm. Plants extraordinarily sensitive to SO$_2$ and hydrogen halides are mosses, lichens, and some fungi. More than a hundred years ago W. Nylander recognized the suitability of lichens as indicators of air quality. As little as 1% of the SO$_2$ concentration harmful to higher plants produces in lichens respiratory disturbances, the breakdown of chlorophyll, and inhibition of growth. From the damage observed in samples of bark overgrown with lichen (lichen explants) and from the composition of the natural lichen growth on trees and stones, it is possible to make inferences about the effects upon a habitat of long-term exposure to SO$_2$. In areas that receive maximal SO$_2$ exposure, lichens do not survive; the area becomes a lichen desert. With increasing distance from the SO_2-emitters, resistant crustaceous lichens begin to appear; these are followed by lichen communities comprising many species. Only in the undisturbed regions, however, is there again a luxuriant growth of lichens on tree trunks and rock surfaces (Fig. 4.28).

4.6 Mineral Balance and Circulation in Plant Communities

4.6.1 The Mineral Balance of a Plant Community

The minerals and the carbon in vegetation are delicately balanced against one another. The uptake of minerals regulates the increase in plant mass, and carbon assimilation makes available the material in which the minerals are incorporated. Using the production Eq. (3.16), therefore, one can determine the yearly mineral uptake by vegetation if the content and composition of the ash of the plant mass are known. A certain fraction of the *total quantity of minerals absorbed* (M_{abs}) by the vegetation per area of ground in the course of a year is lost during the same year in the original mineral form (*recretion*) and is washed away from the shoots by precipitation; this amount is called M_r. The remainder, M_i, is *incorporated* in the plant mass.

$$M_{abs} = M_i + M_r. \tag{4.1}$$

The minerals fixed in the phytomass are apportioned, as in the productivity Eq. (3.17), to the *increase in biomass* (positive ΔB) that occurs during the year and to the annual *losses* of dry matter as *detritus* (L) and due to *grazing* by consumers (G).

$$M_i = \Delta M_B + M_L + M_G. \tag{4.2}$$

The terms in this equation represent the content of minerals or the ash content of the dry matter. The various mineral elements in a plant are nonuniformly distributed, accumulated, fixed, and eliminated (Fig. 4.29); moreover, mineral concentration depends on the state of development of the vegetation. For these reasons the mineral balance of a stand of plants cannot be found simply by multiplying the terms of the production equation by the average ash content of the plant mass. As in deriving carbon balance, the computation must be based on the separate contributions of the individual organs at the different sampling times.

The fraction of the total mineral nutrients lost as M_r and M_L can be expressed by the "*recycling coefficient*" k_M introduced by Ulrich (1968):

$$k_M = \frac{M_L + M_r}{M_{abs}}. \tag{4.3}$$

In a spruce forest in West Germany, 93% of the K taken up, over 80% of the Ca and Mg, and over 70% of the N and P were found to be returned to the soil in the same year. In a tropical rainforest in Panama the turnover factor for Mg was 100% ($^1/_3$ of this by leaching from the crown); that for Ca was 97%, and that for K, 86% (half by leaching).

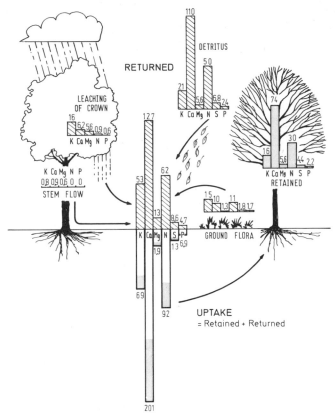

Fig. 4.29. The cycling of minerals in a mixed forest of oak, hornbeam and beech in Belgium. All quantities are given in units of kg · ha⁻¹ · yr⁻¹. The yearly uptake of minerals by the stand was calculated by summation of the quantities retained in the phytomass (*dotted areas*) and those returned to the soil (*cross-hatched*) by way of washing and leaching of the canopy, stem flow, and the litter from trees and ground flora (cf. the carbon turnover of the same stand in Fig. 3.66). After Duvigneaud and Denaeyer-de Smet (1970). For examples of other forests see Table 4.12; for sclerophyll shrub, Lossaint and Rapp (1971); for ericaceous heaths, Gimingham (1972), Wielgolaski et al. (1975) and Smeets (1977); for subshrub steppes, Margaris (1976); for savannas, Egunjobi (1974); for temperate grassland, Innis (1978) and Perkins (1978); for reeds and swamps, Dykyjová and Hradečká (1976) and Dykyjová and Květ (1978); for tundra, Rodin and Bazilevich (1965), Bunnell et al. (1975), Chapin et al. (1975) and Babb and Whitfield (1977); for deserts, West and Skujinš (1977). For accumulation of anthropogenic sulphur and cycling in a forest ecosystem see Johnson et al. (1982)

Values for an evergreen and a deciduous forest are given in Table 4.12; again potassium leads the list, followed by nitrogen and phosphorus.

In returning the greater part of its mineral uptake to the soil every year with the fallen leaves, the plant cover makes an important contribution to the circulation of minerals. These are withdrawn from deep layers of the soil by the roots, raised above the surface of the ground by the plant, and then sent back to a higher soil level. The trees in particular, because of their extensive root systems, raise nutrient minerals that have

Table 4.12. Mineral content and turnover in forests. Units of mineral content are kg · ha⁻¹. For examples of mineral balances in temperate-zone forests see Ovington (1959a,b), Ulrich and Meyer (1973), Woodwell et al. (1975), and Likens et al. (1977; includes surveys of data); in boreal coniferous forests, Rodin and Bazilevich (1965); in tropical forests, Odum and Pigeon (1970), Golley et al. (1975), Klinge (1976), and Likens et al. (1977)

Stand	Deciduous oak-beech-hornbeam mixed forest with undergrowth, Belgium (Virelles), 30—75 years old					Evergreen oak forest, southern France (Rouquet), ca. 150 years old				
Authors	Duvigneaud et al. (1969)					Rapp (1969, 1971)				
Phytomass of stand	156 t dry matter · ha⁻¹					304 t dry matter · ha⁻¹				
Annual net primary production	14.4 t dry matter · ha⁻¹					7 t dry matter · ha⁻¹				
	N	P	K	Ca	Total[a]	N	P	K	Ca	Total[a]
Mineral content of phytomass above ground	406	32	245	868	1632	763	224	626	3853	5505
Amount fixed in new growth (M_B) annually	30	2.2	16	74	127.8	13.2	2.6	8.9	42.7	68.3
Mineral content of the annual loss (M_L)	61	4.1	36	120	228.0	32.8	2.8	16.2	63.9	120.3
Mineral incorporation $M_i = M_B + M_L$	**91**	**6.3**	**52.0**	**194.0**	**355.7**	**46.0**	**5.4**	**25.1**	**106.6**	**188.6**
Leaching (M_r)	0.9	0.6	17.0	7.1	31.8	0.5	0.8	25.7	19.4	48.7
Annual adsorption of minerals $M_{abs} = M_i + M_r$	**91.9**	**6.9**	**69.0**	**201.1**	**387.5**	**46.5**	**5.4**	**50.8**	**126.0**	**237.3**
Mineral cycling $k_M = \dfrac{M_L + M_r}{M_{abs}}$	0.68	0.68	0.77	0.63	0.67	0.72	0.67	0.83	0.66	0.71

[a] Total amount of mineral in the dry matter; this value is greater than the sum of N, P, K and Ca, since it includes the other elements in the ash.

sunk quite deep into the ground; once the litter from the trees has been broken down, the minerals are available to plants of the herbaceous stratum with shallow roots.

4.6.2 The Circulation of Mineral Nutrients Between Plants and Soil

Under natural conditions, the cycling of minerals between the plants and the soil is, to a great extent, a closed system. The mineral nutrients taken up by the plants are returned to the soil either directly by recretion or as components of organic material; there they are mineralized by animals and microbes and adsorbed by the soil colloids. The mechanism for this recycling of minerals is a crucial part of the circulation; if the organic wastes (of which there are more as the productivity of the plant canopy increases) were removed, the store of nutrients represented by the incorporated minerals would be lost. Recognition of this fact is the basis of well-planned fertilization; the nutrient depletion brought about by harvesting of crops or removal of litter must be compensated by the provision of fertilizers in amounts of the same order of magnitude as the annual loss.

However, plants are not entirely dependent on the minerals supplied by *litter remineralization*. Weathering brings to the soil and the plants quantities of mineral nutrients that are hard to estimate, but no doubt significant. In raw soils it is primarily by weathering that the nutrient requirements of the plants are met. Minerals are moved into the root zone by groundwater and by water rising by capillarity. Precipitated water carries inorganic materials contained in the atmosphere as gases (SO_2, NH_3, and gases containing nitrogen oxides), dust, fog or aerosols. These atmospheric components are also captured directly by the plant cover.

Mineralization and the production of organic matter must keep pace with one another. Where the decomposition of litter and humus proceeds too slowly, the growth rate of the plants usually slows until it is in equilibrium with the rate of mineralization. If the minerals bound in organic matter are liberated rapidly and returned in abundance, the primary producers are better supplied with nutrient ions and can build up a greater mass. The remarkable productivity of tropical rain forests is thus determined by the rapid mineral turnover there. The constant favorable temperature and the high humidity promote decomposition by microorganisms at an extraordinary rate, so that the nutrient elements in the soil are only briefly tied up in organic compounds and soon are made available to the plants again in inorganic form. Excessive liberation of the inorganic elements or a reduction in withdrawal by the vegetation (for example, after the system has been disturbed by fire or the actions of man) can disturb the mineral balance of the ecosystem. Nutrient elements in the mineralized state are not only more accessible to the plants; they are also more mobile in the soil and thus can be more easily *leached out*. Therefore the organically bound minerals in the biomass and in the soil are important in that they constitute relatively secure stores.

5 Water Relations

Life evolved in water, and water remains the essential medium in which biochemical processes take place. Protoplasm displays the signs of life only when provided with water — if it dries out, it does not necessarily die, but it must at least enter the anabiotic state, in which vital processes are suspended.

Plants are composed mainly of water. On the average, protoplasm contains 85—90% water, and even the lipid-rich cell organelles such as chloroplasts and mitochondria are 50% water. The water content of fruits is particularly high (85—95% of the fresh weight), as is that of soft leaves (80—90%) and roots (70—95%). Freshly cut wood contains about 50% water. The parts of plants having the least water are ripe seeds (usually 10—15%); some seeds with large stores of fat contain only 5—7% water.

5.1 Poikilohydric and Homoiohydric Plants

In land plants, the assimilative parts of which are in contact with air and continually lose water by evaporation, the establishment of suitable *water relations* is the first requirement. Depending on whether or not they can compensate for short-term fluctuations in water supply and rate of evaporation, terrestrial plants may (according to H. Walter) be distinguished as poikilohydric or homoiohydric. Like lifeless protein gels, poikilohydric plants match their water content with the humidity of their surroundings. Bacteria, blue-green algae, lower green algae in the order Protococcales, fungi and lichens all have small cells that lack a central vacuole; when they dry out, these cells shrink very uniformly, without disturbance of the protoplasmic fine structure — the plant remains viable. As the water content decreases, the vital processes — e.g., photosynthesis and respiration (cf. Chap. 3.2.3.3) — are gradually suppressed. When sufficient water has been imbibed, such plants resume normal metabolic activity and growth. Poikilohydric plants thus have an advantage wherever there are frequently alternating periods of dryness and moisture. The humidity threshold for activity is species-specific and determines the preferred range of distribution of the various species. Soil bacteria and fungi, in general, are active only when the relative humidity (RH) is above 80—95%, but many molds can germinate and grow at RH between 75% and 85%. Yeasts require 88% or higher, and there are fungi (for example, species of *Xeromyces*) that begin to grow at RH as low as 60% (Fig. 5.1).

Poikilohydric forms are found not only among thallophytes, but also among mosses of dry habitats, certain vascular cryptogams, and a very few angiosperms. Pollen grains, and embryos in seeds, are also poikilohydric.

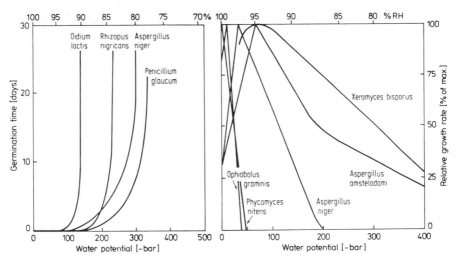

Fig. 5.1. Germination and linear growth rate of hyphae of fungi, as a function of the water potential of the habitat. After Heintzeler (1939) and various authors cited by Griffin (1972)

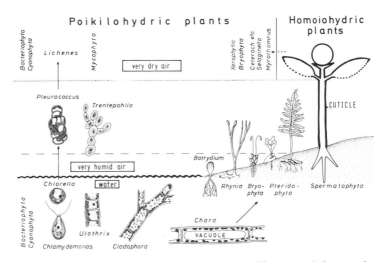

Fig. 5.2. Evolution of the water economy of plants to terrestrial life. *From left to right*: Transition from aquatic lower algae with non-vacuolated cells to primarily poikilohydric aerial algae; development of vacuole in aquatic green algae and Characeae; transition from vacuolated thallophytes to homoiohydric vascular plants (hygrophytic mosses are still restricted to habitats with high air humidity, and in dry habitats become secondarily poikilohydric; there are also secondarily poikilohydric forms among the pteridophytes and angiosperms, but not among gymnosperms). Most vascular plants, because they are equipped with a cuticle that limits its transpiration and because their cells are considerably vacuolated, are homoiohydric. From Walter (1967)

Homoiohydric plants are descended from green algae with vacuolated cells; a *large central vacuole* is the common characteristic of all homoiohydric plants. Because the water content is stabilized within limits by the water stored in the vacuole, the protoplasm is less affected by fluctuating external conditions. However, the presence of a large vacuole also results in the loss of the cell's ability to tolerate exsiccation. Thus one finds the predecessors of homoiohydric land plants pressed close to wet soil or living in permanently moist habitats (Fig. 5.2). Only with the evolution of a protective *cuticle* to slow down evaporation, and of stomata to regulate transpiration, were plants able to control their water economy adequately; with these adaptations, and an extensive system of roots, the protoplasm can be maintained in steady activity despite sudden changes in humidity. Thus the plants were sufficiently productive to form a closed cover over broad areas — and ultimately the enormous phytomass now clothing the continents.

5.2 Water Relations of the Plant Cell

5.2.1 The Water in the Cell

The water in plant cells occurs in several forms: It is a chemically bound constituent of protoplasm; as water of hydration, it is associated with ions, dissolved organic substances and macromolecules, filling the gaps between fine structures of the protoplasm and of the cell wall; it is stored as a reserve supply in cell compartments and vacuoles; finally, as interstitial water it serves as a transport medium in the spaces between cells and in the conducting elements of the vessel and sieve-tube systems.

Water of Hydration

In accordance with their dipole nature, water molecules aggregate and accumulate at charged surfaces in flexible arrays (known as structured water). Strongly charged ions of about the same size as water molecules bind the latter more firmly, the greater their charge and the smaller the radius of the ion. The sodium ion, with the same charge but half the surface area of the potassium ion, has twice the "density" of charge and accordingly almost four times as thick an envelope of hydration. Water molecules are bound tightly at the surfaces of the ions by electrostatic forces. A similar situation occurs at the surfaces of protein molecules and polysaccharides. Water molecules become associated with polar groups (hydroxyl, carboxyl and amino groups) and form several layers of structured water; the water molecules are more readily displaceable the further they are from the polar group.

Water of hydration accounts for only 5—10% of the total cell water, but this amount is absolutely essential to life. Only a slight decrease in the content of water of hydration results in severe alterations in protoplasmic structure and in the death of the cell. Most of the water in the protoplasm fills the spaces in the cytoplasm and cell wall, but it is not entirely free to move; its mobility is limited by capillary forces, and it is retained by dissolved substances. The cell walls of plants hold water with a pressure of 15—150 bar, depending on the density with which the fibers are packed. The surface forces holding water to structural elements in a matrix (cell wall, plasma colloids) can be expressed in terms of the "matric" pressure τ.

Stored Water

The most easily translocated water is that in those cell compartments specialized as reservoirs for solutions. But even this water is not completely mobile; in addition to hydration and matric effects, dissolved substances such as sugars, glycosides, organic acids and ions affect the net diffusion of water, through semipermeable membranes, from one solution to another. This effect is described in terms of the *osmotic pressure* of each solution.

If two solutions (for example, cell sap and pure water) having different concentrations of various solutes are separated by a membrane permeable only to water, there is a tendency for net flow of water to occur into the solution having the greater solute concentration. This is not paradoxical, and can be understood in the sense that the water diffuses *down* its *own* concentration gradient. If now the more concentrated fluid is confined (as by the cell wall enclosing the protoplast), the inflow of water causes the pressure in the cell to rise; it rises until the pressure difference just counterbalances the tendency of water to flow in. Although this equilibrium pressure difference depends upon the concentrations of the *two* solutions, one speaks of the osmotic pressure of a single solution, implying the pressure that will just counterbalance the water influx which would occur were it in communication with pure water, through a membrane permeable to water.

The osmotic pressure of a solution increases with absolute temperature and with the concentration n of dissolved particles (the molarity in terms of molecules and ions that do not further dissociate). Van't Hoff pointed out in 1855 that, at least for dilute solutions, the osmotic pressure π is given approximately by

$$\pi = nRT. \tag{5.1}$$

Here R is the universal gas constant ($8.3143\ \mathrm{J \cdot K^{-1} \cdot mol^{-1}}$), so that a 1 M solution at standard conditions has an osmotic pressure of about 22 bar.

Solutions of dissociating substances contain more independent particles than their molarity indicates, so that for electrolytes the equation includes another factor taking into account the degree of dissociation. Conversely, macromolecular substances can be prominent constituents, in terms of weight, without raising the osmotic pressure appreciably. Through the polymerization of small molecules (with marked osmotic effect) to macromolecules — for example, the conversion of sugar to starches — and the reversal of this procedure by hydrolysis, the cell can rapidly alter its osmotic pressure; as a result, the net influx of water can be regulated.

5.2.2 The Water Potential of Plant Cells

The binding of water to macromolecular structures and dissolved substances results in a decreased availability for chemical reactions and for use as a solvent. It is the

Table 5.1. Values of relative humidity (in the air above a solution) and the osmotic pressure (in bar) of the solution, when the two phases are enclosed and allowed to equilibrate at 20° C. Recomputed from Walter (1931)

bar	% RH	bar	% RH
0	100	97.9	93.0
6.7	99.5	112	92.0
13.5	99.0	126	91.0
20.3	98.5	141	90.0
27.2	98.0	301	80.0
34.1	97.5	481	70.0
41.0	97.0	687	60.0
55.0	96.0	933	50.0
69.1	95.0	∞	0
83.2	94.0		

availability of water, rather than its total quantity, that influences the biochemical activity of protoplasm.

The thermodynamic state of the water in a cell can be compared with that of pure water, and the difference expressed in terms of potential energy. The water potential Ψ, as defined by Slatyer and Taylor (1960), is equal to the difference in free energy per unit volume between matrically bound, or pressurized, or osmotically constrained water and that of pure water. The dimensions of water potential are energy per unit mass or per unit volume ($J \cdot kg^{-1}$). Because a dyne-cm is an erg, force per unit area and energy per unit volume have the same dimensions, the conversion factor being approximately $100 \ J \cdot kg^{-1} = 1$ bar.

The increasing acceptance of a terminology in which potential terms are used consistently and exclusively has done a great deal to eliminate sources of confusion in the literature. In the past water relations have been discussed in terms of osmotic pressure, diffusion pressure deficit, suction pressure, hydrostatic pressure, the pressure exerted by an enclosing structure, and the depression of water potential — all of which are defined in such a way as to represent positive values. Although the use of positive variables has been considered an advantage, this has been more than offset by the multiplicity of terms and of the symbols used to represent them. Here we adopt a terminology based entirely on potential, in which the water potential Ψ (negative) expresses the net effect of the osmotic potential Ψ_π (negative), the matric potential Ψ_τ (negative) and the pressure potential Ψ_p (positive). The relationships between these quantities will be discussed in the following sections.

5.2.3 Water Potential and the Cellular Translocation of Water

Between sites differing in water potential there is a *potential difference* $\Delta\Psi$. This situation is analogous to that in an electrical system, in which different points are at dif-

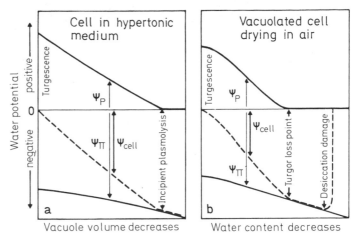

Fig. 5.3a and b. Water-potential diagrams for vacuolated plant cells in a hypertonic medium (**a**) and for desiccating vacuolated cells in air (**b**). As water loss procedes the pressure potential Ψ_P falls from positive values to zero, and the osmotic potential Ψ_π becomes more negative. In the state of incipient plasmolysis or at the turgor loss point Ψ_{cell} corresponds to the osmotic potential, and if more water is lost the system breaks down as a result of damage to the biomembranes. Diagrams based on data of Barrs (1968), Kyriakopoulos and Larcher (1976) and Pospišilová (1975). For detailed information on water in tissue and cells see Tyree and Jarvis (1982)

ferent voltages. Just as there is a tendency for current to flow from the higher-voltage to the lower-voltage points, in living cells there is a tendency toward the reduction of water-potential differences. In the cell this occurs by the transport of water or other substances; *water transport* always proceeds in the direction toward the system at lower water potential. A state of equilibrium within the cell, and between the cells and their surroundings, is usually rapidly achieved (as long as there is no obstacle to diffusion).

The water balance of cells is of course intimately related to the availability of water in their environment. For a cell in air, as Table 5.1 indicates, there is an equilibrium between the humidity and the water potential of the cell; if the relative humidity of the air is less than the given equilibrium value, water will be lost and the water potential will become more negative. Within a cell, too, water shifts from one cell compartment to another until an equilibrium is reached, at which point the water potentials of the different compartments are equal. The processes underlying osmotic water transport can be most simply explained in terms of the model of the single cell in hypotonic and hypertonic media that has been derived from *experiments* in plant physiology (Fig. 5.3). Note, however, that this situation occurs *in nature* only in the case of marine plants in tide pools that are drying up or becoming diluted with fresh water.

The Single Cell in a Hypotonic Medium. If a cell with a non-rigid wall is placed in a hypotonic medium, water enters the cell; it flows through the protoplasm into the cell sap, and as a result the volume of the vacuole increases considerably. The expansion of the vacuole presses the plasma against the cell wall, which stretches elastically and exerts a counterpressure (the turgor pressure P). The tendency for water to enter the cell is determined by the difference in water potential between the cell contents and the external medium, the water potential of the cell being given by

$$\Psi_{cell} = \Psi_\pi + \Psi_P . \tag{5.2}$$

As water enters the cell, Ψ_π rises (becomes less negative, as the cell solution becomes diluted) and Ψ_P, due to the restoring force of the cell wall, also increases. Ψ_{cell} thus becomes increasingly less negative, until it equals the water potential of the hypotonic medium.

If the cell is in pure water, equilibrium will be reached only when Ψ_{cell} is zero. Under these conditions, the positive term in Eq. (5.2), Ψ_P, must just balance the negative term; i.e.,

$$\Psi_\pi + \Psi_P = 0, \text{ and } \pi = P . \tag{5.3}$$

The water potential equation holds for the cell as a whole, the protoplasm behaving like a semipermeable membrane, so that the entire system functions as an osmometer. Separate potentials can also be identified for the components of the system—the vacuole, the protoplasm and the cell wall. Whenever the water content changes, these components immediately adjust to a new state of equilibrium. That is,

$$\Psi_{vacuole} = \Psi_{\pi \ vac} + \Psi_P$$

$$\Psi_{protoplasm} = \Psi_{\tau \ ppl} + \Psi_{\pi \ ppl} + \Psi_P$$

$$\Psi_{cell \ wall} = \Psi_{\tau \ cw}$$

$$\text{at equilibrium, } \Psi_{vacuole} = \Psi_{protoplasm} = \Psi_{cell \ wall}. \tag{5.4}$$

The Single Cell in a Hypertonic Medium; Incipient Plasmolysis. If a cell is placed in a solution more concentrated than that in the vacuole, water is drawn out of the cell. As a result, the volume of the vacuole is reduced and the elastic cell wall is less distended, exerting less pressure on the protoplasts. Finally the volume of the cell shrinks to a limiting value, beyond which the cell wall can no longer follow. At that point the protoplast begins to pull away from the cell wall; this stage is called *incipient plasmolysis*. In the state of incipient plasmolysis, the osmotic pressure is unopposed by the tension of the cell wall:

$$P = 0, \text{ so that } \Psi_{cell} = \Psi_\pi . \tag{5.5}$$

Tissues in Aqueous Media. When cells are not isolated, but form a tissue, their expansion is facilitated or opposed by that of their neighbors; i.e., P in the equation for a given cell is modified by external factors. If the adjacent cells are pushing against the cell considered, P is increased and equilibrium is reached when the water content is still low (and the volume of the vacuole is relatively small). This is important in tissues with delicate walls, for under such conditions they can maintain suitable turgidity without excessive filling with water. If, on the other hand, the cell walls form a rigid surface, they would not readily follow the shrinking protoplasts. Under these conditions P ceases to counteract Ψ_{cell} when only a moderate amount of water has been withdrawn.

Cells of Land Plants in Their Natural Surroundings. The cells of land plants lose water by evaporation from the surfaces of the cell walls. As the cell wall dries out, water flows into it from the protoplasm and, in homoiohydric plants, is replaced by water from the vacuole; the cells begin to wilt. But plasmolysis does not occur, since the cell walls are impermeable to air. The walls follow the protoplasts; they are drawn inward and can even fold up in the process. As turgor diminishes there is a rapidly increasing tendency for water to be drawn into the cell until the turgor loss point is reached; from then on this tendency increases in proportion to the increase in cell-sap concentration. Finally, if desiccation is so extreme that the biomembranes are damaged, the osmotic system breaks down and no more water can be drawn in.

5.3 Absorption, Transpiration, and Water Balance in the Plant

The shoot systems of land plants stand in the air, and steadily lose water that must be replaced from the soil. Transpiration, water uptake and conduction of water from the roots to the transpiring surfaces are inseparably linked processes in water balance. The vapor-pressure deficit of the air (the saturated vapor pressure minus the actual vapor pressure) is the driving force for evaporation, and the water in the soil is the crucial quantity in water supply.

5.3.1 Water Uptake

Plants can absorb water over their entire surface, but the greater part of the water supply comes from the soil. In the higher plants water uptake occurs by way of the roots, the specialized organ of absorption. Lower plants are rootless and thus dependent upon direct water uptake via organs above ground.

5.3.1.1 Direct Water Uptake by Thalli and Shoots

Thallophytes draw water by capillary action from damp substrates and from their surfaces after wetting by rain, dew and fog; in the process they swell. When saturated with water gelatinous lichens, peat moss and the fruit bodies of fungi contain up to fifteen times as much water as in the dry state; most mosses contain 3—7 times as much, and lichens 2—3 times as much as when they are dry. Water is taken up rapidly, so that the thalli are soaked in a few minutes and maximally swollen in half an hour. Bacteria, lower fungi, fungal hyphae, and some algae and lichens can also withdraw water from *humid air*; they, too, swell as a result. But they never take up as much water in this way as when their surfaces are wet, and as a rule several days are required to match their moisture content to that of the surroundings.

Vascular plants are protected against evaporative water loss by a cuticle. To the same degree that the cuticle slows the loss of water through the surface of the shoot, it interferes with the entry of water when the surface is wet. Therefore appreciable direct water intake by shoots occurs only, if ever, at specialized parts of the plant such as *hy-*

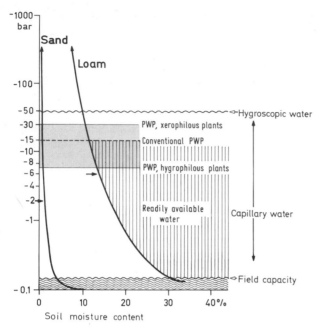

Fig. 5.4. Diagram of the dependence of the water potential of a sandy soil and a loam soil upon the water content of the soil. Depending on pore size, the water potential at field capacity is 0.05 bar (sandy soil) to 0.15 bar (loam). Conventional limiting values: water is exclusively hygroscopically bound at values of Ψ_{soil} of -50 bar and below; water content at field capacity is considered to correspond to $\Psi_{soil} = -0.15$ bar, and the permanent wilting percentage (*PWP*) to correspond to $\Psi_{soil} = -15$ bar. The readily available water depends upon the specific *PWP* of the plants growing on this soil. The *black arrows* are referred to in the text. After Kramer (1949), Laatsch (1954), Slatyer (1967), Rutter (1975). For water uptake and flow in roots see Weatherley (1982)

dathodes (water-permeable places in the epidermis) and non-cutinized attachment points of wettable hairs. In epiphytic Bromeliaceae, a significant part of the water supply enters through *scales* specialized for imbibition.

5.3.1.2 Water Uptake from the Soil

The Water in the Soil

Water infiltrates the soil following precipitation and gradually percolates deeper, down to the water table. In highly permeable soils the rate of percolation is several meters per year, in loamy soils 1–2 m per year, and in very dense soils only a few decimeters per year. Some of the infiltrating water, however, is retained and stored in the pore spaces of the soil. The relative amounts of water that are retained as *capillary water* in the upper layers of the soil and that sink through them as *gravitational water* depend on the nature of the soil and the distribution of pore sizes within it. Pores up to about 10 μm in diameter hold water by capillary action, whereas coarser pores (over 60 μm) let it rapidly pass through.

214

The Water-Storage Capacity of the Soil. The water content at saturation of the soils in their natural locations, after the gravitational water has percolated through, is called the *field capacity* and is expressed in g H_2O per 100 ml of soil. After long periods of rain, and when the snow has just melted in the spring, gravitational water can remain in the upper layers of the soil and thus be available to the roots of plants.

Fine-grained soils, as well as soils with a high colloid content and rich in organic substance, store more water than coarse-grained soils. The field capacity thus increases according to the following sequence: sand, loam, clay, muck. A large water-storage capacity is of course advantageous to the plants, enabling them to survive periods of drought.

How Water is Held by the Soil. The water that remains in the soil after the passage of gravitational water is held in the pore space by *capillary action*; it is also attached to the soil colloids by *surface forces* and (especially in saline soils) can be *osmotically* bound to ions. Thus the soil water, like the water within a plant, is not entirely free. When the soil is not saturated with water, the *water potential* becomes increasingly negative. The soil's tendency to hold water can be described by the water potential equation and be given in bar or $J \cdot kg^{-1}$. In most soils that part of the total water potential represented by the osmotic pressure of the soil solution is negligibly small, as is P (the hydrostatic pressure of the water in the pore space). The crucial component of the soil's water potential is the *matric potential* τ — the force with which the water is held by capillary action and adsorption to colloids. The matric potential is particularly large in soils with fine pores. Gardner (1968) has described the capillary component of the matric potential by the formula

$$\Psi_{cap} = -\frac{4\sigma}{d} \approx -\frac{290}{d} \; [J \cdot kg^{-1}] \tag{5.6}$$

where σ is the surface tension of the water and d is the pore diameter (in μm). The force with which the soil water is held increases greatly as, during drying out, the large-diameter pores are emptied and capillary water remains only in the finer (less than 0.2 μm) pores (Fig. 5.4). In sandy soils with a coarse granular structure the transition is particularly sharp, whereas in loamy and clay soils, in which there is a range of intermediate pore sizes between the average and the smallest, the water potential changes less abruptly.

Water Uptake by Roots

A plant withdraws water from the soil only as long as the water potential of its fine roots is more negative than that of the soil solution. The rate of water uptake is greater, the larger the absorbing surface of the root system and the more readily roots can withdraw water from the soil. According to the formula of W. R. Gardner (1968),

$$W_{abs} = A \cdot \frac{\Psi_{soil} - \Psi_{root}}{\Sigma r}. \tag{5.7}$$

That is, the amount of water absorbed by the roots per unit time (W_{abs}) is proportional to the exchange surface area (A) in the region penetrated by the roots (*active root area per unit volume of soil times volume of soil occupied by roots*) and the *potential difference* between root and soil; it is inversely proportional to the *resistances* (r) to water transport within the soil and in the passage from soil into plant. The active root area of herbaceous crops amounts to about $1 \text{ cm}^2 \cdot \text{cm}^{-3}$; that of woody plants is of the order of $0.1 \text{ cm}^2 \cdot \text{cm}^{-3}$.

The cell-sap concentration in *roots* is usually enough to give a *water potential* of only a few bars; this, however, is sufficient to withdraw the greater part of the capillary water from most soils. This effect can be seen in Fig. 5.4; with Ψ_{root} of only -2 bar the roots withdraw more than $^2/_3$ of the water storable in a sandy soil; a loam soil, which holds water more strongly because of the fineness of its pores, gives up half its capillary water content to roots with a water potential of only -6 bar. To a limited extent, some plants can increase the difference between their water potential and that of the soil solution still further, so as to obtain even more water. Among the herbaceous plants of humid regions, those restricted to permanently moist sites (hygrophytes) can lower the water potential of the roots to at most about -10 bar; plants growing in dry areas (xerophytes) to as much as -60 bar and more. Crop plants achieve -10 to -20 bar; -30 bar is considered the limit for forest trees. If the water potential of the soil decreases further water can be taken from the soil if there is a *replacement* by influx from the regions of the soil in which there are no roots. In sandy soils, with large pores, the columns of water held up by capillarity break under slight tension, so that the *hydraulic conductivity* is readily interrupted. In clay soils, with very fine capillaries, water is replaced even if the potential is low, but the movement of water is very slow and takes place only over short distances (a few mm to cm). When the "readily available water" in the immediate surroundings of the roots is exhausted, the only possibility remaining to the plant is to follow the water by *root growth* and to enlarge the active surface area of the roots. The root system of a plant moves constantly in the search for water. With advancing desiccation of the soil, parts of a root system can die and dry up, while in other places the root is growing out for many meters and branching profusely. The capacity to do this is especially pronounced in plants growing in dry regions.

Wilting Percentage and the Available Water in the Soil

When a plant can no longer drive its water potential below that of the soil it wilts permanently — it does not recover even at night or if protected from evaporation (for example, by covering with a nylon bag). This level is called the *permanent wilting point* (PWP; Briggs and Shantz, 1913). In agriculture and soil science, a soil water potential of -15 bar is taken as the conventional norm for the PWP. Where ecological problems are concerned, the permanent wilting range for the particular species under study must be determined; this can lie between -10 and -40 bar. For this reason the amount of *water available to the plants* cannot be computed simply from the difference between soil water content at field capacity and at the permanent wilting level; it must be determined in each case from the water-potential difference between soil and root [see Eq. (5.7)]. An additional reason is that the withdrawal of the available water from the soil becomes progressively more difficult as soil water potential falls toward the wilting point.

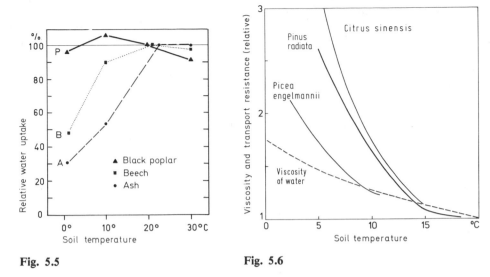

Fig. 5.5

Fig. 5.6

Fig. 5.5. Temperature dependence of water uptake by the roots of *Populus nigra*, *Fagus sylvatica* and *Fraxinus excelsior*. After Döring (1935)

Fig. 5.6. Relative resistance to water transport into and through the plant, and the relative viscosity of the water (as compared with that at 20° C), as soil temperature decreases. In orange trees and Monterey pines reduced root permeability begins to limit water uptake a little below 15° C, whereas in Engelmann spruce from 3000 m above sea level this does not occur until the temperature has fallen to between 5° and 10° C. After Elfving et al. (1972) and Kaufmann (1975, 1977). For classical experiments see Kramer (1940, 1942); low temperature effects on water transfer and uptake are discussed by Dalton and Gardner (1978)

Water Uptake and Soil Temperature

The soil temperature influences water uptake by plants in that both the capacity of the roots to absorb and the resistance to movement of water through the soil and into the roots are temperature-dependent. Plants can extract water from warm soils more readily than from cold soils. At low temperatures the permeability of protoplasm to water falls, and the rate of root growth is reduced (cf. p. 29), a most critical process in the plant's exploitation of water. In the temperature-dependence of water uptake one can discern adaptation of the plants to the soil temperature prevailing in their habitats (Figs. 5.5 and 5.6). Temperate-zone species that begin development early are, as a rule, less hampered by low soil temperatures than are species developing later in the season. In beans, tomatoes, cucumbers, squashes and other plants of warmer countries, water uptake is reduced at 10°—15° C and suspended at temperatures below 5° C, whereas tundra and alpine plants and some forest trees can continue to take up water from ground in which the temperature is barely above 0° C or even from ground that is partially frozen. Below −1° C all the capillary water in the soil is frozen [r in Eq. (5.7) is infinitely large], and no water at all can enter the plants.

5.3.2 The Translocation of Water

5.3.2.1 The Path of Water in the Plant

Within the plant water moves along the water-potential gradient, by *diffusion* from cell to cell (short-distance transport) and by *conduction* through the xylem (long-distance transport). The *diffusion* through the tissues results partly from *osmotic* effects, but is primarily due to *capillary action*, the cell walls acting like wicks. In this way the water passes through the parenchyma of the root cortex until it reaches the endodermis. The root cortex serves as a reservoir, compensating for short-term fluctuations in the supply of water from the soil. In the *endodermis* apoplasmatic transport is blocked by the inclusion of hydrophobic or woody elements in the cell walls. All the inflowing water is channeled to particular sites in the endodermis through which it can pass. In the *central cylinder* of the root the water enters the long-distance transport system; from this point on, it moves through the plant by *conduction*. In the parenchyma cells of the central cylinder arises the *root pressure* generated when water (together with nutrient ions) flows at an accelerated rate into the vascular system. The system of *conducting vessels* is specialized for rapid translocation and distribution, most of the water moving by mass flow through the lumens of the *vascular elements*. It is possible also to demonstrate water transport in cell walls in the long-distance system, but the amounts of water translocated in this way are insignificant. In the leaves the xylem ducts divide into fine branches, and through the tracheids at their tips the water passes to the parenchyma around the veins; from the parenchyma it is displaced to the mesophyll cells, again by diffusion.

5.3.2.2 The Plant in the Water-Potential Gradient Between Soil and Atmosphere

The plant bridges the steep *water-potential gradient between soil and air* (Fig. 5.7). Because the shoot is exposed to the vapor-pressure deficit of the air (that is, to a low water potential), a flow of water through the plant is set in motion.

The rules governing *water transport* are analogous to those for the flow of electrical current, described by Ohm's Law. Analog circuit diagrams are thus a useful way to represent the relationships in the soil-plant-atmosphere system. The *potential gradient* in the continuum soil-plant-atmosphere is the *driving force for water transport* through the plant. Owing to the various resistances to hydraulic conductivity within the plant, there develops a gradient in water potential from the leaf surface through the conduction system to the root (Fig. 5.8).

B. Huber, H. Gradmann and T. H. van den Honert have expressed the relationship between *water flow* (J), the *potential gradient* $\Delta \Psi$, and the *hydraulic and transition resistances* (r) in the following form:

$$J = \frac{\Delta \Psi}{\Sigma r}.$$

The water potential Ψ_z at a particular location in the plant is lower, the lower the *water potential in the soil*, the more effective the *force of gravity* (low Ψ_g in tall trees or vines),

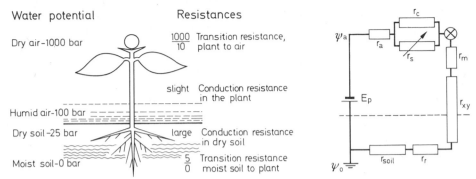

Water potential

Resistances

Dry air-1000 bar

$\frac{1000}{10}$ Transition resistance, plant to air

slight Conduction resistance in the plant

Humid air-100 bar

Dry soil-25 bar

large Conduction resistance in dry soil

Moist soil-0 bar

$\frac{5}{0}$ Transition resistance moist soil to plant

Fig. 5.7. Water-potential gradients and transport resistances between soil, plant and atmosphere. *Left*, Order-of-magnitude estimates of water potential and resistances. The sharpest potential gradient is that between the shoot surface and dry air. This is also the site of the largest transport resistance; the latter is associated with the high energy requirement for the evaporation of water and with the cuticular resistance to diffusion. After Kausch (1955). *Right*, Resistances to water conduction as represented in a circuit diagram. E_p, potential evaporation; Ψ_0, water potential of the liquid phase in the soil; Ψ_a, water potential of the atmosphere; r_{soil}, resistance to diffusion through the soil; r_r, transport resistance in the secondary roots and root cortex; r_{xy}, conduction resistance in the xylem ducts of roots, shoot, leaf petioles and veins; r_m, transport resistance in the mesophyll; r_c, cuticular resistance (very high); r_s, stomatal resistance (variable); r_a, boundary layer resistance; $\otimes$, transition from liquid to vapor phase. After Cowan (1965), Boyer (1974), Kreeb (1974)

Fig. 5.8. Water-potential gradients in an ivy plant. The water-potential depression is always least in the region surrounding the parts of the plant conducting the transpiration stream: in the roots it is greatest in the rhizoderm, where the water is drawn in; above ground, it is greatest in the tissue that is transpiring most vigorously (the epidermis of the shoot axis and the mesophyll of the leaves). Within the plant as whole, it rises from the bottom to the top and along the petiole to the leaf. From Ursprung and Blum (1918). Examples of water-potential gradients in *Ulmus parvifolia* are given by Wiebe et al. (1970); in *Eucalyptus camaldulensis*, by Manohar (1977); in maize, by Turner (1975)

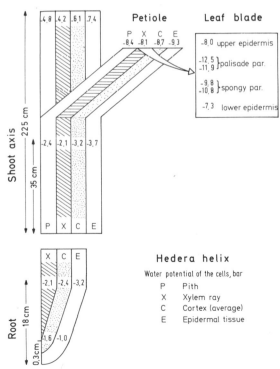

219

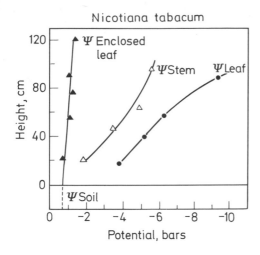

Nicotiana tabacum

Fig. 5.9. Vertical gradients of leaf and stem water potential in tobacco plants transpiring at a high rate and in plants with transpiration restricted by enclosing leaves in polyethylene bags (Begg and Turner, 1970)

the greater the *hydraulic resistances* r between the soil and the reference point z in the shoot, and the more *water flows through the plant* (sum of the partial fluxes f_i). A formula expressing these relationships has been given by Richter (1973):

$$\Psi_z = \Psi_{soil} + \Psi_g + \Sigma^z_{soil}\ f_i \cdot r_i . \tag{5.9}$$

From the two Eqs. (5.8) and (5.9) it is evident that the plant would be expected to exhibit a *steep gradient in water potential* only when large quantities of water are flowing through it — that is, when conditions demand *intensive transpiration* (Fig. 5.9). In this situation the cohesive force acting on the conducting system is large. This effect is of practical value, because by *determining the xylem water potential* one can estimate the water-potential state of the entire plant: When a leaf or twig is cut off, the column of water in the xylem is broken and the water in the vessels is drawn back into the tissues. If the twig, immediately after it is cut off, is introduced into a pressure chamber like that developed by Scholander et al. (1965), the external pressure can gradually be increased until the water potential in the cells is compensated and the water moves back into its original position in the vessels. The extra pressure applied, which can be read off from a manometer, corresponds to the negative pressure previously existing in the twig, a function of the water potential of the cells and the cohesive tension of the water in the xylem ducts of this part of the plant.

5.3.2.3 The Rate of Sap Flow

The *amount of water* moved through the system per unit time is greater, the larger the cross-sectional area of the vessels (the "conducting area"). The physiology and ecology of water conduction were emphasized in the research of H. Dixon, J. Böhm, and B. Huber, and it is their work that has laid the foundation for our present-day understanding of these factors in water balance.

The area of the water-conducting system in a shoot axis or the petiole of a leaf is the sum of the *cross-sectional areas of all the xylem elements*. The conducting area is usually ex-

Table 5.2. Specific conductivity of various tissues and maximum sap-flow velocity. Data taken from Berger (1931), Huber (1956), Zimmermann and Brown (1971), Baxter and West (1977), and Raven (1977). Sap flow velocities of various forest trees are reported by Hinckley et al. (1978)

A. Specific conductivity ($cm^2 \cdot s^{-1} \cdot MPa^{-1}$)

Parenchyma	$10^{-7}-10^{-5}$
Conifer wood, axial	5—10
Ericaceous dwarf shrubs	2—10
Microporous angiosperm wood	5—50
Macroporous and cycloporous angiosperm wood	50—300
Liana wood	300—500
Woody roots	200—1500
Fibrous roots	1—2
Conducting bundles of herbaceous plants	30—60 (250 max.)

B. Maximum velocity of sap flow ($m \cdot h^{-1}$)

Evergreen conifers	1—2
Mediterranean sclerophylls	0.4—1.5
Deciduous diffuse-porous trees	1—6
Ring-porous broad-leaved trees	4—44
Herbs	10—60
Lianas	150

pressed with reference to the mass of the plant parts supplied (i.e., as a *relative conducting area*); for example, in a petiole the conducting area is given as area per unit fresh weight of the leaf, and in a stem as area per total weight of the shoot. The relative conducting area is a measure of the ease with which the individual shoot components of a plant can be supplied with water. It is large in plants that transpire very strongly; some desert plants have relative conducting areas of $2-3$ $mm^2 \cdot g^{-1}$, and in dwarf shrubs of the Ericaceae, Mediterranean shrubs, steppe plants, and herbs of sunny habitats values between 1 and 2 $mm^2 \cdot g^{-1}$ are found. Most woody plants and sciophytes have conducting areas less than 0.5 $mm^2 \cdot g^{-1}$. Water plants, as well as succulents, have especially small conducting areas. Within a plant, too, the relative conducting area varies. In trees it increases from the bottom upward, so that the shoots at the treetop are at an advantage. In this way the plant compensates for the longer distance over which water must be conducted.

Specific Resistance to Water Flow. The sap flow must overcome filtration resistances in the transverse walls crossing the conducting elements at certain intervals, as well as the friction in narrow vessels. The specific resistance of xylem elements in the shoot axis — or its inverse, the *specific xylem conductivity* — is a measure important in characterizing the water translocation system of a plant. The specific xylem conductivity of conifers is half that of evergreen foliage trees, and the latter in turn is half that of deci-

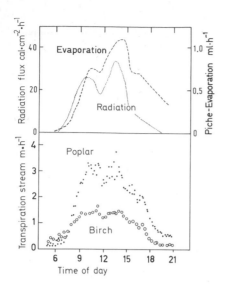

Fig. 5.10. Daily variation in sap-flow velocity in poplars and birches. When radiation and low humidity (high evaporation rate as measured by Piche evaporimeter) promote transpiration through leaves, the water is translocated more rapidly through the conducting elements. After Klemm as cited by Polster (1967)

duous trees (Table 5.2). The roots, with their large-diameter vessels, conduct water particularly well, as do lianas.

Sap-Flow Velocity. The *maximum possible velocity* depends upon the structure of the conducting system concerned (particularly as it affects resistance to flow) and varies among different plant parts and different types of plants (Table 5.2). As long as water uptake by the roots is not hampered, the rate of flow increases with the main driving force, the rate of evaporation (Fig. 5.10). Sap ascent adjusts extremely rapidly to the rate of transpiration, reflecting even short-term fluctuations; thus one can use measurements of flow velocity in tree trunks to infer the progress of transpiration in the entire crown of the tree. In larger trees movement of water begins in the morning at the very top of the crown and at the tips of the branches, drawing up the column of water extending to the base of the trunk. Then the sap begins to flow rapidly, and soon after sunrise it attains its maximum rate. In the evening flow becomes slower, but until late at night there can be a slow influx to the trunks, so that their water reserves are replenished.

5.3.3 Water Loss from Plants

Plants lose water through evaporation (*transpiration*) and occasionally also, to a slight extent, in liquid form (*guttation*). The quantitative contribution of guttation to water balance is negligible, so that in the following discussion, when water loss is mentioned, it is always transpiration that is meant.

5.3.3.1 Evaporation from Moist Surfaces

As a physical process, transpiration by plants follows the rules governing the *evaporation* of water from moist surfaces. An exposed water surface gives off more water vapor per unit time and area, the steeper the *gradient of vapor pressure* between surface and air. A vapor-pressure gradient arises when the water-vapor content of the air at the evaporating surface is greater than that at some distance from this surface. This is always the case when the evaporating surface is adequately supplied with water and is warmer than the air. Strong irradiation *warms the surface* and thus leads to a sharper vapor pressure gradient and to more rapid evaporation. In this way transpiration (and thus movement of water through the plant) occurs even under conditions of high humidity — in fact, even when the air is saturated with water. This is an important factor in the transport of minerals, particularly in humid tropical regions.

Evaporation under conditions of unlimited water supply is called *potential evaporation* (E_p). Potential evaporation is determined by measuring the weight of water evaporated under standard conditions, or is computed from known values for energy balance and mass air exchange. Formulas for applications in agrarian meteorology have been developed by C. W. Thornthwaite and H. L. Penman. In the area of plant ecology standardized evaporating devices (evaporimeters, or more correctly *atmometers*) are used to estimate potential evaporation. Atmometer measurements are useful for relative comparisons, but are not suitable for analysis of the physical process of evaporation. In arid subtropical regions potential evaporation reaches $10-15$ mm $\cdot$ d^{-1} (that is, $10-15$ l of water per m^2 surface area per day), in regions with Mediterranean climate during the dry season $5-6$ mm $\cdot$ d^{-1}, and in the equatorial zone $3-4$ mm $\cdot$ d^{-1}. In the temperate zone, potential evaporation on fine summer days can amount to as much as 4 mm $\cdot$ d^{-1}, whereas the average over the growing season is around 2 mm $\cdot$ d^{-1} and in winter, only $0.1-0.2$ mm $\cdot$ d^{-1}.

The *actual evaporation* from moist surfaces (soil, the cell walls of plants) is usually less than the potential evaporation, because the water is almost never replenished as rapidly as it is lost.

5.3.3.2 The Site of Transpiration in the Plant

Water evaporates from the entire outer surface of a plant, and from all interior surfaces that come into contact with air. In thallophytes the outer surfaces of the thallus are involved, whereas in vascular plants external transpiration occurs by way of the cutinized epidermis (*cuticular transpiration*) and suberized surfaces (*peridermal transpiration*). Within the organs of the plant water evaporates from the surfaces of cells bordering on intercellular air space. In this case the water first is converted from the fluid phase to the vapor phase, and then the water vapor escapes through the stomata (*stomatal transpiration*). From the surface of the plant the vapor diffuses into the adjacent layer of air (the boundary layer, cf. Chap. 3.2.1.1) and thence into the open air. The movement of water vapor from the evaporating surface to the open air is brought about by diffusion.

5.3.3.3 Transpiration as a Diffusion Process

Transpiration can be considered as a process of diffusion. Thus *transpiration intensity* is proportional to the difference between the *concentration of water vapor* at the

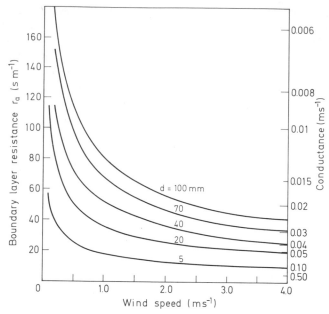

Fig. 5.11. Boundary-layer resistance over leaves of different sizes, as a function of wind velocity. *d* indicates the leaf dimension parallel to the direction of air movement. From Grace (1977). See also the classical study of the effect of wind on transpiration by Gäumann and Jaag (1939)

evaporating surfaces and the water-vapor content of the atmosphere (g $H_2O \cdot m^{-3}$), and is inversely proportional to the sum of the *diffusion resistances* Σr:

$$Tr = \frac{\Delta C}{\Sigma r}. \tag{5.10}$$

Transpiration intensity is expressed in terms of water loss (mg H_2O) per unit time (usually minutes or hours) and unit surface area (usually dm_2^2, the total leaf area — i.e., both upper and lower sides). It can be determined directly — for example, gravimetrically by measuring the water lost during a certain time from plants or parts of plants — but it can also be computed by Eq. (5.10) if the humidity gradient ΔC (between the transpiring surface and the surrounding air) and the leaf diffusion resistance Σr are known. In *leaves* the diffusion resistance to water vapor is given by the boundary-layer resistance and the cuticular and stomatal diffusion resistances. At *surfaces lacking stomata* the diffusion of water vapor is limited by the boundary-layer resistance and by cuticular or peridermal diffusion resistances.

The *boundary-layer resistance* r_a to water-vapor transport, like that to CO_2 exchange, is very much dependent on the size, shape and surface properties of the leaves, as well as on wind velocity. Small leaves (e.g., conifer needles) rarely have r_a greater than 1 $s \cdot cm^{-1}$, even in still air; in large leaves such as those of banana plants the boundary-layer resistance in still air can be 3 $s \cdot cm^{-1}$ or more. As air movement increases these differences are progressively reduced (Fig. 5.11). At wind velocities of about 2 $m \cdot s^{-1}$

or more the boundary-layer resistance is in general less than 0.5 s $\cdot$ cm^{-1} and is no longer an appreciable factor as compared with stomatal diffusion resistance.

The *stomatal diffusion resistance* r_s depends on the degree of opening of the stomata. The *minimal stomatal diffusion resistance* is determined by anatomical features such as the size, structure, arrangement and density of the stomata of the species concerned; it can thus be used to characterize the maximal transpiration capacity of the species. In the great majority of herbaceous plants the *minimal leaf diffusion resistance* (referred to the total surface area, both upper and lower sides of the leaf) lies between 0.7 and 5 s $\cdot$ cm^{-1}; in herbaceous crops, including cultivated grasses, it ranges from 0.7 to 4.5 s $\cdot$ cm^{-1}. The corresponding values for wild grasses are $1.2-6.6$ s $\cdot$ cm^{-1}, for herbs of sunny habitats $0.8-3.3$ s $\cdot$ cm^{-1}, and for shade herbs $3-8$ s $\cdot$ cm^{-1}. Broad-leaved woody plants, both deciduous and evergreen, have minimal leaf diffusion resistances in the range $1.6-8$ s $\cdot$ cm^{-1}; fruit trees account for the lower part of this range ($1.6-3$ s $\cdot$ cm^{-1}). The greatest diffusion resistances with open stomata are found in xerophytes ($2-10$ s $\cdot$ cm^{-1}), succulents and conifers (both $3-8$ s $\cdot$ cm^{-1}). In conifer needles this low value may well be associated with the spongy wax plug in the stomatal chamber, which reduces water-vapor diffusion to one-third of that otherwise possible. A review of the available data has been given by Körner et al. (1979).

5.3.3.4 Transpiration Capacity, and Maximal Transpiration Under Field Conditions

Transpiration capacity is an expression of the specific intensity of stomatal and cuticular water loss from leaves or other plant surfaces under *defined conditions of evaporation*. Transpiration capacity depends primarily on morphological or anatomical characteristics of the leaf. It can be computed from a given, standardized ΔC_{H_2O} (difference between the concentrations of water vapor at the plant surface and in the atmosphere) and the minimal leaf diffusion resistance, or it can be measured directly under standard conditions. Transpiration capacity can also be expressed by the specific *relative transpiration* E/E_p, as suggested by B. E. Livingstone; here E is the amount of water transpired and E_p is the potential evaporation at the same time and place. The *maximal transpiration* is the average peak transpiration rate measured *in the natural habitat*. In order to evaluate specific water consumption it is important to know the transpiration capacity; data on maximal transpiration, like that given in Table 5.3, are useful in estimating the water consumption typical of a particular habitat.

All these measures express water loss per unit area of the transpiring leaves. If the total surface area or the mass of the foliage of a plant is known, they can be used to estimate the water consumption of the entire plant (Table 5.4). In the case of trees, extrapolation from the transpiration of a twig to that of the whole crown provides only a rough estimate, because the movement of water vapor away from the transpiring surfaces is impeded by the high degree of overlapping within the crown. To avoid such errors, it is preferable to use procedures for more direct measurement of overall transpiration — for example, by weighing pot plants or by calculations based on measurements of water flow through the xylem.

Table 5.3. Maximal transpiration and average cuticular transpiration from the leaves of morphologically and ecologically different plant types and from the shoot surface of leafless succulents under the evaporation demand in the habitat. These are typical values, drawn from the data in original works of many authors. Transpiration intensity is referred to the total surface area (in the case of leaves, upper and lower surface) and given in mg $H_2O \cdot dm_2^{-2} \cdot h^{-1}$

Plant type	Transpiration with stomata open	Cuticular transpiration	
		after stomatal closure	as % of total
Hydrophytes	1800—4000		60—70
Herbaceous dicotyledons			
Sunny habitats	1700—2500	(100) 200—300	(6) 10—20
Shady habitats	500—1000	50—250	10—25 (30)
Mountain habitats	800—1600		
Grasses			
Tundra	70—120		
Reeds and meadows	1000—1500		
Dry habitats	(600) 1500 (3000)	250—300	(5) 15—25 (40)
Halophytes	400—800 (1500)		
Dwarf shrubs			
Tundra	50—150		
Heath	600—1000	50—60	5—10
Evergreen conifers	450—550	12—15	ca. 3
Winter-deciduous forest trees, temperate zone			
Light-adapted species	(500) 800—1200	90—110	10—20
Shade-adapted species	(250) 400—700	(30) 60—100	12—18
Temperate fruit trees			
Pomoidea	700—1000	120—160	10—20
Prunoidea	400—700	80—110	15—20
Grapevines	400—500	80—90	17—24
Tropical forest trees			
Sunny sites	100—600		
Shady sites	60—120		
Mangrove trees	200—600		
Palms	400—600 (900)		
Sclerophylls	(200) 500—1000 (1400)	(10) 50—100	(4) 7—15
Shrubs of subtropical deserts	900—3400	30—100	3—10
Succulents [a]			
Leafy succulents	300—600	5—12	1—2
Cacti	200—600	3—10	1—2

[a] P.S. Nobel (pers. comm.).

Table 5.4. Water consumption by trees of various size in the field; for transpiration of tropical rain forest trees in relation to stem size see Jordan and Kline (1977)

Plant species	Height (m)	Age (yrs)	Leaf area (m², both sides)	Foliage mass (kg dry matter)	Leaf number	Transpired water (l per tree) Maximum per day	Transpired water (l per tree) Average per year or season	Region	References
Betula pendula	12	28	115	4.6		140		Denmark	a
Populus canadensis	13	18	135	5.4		128		Denmark	a
Quercus petraea	12	25	90	3.5		73		Denmark	a
Fraxinus excelsior	15	27	60	2.4		47		Denmark	a
Alnus incana	11	19	116	3.7		47		Denmark	a
Fagus sylvatica	12	27	100	2.8		32		Denmark	a
Picea abies	13	33		10.9		27		Denmark	a
Betula pendula	15	21		3.0			5307	Southwest Germany	c
Fagus sylvatica	14	21		2.3			3801	Southwest Germany	c
Pinus sylvestris	15	21		4.4			3037	Southwest Germany	c
Pseudotsuga menziesii	15	21		6.1			2368	Southwest Germany	c
Pseudotsuga menziesii	28	45	400				23,760	USA (Washington)	d
Salix alba	3.5		10		25,000	9.1	800	Southwest Germany	b
Alnus glutinosa	2.5		8		2500	3.7	350	Southwest Germany	b
Fraxinus excelsior	3		7		500	3.6	340	Southwest Germany	b
Acer platanoides	2.5		6		900		160	Southwest Germany	b
Acer pseudoplatanus	2.5		3		300		130	Southwest Germany	b
Elaeis guineensis	5	11	4		32	30	23,300	Zaire	e

References: (a) Ladefoged, 1963; (b) Braun, 1974, 1977; (c) Künstle und Mitscherlich, 1977; (d) Fritschen et al., 1977; (e) Ringoet, 1952.

Stomatal Transpiration

The *specific transpiration capacity* of leaves is greatest in swamp and floating plants. Herbaceous species in sunny habitats transpire most strongly, shade herbs only half as strongly, and trees and sclerophyll plants more weakly still. Plant organs also vary in the rate at which water is lost from the stomata; the stems give off considerably less water than the leaves, as do the flowers (which have very few stomata). The *maximal rates* of transpiration of plants in the field follow fairly closely the specific transpiration capacities. The only conspicuous departures from this rule are the trees of tropical forests and desert plants. Because of the high air humidity in their habitats, the plants of tropical rain forests and particularly elfin forests transpire at moderate intensities. By contrast, desert plants exposed to intense radiation and dry air evaporate off a great deal of water as long as their stomata are open, even if their transpiration capacity is low.

Cuticular Transpiration

Cuticular transpiration can be regarded as diffusion through a hydrophobic medium, since the water molecules must pass through the cutinized layers of the outer wall of the epidermis and through the cuticle itself. The *cuticular diffusion resistance* is usually very high; it varies in different species with arrangement, density, and number of the cutin and wax lamellae embedded in the outer wall of the epidermis, and with the thickness of the cuticle. The conditions under which the individual plant has grown greatly affect the development of the leaf surface; plants adapted to dry air and soil have leaves with thicker layers of cuticle and a thicker wax coating than those that developed in higher humidity. In hygromorphic leaves the cuticular diffusion resistance is $40-100$ s $\cdot$ cm^{-1}, but in leaves and needles with massive protection against transpiration it can reach values around 400 s $\cdot$ cm^{-1}. When the outer epidermis dries out and shrinks the hydrophobic layers are brought closer together, and as a result the cuticular diffusion resistance can double. At low temperatures, too, the cuticular diffusion resistance increases. Cuticular protection against transpiration is very effective. The specific *effectiveness of stomatal closing* can be expressed by the *modulation amplitude of transpiration* — that is, the ratio of unrestricted total transpiration to cuticular transpiration. Even in plants of shady and humid habitats, cuticular transpiration amounts to little more than a third of the total; in sclerophylls, evergreen conifer needles and desert shrubs it is $3-10\%$, and in succulents only $1-2\%$. This means that in those plants best protected against transpiration the cuticular transpiration capacity is reduced to $0.1-0.05\%$ of the potential evaporation from a moist surface. These proportions have been derived from measurements made under the conditions of evaporation in the habitat. When particular plant species, varieties and states of adaptation are to be characterized, the effectiveness of stomatal closing is better expressed by a *comparison of the diffusion resistance* when the stomata are maximally opened with that when they are completely closed. Such a comparison shows that the maximal leaf diffusion resistance (stomata closed) in broad-leaved trees is $30-40$ times the minimal diffusion resistance (stomata fully open). In herbaceous dicotyledons leaf diffusion resistance is increased by a factor of $10-50$, and in grasses by a factor of $5-15$, when the stomata close.

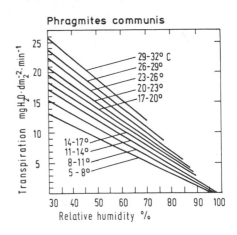

Fig. 5.12. Unrestricted transpiration of the leaves of reeds in relation to the relative humidity and the air temperature. After Tuschl as cited by Burian (1973). Further examples are given by Hall and Kaufmann (1975)

Peridermal Transpiration

The loss of water through the surfaces of *suberized* shoot axes is comparable in magnitude to cuticular transpiration; the amount of such loss, in any case very small, depends upon the species-specific structure of the periderm, the permeability of the lenticels, and the presence or absence of cracks in the bark. For this reason the trunks of oaks, maples, and pines give off more water than those of spruce, beech, and birch trees, with their smoother and denser bark. In trees of the temperate zone peridermal transpiration in summer is of the order of magnitude of $10-100$ mg $\cdot$ dm^{-2} $\cdot$ h^{-1}, about 1% of the potential evaporation. No one has yet made a comparative study of peridermal transpiration in smooth-barked trees of the wet tropics and thick-barked woody plants of dry regions.

5.3.3.5 Variability in Rate of Transpiration

Dependence of Transpiration on External Factors

External factors affect transpiration to the extent that they alter the *steepness of the water-potential gradient* between plant surface and surrounding air. The rate of transpiration rises with decreasing *humidity* and increasing *temperature* (Fig. 5.12), and is greater when the *air is moving*. Within closed stands of plants, dense crowns of trees, and other closely packed vegetation humidity is higher than outside and convection is slower, so that the rate of transpiration is reduced.

Physiological Control of Transpiration

Transpiration is strictly dependent on the physical conditions affecting evaporation only as long as the degree of stomatal opening does not change — that is, as long as the stomata remain either open to a *fixed* degree or firmly closed. Only then is the amount of water lost proportional to the evaporative power of the air. The ability of a plant to regulate stomatal opening enables it to adjust the rate of transpiration to the requirements of its water balance. Changes in stomatal opening can be determined *porometrically*. By diffusion porometry one can obtain a quantitative measure of the stomatal diffusion resistance, the most precise definition of the degree of stomatal opening. It is useful to describe stomatal regulatory processes directly in terms of a change of the stomatal diffusion resistance (or conductivity) rather than as a change in rate of transpiration.

The *response threshold*, the *rapidity*, and the *effectiveness* of stomatal regulation vary among species and with the degree of adaptation to the habitat. Trees and herbaceous sciophytes narrow their stomatal openings even when there is only a slight water deficiency, and briskly complete the process of closing. Herbs of sunny habitats restrict their stomatal transpiration only under much drier conditions, and even then closing is slow (cf. Fig. 5.36). There are differences in the way stomata behave in different plant species and even in different individuals of the same species. In fact, even within a single plant, leaves vary appreciably in this respect, depending on their form and their position on the shoot.

5.3.3.6 Transpiration and CO_2 Exchange

As already emphasized, transpiration and CO_2 acquisition by a plant are linked by the stomata, through which both water vapor and CO_2 diffuse. In order to take up CO_2 the plant must give off water, and when water loss must be reduced the influx of CO_2 is reduced as well.

The chief problem associated with gas exchange, as O. Stocker has put it, is that of "tacking between thirst and starvation". For the management of plantations in agriculture and forestry, where the target is the greatest possible productivity, it is important to know the relationship between water consumption and productivity. As an expression of this relationship Hellriegel (1883) introduced the term *Transpirationskoeffizient* (transpiration ratio), the amount of water in liters consumed by a plant or a stand during the growing season per kg dry matter produced (Table 5.5). One can also compute the number of grams of dry matter synthesized per liter of evaporated water ("*efficiency of transpiration*"; Maximov, 1923). The water required per unit of dry matter formed varies among different species and varieties and is very much dependent on the state of development, the density of the stand and the environmental conditions — the water supply and the rate of evaporation, in particular. Knowing the specific transpiration ratios of crop plants, the grower can select species and varieties appropriate to the situation and, in dry areas, can accurately adjust the amounts of water used for irrigation.

A measure that has been employed since its introduction by Polster (1950) to analyze the influence of environmental factors and the structural and functional characteristics of certain plants upon the optimization of dry-matter production and water consumption is the *photosynthesis/transpiration (F/E) ratio*. Holmgren et al. (1965) have proposed that the specific diffusion behavior of a plant be described by the ratio of the sums of all diffusion resistances to water vapor and to carbon dioxide, $\Sigma r_{H_2O}/\Sigma r'_{CO_2}$. Recently the expression *water-use efficiency* ω has been used in place of all these ratios. But it must not be overlooked that the relationships are quite different in the different expressions; the transpiration ratio in the sense of Hellriegel and the "efficiency of transpiration" are measures of the dry-matter production and water consumption of plants or stands over periods that are ordinarily quite long (weeks or entire seasons),

230

Table 5.5. Average transpiration ratios for productivity ($1/\omega_p$, in liters of transpired water per kg dry matter produced). Data taken from Maximov (1923), Shantz and Piemeisel (1927), Ringoet (1952), Joshi et al. (1965), Polster (1950, 1967), Black (1971), Nobel (1977a), Caldwell et al. (1977), and André et al. (1978)

Herbaceous C_3 plants	
Grain	500–650
Legumes	700–800
Potatoes and beets	400–650
Sunflowers (young)	280
Sunflowers (flowering)	670
Woody plants	
Tropical foliage trees (crop plants)	600–900
Temperate-zone foliage trees	200–350
Conifers	200–300
Oil palms	ca. 300
C_4 plants	220–350
Maize species in field experiments:	
INRA 260 (early variety)	320
Pioneer 3567 (late variety)	266
Maize in growth chamber:	
INRA F7 $\times$ F2	136
CAM plants	50–100

whereas the E/F or F/E ratios refer only to gas exchange in assimilating organs over short periods of time (hours or days). The ratio of diffusion resistances, in contrast to the other two ratios, does not take into account the influence of variations in the concentration gradients of H_2O and CO_2 that depend on environmental factors. Here we propose, for the sake of clarity, to use the terms "transpiration ratio" and "water-use efficiency" in the broadest sense but to make them more precise by stating the frame of reference. In general it holds that the transpiration ratio and the water-use efficiency are inversely proportional (transpiration ratio $= 1/\omega$).

The *water-use efficiency of productivity* ω_p is defined as

$$\omega_p = \frac{\text{Dry-matter production}}{\text{Water consumption}} \; [\text{g DM} \cdot \text{l}^{-1} \, H_2O]. \tag{5.11}$$

Dry-matter production and water consumption can be expressed with reference to a single plant or to a stand; in the latter case dry-matter production is that given by Eq. (3.14), and the value for water consumption is the overall evapotranspiration. The *water-use efficiency of photosynthesis* ω_F is defined as

$$\omega_F = \frac{F}{E} \; [\text{mg } CO_2 \cdot \text{g}^{-1} \, H_2O]. \tag{5.12}$$

The formula for the ratio of diffusion resistances ω_r is

$$\omega_r = \frac{\Sigma r_{H_2O}}{\Sigma r'_{CO_2}}. \tag{5.13}$$

With the stomata maximally opened, ω_r in most C_3 plants is 0.1–0.4, in some trees 0.5, and in C_4 plants as great as 0.8 (for a survey of the data see Fischer and Turner, 1978, and Körner et al., 1979).

Conversion of ω_F to ω_p, using the conversion factors for CO_2 in the dry matter (see p. XVII), can be useful as an estimator but is not quite accurate, because the dry-matter production of a plant does not depend only on the intensity of CO_2 exchange; it is affected much more by the CO_2-exchange balance and the carbon allocation pattern typical of the species.

ω_F and ω_r are interconvertible if the concentration gradients for CO_2 and H_2O are known. The rates of diffusion of the two substances, however, are not identical. The

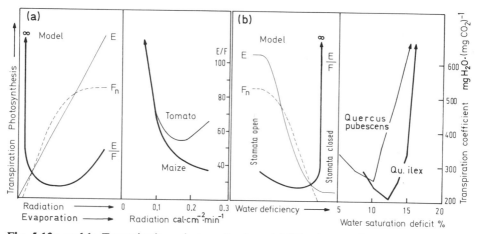

Fig. 5.13a and b. Transpiration, photosynthesis and E/F ratio as a function of insolation, evaporative power of the air and water supply. **a** As the incident radiation increases, the stomata open and the evaporative power of the air becomes greater. Transpiration steadily increases, but photosynthesis increases only up to the saturation level, which is reached earlier in C_3 plants (tomato) than in C_4 plants (maize). At intensities of illumination a little below the light-saturation region for photosynthesis, the E/F ratio falls to a minimum. The "model" curve (valid for C_3 plants) is taken from de Wit (1958), and the E/F ratios for tomato and maize from Barrs (1969). **b** Limitation of E and F with increased water deficiency and the associated closing of stomata. The transpiration coefficient is particularly low when the stomatal pores have already narrowed somewhat, at the beginning of the closure brought about by desiccation. In this respect there are clear differences between species (compare the curves for the deciduous oak *Quercus pubescens* and the evergreen oak *Quercus ilex*). When the stomata are closed the E/F ratio approaches infinity, because photosynthesis is entirely suppressed while the cuticular component of transpiration continues to operate. The model curve is original, the data taken from Larcher (1960). The influence of temperature is demonstrated by Wuenscher and Kozlowski (1971) and age-dependence by André et al. (1978); for an analysis of the dependence of water use efficiency of rice on external factors see Sugimoto (1973)

CO_2 concentration difference between the outside air and the chloroplasts is much less than the water-vapor pressure difference between the interior of the leaf and the atmosphere, as long as the outside air is not saturated with water. At a temperature of 20° C and a relative humidity of 50%, the water-vapor gradient is about 20 times as sharp as the CO_2 gradient. For this reason alone, the evaporation of water proceeds much more rapidly than the uptake of CO_2. Moreover, the water molecules, being smaller, diffuse 1.5 times as fast as the larger CO_2 molecules, given identical gradients. There is also a fundamental difference with respect to the diffusion pathways. The route is longer for CO_2, which must enter the chloroplasts, and there is an additional obstacle in that CO_2 movement in solution occurs exceedingly slowly. The *ratio of transpiration to photosynthesis* (*E/F*) is therefore always changed whenever any of the factors affecting diffusion is altered.

When the *stomata are open*, CO_2 uptake is limited more severely than transpiration by the diffusion resistances inside the leaf (above all by the carboxylation resistance). When the *stomata are closed*, CO_2 uptake is blocked, but water continues to escape through the cuticle so that the *E/F* ratio, for practical purposes, approaches infinity. The most favorable compromise between water consumption and CO_2 uptake is achieved when the stomata are partially open. This is evident not only in the light-dependence relation but also in the behavior during progressive desiccation; as the plant begins to dry out, the *E/F* ratio is lowest when both exchange processes are somewhat restricted (Figs. 5.13 and 5.14). In the *natural habitat*, the expenditure of water necessary to obtain the CO_2 for photosynthesis is least in the early hours of the morning, since photosynthesis gets off to a more rapid start than does transpiration. As warming by the sun progresses and humidity decreases, the rate of water loss rises more rapidly than that of CO_2 uptake, and the *E/F* ratio increases.

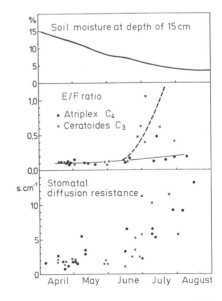

Fig. 5.14. Changes in the E/F ratio and in the stomatal diffusion resistance, measured in the morning, of leaves of the C_3 plant *Ceratoides lanata* and the C_4 plant *Atriplex confertifolia*, during progressive drying of the soil in a cool-temperate saline desert (Utah). At first, when there is adequate water, the E/F ratio is similar in the two plants; during the summer drought stomatal diffusion resistance increases steadily in both species. Because of its greater carboxylation capacity the C_4 plant exhibits a smaller effect on photosynthesis than does the C_3 plant, in which CO_2 uptake is markedly reduced beginning in June. Accordingly, the E/F ratio increases much more rapidly in *Ceratoides* than in *Atriplex*. Modified from Caldwell et al. (1977)

5.3.4 The Water Balance of a Plant

5.3.4.1 Water Balance: A Shifting Equilibrium

The water balance of a plant is given by the *difference* between the rates of water intake and water loss:

$$\text{Water balance} = \text{water absorption} - \text{transpiration}. \qquad (5.14)$$

Here transpiration is viewed as a measure of the expenditure of water and not as a physical process; therefore it is expressed in terms of the amount of water lost per unit mass (usually fresh weight) and not per unit area. It would be better, however, to compare absorption and transpiration with the water content of the plant, expressing the *water turnover* as mg water evaporated per g water content. The *water turnover rate* gives the percentage of the water present in the plant (in the leaf) which is lost in a given period of time (minute, hour, day) and must be replaced if the water balance is to be maintained.

A satisfactory *water balance* can be maintained only if the rates of uptake, conduction and loss of water are suitably adjusted. The balance becomes *negative* as soon as the supply of water no longer meets the transpiration requirements. If the stomata narrow as a result of this deficit, so that transpiration is decreased while uptake continues as before, the *balance is restored* after a transient overshoot to positive values. Thus the water balance of a plant oscillates continually between positive and negative deviations. It is useful to distinguish the short-term oscillations from the long-term disturbances of this equilibrium. *Short-term fluctuations* reflect the interplay of the various water-regulatory mechanisms, particularly the changes in stomatal aperture (Fig. 5.15). There are more marked departures from equilibrium in the *course of a day*, particularly in the alternation between day and night. In the daytime, in a natural habitat, the water balance is almost always negative; the water content of the plant is not restored until evening or during the night (and then only if there are sufficient water reserves in the soil). During dry periods the water content is not entirely restored overnight, so that a deficit accumulates from day to day, until the next rainfall; similarly, there are *seasonal fluctuations* in water balance.

5.3.4.2 Determination of Water Balance

Water balance can be *computed directly* from quantitative determinations of water uptake and transpiration. One can also make an *indirect estimate* of the water balance through its effect upon water content or water potential of the plant. A negative balance always eventually produces a decrease in turgidity and water potential of the tissues. These changes appear first in the leaves, which are the site of most intensive evaporation and moreover are the furthest removed from the roots.

Change in Water Content as an Indicator of Water Balance. A water deficit can be demonstrated by repeated measurements of the water content of leaves and other parts of the shoot. The actual water content at any given time (W_{act}) must be given with respect to a standard measure — for example, the water content of the leaves under

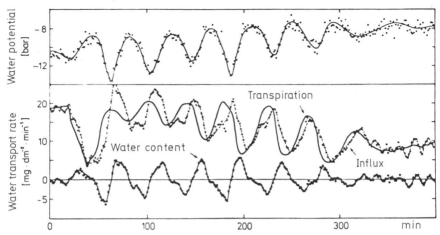

Fig. 5.15. Short-term fluctuations in water turnover, water balance and water potential of cotton leaves. During the phase of rapid transpiration the water content of the leaf falls and its water potential is reduced. The amount of water passing through the petiole (influx) follows a curve 180° out of phase with that for water potential. The short-term fluctuations in transpiration are brought about by oscillation of the stomatal aperture. After Lang et al. (1969). See also Cowan (1972) and Meidner and Sheriff (1976)

conditions of saturation (W_s). The water *content* at any particular time of observation can be expressed as the "*relative water content*" (RWC; Barrs, 1968), a percentage of the water content at saturation:

$$RWC = \frac{W_{act}}{W_s} \cdot 100 \ [\%] \ . \tag{5.15}$$

A measure of water *deficiency* is the *water saturation deficit* (WSD; Stocker, 1929). The water saturation deficit indicates how much water a tissue lacks as compared with complete saturation.

$$WSD = \frac{W_s - W_{act}}{W_s} \cdot 100 \ . \tag{5.16}$$

The water saturation deficit increases as a result of maintained negative water balance, and the relative water content under the same conditions decreases.

Changes in Water Potential. Fluctuations in water content necessarily affect the concentration of the cell sap and the water potential of the cells. The *osmotic pressure*, as a component of the water potential of the cell, provides an indication of changes in the water balance. The osmotic pressure rises as long as water balance is negative. The absolute values of π differ according to plant species, form, state of development and the tissue concerned. The range within which osmotic pressure fluctuates also varies in these respects. There are plants which can support great osmotic stress without damage (*euryhydric* species). In contrast, *stenohydric* species suffer impairment of their vital functions with even slight increases in osmotic pressure. The optimum value (π_{opt}) is

235

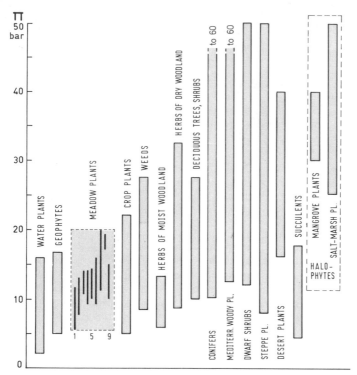

Fig. 5.16. Ranges for the values of osmotic pressure of leaves of ecologically different types of plants (the osmotic spectrum). The sub-ranges (*black bars*) shown for the meadow plants illustrate how to interpret the osmotic range given for each plant group; that is, it is derived from the difference between the lowest and the highest osmotic pressures found among all plants studied in the ecological group. *1, Polygonum bistorta; 2, Taraxacum officinale; 3, Galium mollugo* and *Campanula rotundifolia; 4, Achillea millefolium; 5, Tragopogon pratensis; 6, Poa pratensis; 7, Melandrium album; 8, Cynodon dactylon* and *Lilium perenne; 9, Arrhenatherum elatius.* After Walter (1960)

considered to be the osmotic pressure in a state of optimum water content; the osmotic maximum (π_{max}) is the value found in nature under conditions of extreme drought. To obtain a picture of the water balance in plants of different climatic regions and habitats, one can summarize the full range of variation of π values in the graphic form of "osmotic spectra". In such a diagram (Fig. 5.16) one can see that the plant species fall into ecologically distinct groups.

A measure of changes in water balance that is more sensitive than the osmotic pressure is the leaf water potential (Table 5.6). The immediate result of water deficit is loss of turgor accompanied by a distinct decrease of water potential. It is especially in the range of slight water deficits that $\Delta\Psi$ undergoes greater change than π, so that in this range water potential (usually determined by measuring the xylem water potential) is the most reliable indicator of balance. As a result the range of variation of water potential, the immediate response of the shoot to an inadequate water supply, is always greater than that of the osmotic potential (Fig. 5.17).

Fig. 5.17. Changes in water potential and osmotic potential in the leaves of *Hammada scoparia* during the course of a day at the end of the dry period in the Negev Desert, in one case under the natural drought conditions and in the other with irrigation of the shrubs. After Kappen et al. (1976). The relationships in salt-filtering and nonfiltering mangrove plants are discussed by Miller et al. (1975). For variations during the life cycle of soybeans see Sionit and Kramer (1976); of sorghum and maize, Fereres et al. (1978)

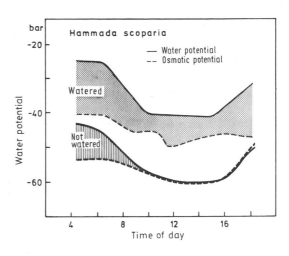

Table 5.6. Minimal water-potential values of assimilative organs in plants of ecologically different groups. From Scholander et al. (1965), Merino et al. (1976), Berger et al. (1978), and the data collected by Richter (1976)

Plant group	Ψ_{min}
Water plants	−12
Swamp plants	−15
Herbaceous crops	−15 to −25
Grasses	−20 to −22 (−45)
Woody plants of the temperate zone	
Deciduous trees and shrubs	−15 to −25
Conifers	−18 to −25 (−60)
Plants of periodically dry regions	
Sclerophylls	−35 to −70
Arid bush	−35 to −80
Garrigue plants	−40 to −80
Desert plants	
Shrubs	−55 to −90 (−160)
Succulents	−18 to −20
Mangroves	−50 to −60
Halophytes	−30 to −55 (−90)

The numbers in parentheses are extreme values.

5.3.5 Water Balance in Different Plant Types

In every climatic zone and habitat, plants with quite diverse water relationships grow side by side. They can be classified in two fundamental categories with respect to water balance, the hydrostable (isohydric) and hydrolabile (anisohydric) types (Fig. 5.18).

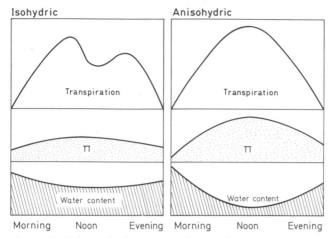

Fig. 5.18. Schematic diagram of the two basic categories of water-balance mechanism as proposed by Berger-Landefeldt (1936). The isohydric type (*left*) avoids pronounced fluctuations in water content and osmotic pressure during the day by stomatal regulation of the rate of water loss. The anisohydric type does not restrict transpiration until it has become very dry; the latter mechanism allows the water balance to stay negative for extended periods, as is reflected in the wide fluctuations osmotic pressure and water content in the course of a day. For characterization of water balance by the amplitude of water-potential fluctuation see Ritchie and Hinckley (1975)

Hydrostable species can, to a great extent, maintain a favorable water content throughout the day, the water balance remaining near zero; their stomata respond with great sensitivity to lack of water, and their root systems as a rule are extensive and efficient. A further factor stabilizing protoplasmic water content is the presence of water reserves in storage organs, roots, the wood and bark of the stems and the leaves. Hydrostable plants include trees, some grasses, shade plants and succulents.

Hydrolabile species are in danger of larger losses of water and greater increases in cell sap concentration. Such vascular plants are always *euryhydric*; their protoplasm must be able to tolerate these rapid and extensive fluctuations in water potential without injury. Many herbs of sunny habitats are hydrolabile, as are the poikilohydric plants.

Trees. In mature trees, with their extensive evaporating surfaces and the long distances water must travel from roots to leaves, it would be especially serious if appreciable water deficits were allowed to develop; incipient water loss must be dealt with from its first appearance. The entire crown of a tree can transpire without restriction only when an adequate water supply is assured. On fine days, at least around noon, trees always encounter difficulties in keeping up with the rate of water loss, so that the guard cells — which in most trees respond even to very small water saturation deficits — temporarily restrict transpiration. Later in the day, when the water content has been restored, the stomata open again and the rate of transpiration increases (Figs. 5.19 and 5.20). The *noon depression of transpiration* is characteristic of trees on clear days. If the water deficit of the tree is not alleviated sufficiently during this period, the afternoon rise in rate

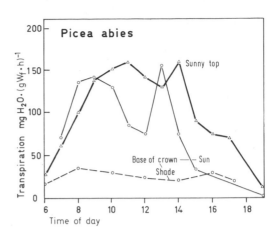

Fig. 5.19. Daily fluctuations in the transpiration of spruce shoots on a sunny August day which had been preceded by dry weather. With an inadequate supply of water, the shoots in the shade at the base of the crown first reduce their water loss, then the twigs in the sun at the lower margin of the crown, and finally the shoots in the sunny top of the crown. After Pisek and Tranquillini (1951)

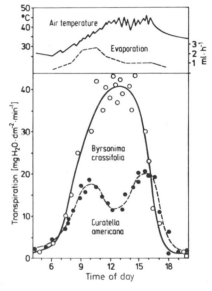

Fig. 5.20. Diurnal fluctuations in transpiration of savanna trees during the dry season. Both species have roots extending far enough into the ground to reach the ground water, but they are of different types with respect to water economy. After Vareschi (1960). Classical examples of changes in transpiration of woody plants in tropical and arid climates during the day are given in the following publications: trees of tropical rainforest, Stocker (1935b); savanna trees, Stocker (1970); oil palms, Ringoet (1952); cocoa trees, Lemée (1956); sclerophyllous shrubs in periodically dry regions, Killian (1931, 1932), Rouschal (1938), Ferri and Labouriau (1952), Grieve (1956), Hellmuth (1971), and Lange and Lange (1963)

of transpiration may be less pronounced or completely absent. Water-conservation measures are not taken in all parts of the crown at once, but occur in a stepwise manner and in an orderly sequence: the first and most prominent *reduction of transpiration* occurs in the shady parts of the crown, where the threshold for stomatal closure is lowest, then in the base of the crown, and finally even the leaves of the tree top limit their evaporation. Correspondingly, the rise in water potential in the sunny tops of trees is sharper than in the shaded parts (Fig. 5.21). The range of *daily fluctuation of osmotic pressure* is very small in trees, rarely exceeding an amplitude of 3 bar; this means that in spite of the increase in the forces drawing water upward during the course of a day the protoplasm itself is not subjected to any great fluctuations in water potential. Not all trees are hydrostable to the same extent. Trees characterized by good maintenance of water

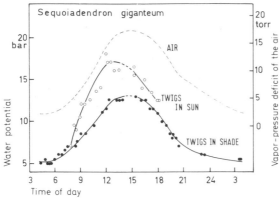

Fig. 5.21. Daily changes in water potential in twigs of the giant sequoia at a height of 7 m (measured with a pressure chamber). As the vapor pressure deficit of the air increases in the morning, the water potential in the twigs becomes more negative, more rapidly in twigs located in the sun than in the shade. With the decline in vapor pressure deficit during afternoon and evening, water balance is restored and the water potential changes in the opposite direction, becoming least negative only a short time before sunrise (Richter et al. 1972). The time courses of water-potential change in leaves at the top and the base of the crowns of cocoa trees under tropical weather conditions have been compared by Alvim et al. (1973). Temporal and spatial variations of Ψ in forest trees are reviewed by Hinckley et al. (1978)

content are conifers, sciophytic species and some heliophytic species such as oaks. There are also trees — for example, species of ash — that do not behave so conservatively; during dry spells the leaves wither prematurely.

Herbaceous Dicotyledons. Among the herbaceous plants one finds all intermediate forms between hydrostability and extreme hydrolability. This *diversity of types* is particularly evident in sunny habitats with a tendency to dryness of the soil; examples are given in Fig. 5.22. When there is a fairly adequate water supply, some of the dicots transpire without restriction throughout the day, at a rate deterined by the evaporative power of the air. Others behave more like trees, depressing the rate of transpiration at midday or even limiting the period of maximal stomatal opening to the early morning and late evening. The extravagant transpiration displayed in the first group can be afforded only by plant species which either can draw upon abundant water reserves through an extensive root system or are able to endure a high degree of desiccation. In the latter case, the osmotic pressure of the cell sap can rise by 3—6 bar during the course of a day. Such species are found frequently among the herbaceous heliophytes.

Grasses and Sedges. These, too, comprise both hydrostable and hydrolabile species. The stomata of hydrostable grasses respond with extreme sensitivity to the first sign of negative water balance. The stomata therefore *close gradually* during the morning so that often no sharp depression is observed around midday (Fig. 5.23). As the leaves age — at the beginning of the dry period in savanna grasses — the stomata lose their mobility, and the grasses gradually lose control of the water balance. Even though the soil is dry, the plants continue to transpire without restraint until the leaves wither.

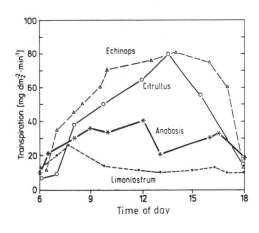

Fig. 5.22. Diurnal fluctuations in transpiration of desert plants in the southern Algerian Sahara at the beginning of the dry season. *Citrullus colocynthis* is a soft-leaved root geophyte with a deep root system. *Echinops spinosus* is a sclerophyllous dwarf shrub, *Anabasis aretioides* is a cushion plant of compact habit, and *Limoniastrum feei* is a succulent capable of colonizing even extremely dry habitats. After Stocker (1974). Many examples of diurnal changes in transpiration of herbaceous plants from various habitats are given by Stocker (1956) and Galazii and Beideman (1975)

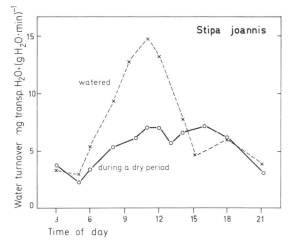

Fig. 5.23. Daily changes in transpiration of the steppe grass *Stipa joannis* during a dry period in July, as compared with the transpiration of artificially watered plants. Here transpiration is given in terms of water turnover (mg transpired water per g water contained in the leaves). After Rychnovská (1965)

5.3.6 Water Balance During Drought

5.3.6.1 Drought as a Stress-Inducing Factor

The term "drought" denotes a *period without precipitation*, during which the water content of the soil is reduced to such an extent that the plants suffer from lack of water.

In dry regions drought is of such regular and prolonged occurrence that annual evaporation exceeds total annual precipitation. The climate in this situation is called *arid*, as opposed to the *humid* climate in regions with surplus precipitation. About $\frac{1}{3}$ of the earth's land area has a rain deficit, and half of this (about 12% of the land area) is so dry that annual precipitation is less than 250 mm, not even a quarter of the potential evaporation. Extensive dry regions are found mainly between 15° and 30° north and south latitude, and on the lee sides of high mountain chains intercepting rain-bearing winds (Fig. 5.24). In lands far from an ocean there is a gradual transition from a humid

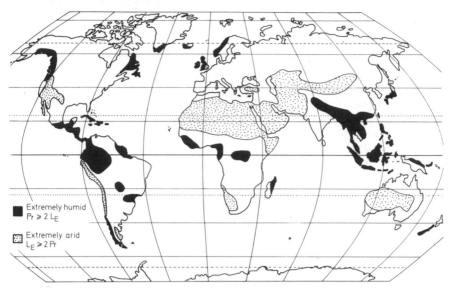

Fig. 5.24. The distribution of extremely humid and extremely arid regions on the earth. Extremely humid: annual precipitation at least twice the amount of water evaporated annually. Extremely arid: annual evaporation at least twice as great as annual precipitation. The demarcation of extremely humid and arid regions is based on the maps of Geiger (1965), giving amount of precipitation and actual evapotranspiration. For precipitation and evaporation maps see Lockwood (1974)

climate, through a semi-arid intermediate region having occasional or periodic dry periods, to an arid region characterized by permanent drought and increased salinity of the soil (Fig. 5.25).

A method used by M. I. Budyko and A. A. Grigorjev to determine the *degree of aridity* is based on the notion that over large areas, and on average over the year, the radiation input (positive radiation balance Q_R; see Chap. 2.2.1.1) is used up to evaporate water. The energy consumed for evaporation can be computed by using the specific heat of evaporation of water λ (see p. XVI). It follows that Q_R is a measure of the potential evaporation in a region. The "Radiational Aridity Index" calculated from the radiation balance relates the annual precipitation Pr, multiplied by λ in order to obtain units of energy, to the radiation balance Q_R as follows:

$$\text{Radiational Aridity Index} = \frac{\lambda \text{Pr}}{Q_R} \, [\text{MJ} \cdot \text{m}^{-2} \cdot \text{yr}^{-1}] . \tag{5.17}$$

Arid regions are characterized by an index greater than 1. For steppes and savannas the aridity index lies between 1 and 2, for semi-deserts between 2 and 3.5, and for deserts over 3.5. Values below 1 indicate a humid climate. In the temperate zone the indices range from 0.3 to 1; in tropical rain forests they are about 0.5 and in the subarctic, about 0.3.

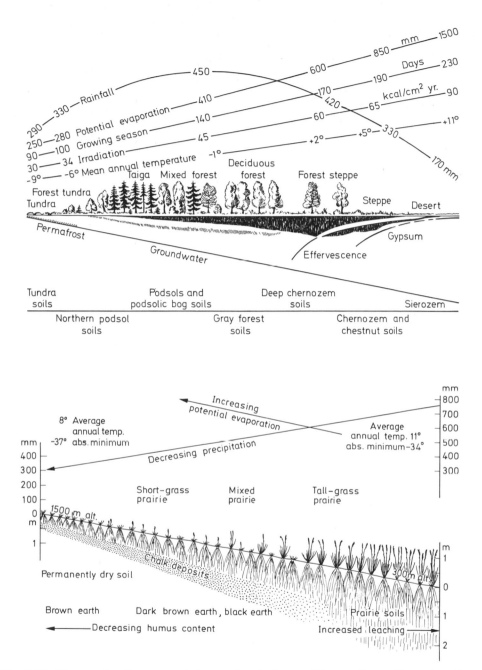

Fig. 5.25. Changes in climate, vegetation and soil along aridity gradients. The point of intersection of the precipitation curve and the curve for potential evaporation marks the boundary between humid and arid climate. *Above*, profile through eastern Europe from northwest to southeast, to the Caspian lowlands. After Vysotskom and Morozov as cited by Walter (1973a). *Below*, east-west profile through the prairie region in the USA that rises from 300 m to the Great Plains at 1500 m above sea level. After Walter (1968)

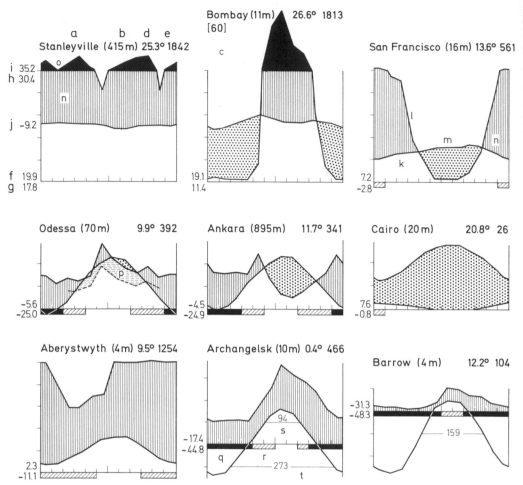

Fig. 5.26. Climate diagrams for *Stanleyville* (Congo, permanently wet equatorial climate); *Bombay* (India, tropical summer-rain climate); *San Francisco* (California, winter-rain region with summer drought); *Odessa* (Black Sea coast, semi-arid steppe climate); *Ankara* (Turkey, Mediterranean climate type with equinoctial rain); *Cairo* (Egypt, subtropical desert climate); *Aberystwyth* (Wales, maritime-temperate climate); *Archangelsk* (taiga zone on the White Sea, cold-temperate climate); *Barrow* (Alaska, arctic tundra climate).

Interpretation of the climate diagrams. *Abscissa*, in the northern hemisphere the months from January to December, in the southern hemisphere from July to June (the warm season is always in the middle of the diagram). *Ordinate*, one subdivision represents 10° C or 20 mm precipitation. *The labels denote: a*, station; *b*, altitude above sea level; *c*, number of years of observation; *d*, mean annual temperature; *e*, mean annual precipitation; *f*, mean daily minimum in the coldest month; *g*, absolute temperature minimum; *h*, mean daily maximum in the warmest month; *i*, absolute temperature maximum; *j*, mean daily temperature fluctuation (tropical stations with diurnal rather than seasonal variation); *k*, curve of the mean monthly temperatures; *l*, curve of mean monthly precipitation; *m*, season of relative drought (*stippled*); *n*, relatively humid season (*vertical shading*); *o*, perhumid season, mean monthly precipitation > 100 mm (scale reduced to 1/10, *black area*); *p*, relatively dry season (precipitation curve

The relationship between annual precipitation and annual evaporation gives only a rough indication of the humid or arid character of an area. As far as the plants growing there are concerned, the important thing is that there should be an assured water supply at the time of greatest need — during the growing season. To provide a picture of *relatively humid and relatively arid seasons*, one can construct climatic diagrams like those suggested by H. Gaussen and Walter (Fig. 5.26). In these, the monthly totals for precipitation and the monthly average air temperature are plotted for the year, the scale being such that *1° C corresponds to 2 mm precipitation*. In such a plot, the temperature curve serves as an indicator of the progressive change in *evaporative power of the air* during the year. That part of the year during which the precipitation curve lies below the temperature curve is a time of drought for the majority of plants that are not irrigated and cannot utilize the groundwater. This procedure offers the advantage that it can be used at stations for which no data on evaporation or radiation are available. Limited precipitation does not in itself bring about aridity; the cold polar regions have little precipitation but are not arid, since there the evaporative power of the air is low.

5.3.6.2 Restriction of Water Consumption During Drought

Plants in humid regions, when no rain falls for days or weeks at a time, so that the water reserves in the soil are depleted and their water balance becomes increasingly unfavorable, progressively reduce their water consumption by *opening their stomata less and for shorter periods*. In Figs. 5.27—5.29 this trend is illustrated; at first transpiration is reduced during the hottest hours of the day, then the afternoon resumption of transpiration is omitted, and finally the stomata open only in the morning. Eventually, while their water content is still adequate, the plants transpire only through the cuticle.

Plants of dry regions, as a rule, have roots that grow deep into the soil or have water-storing tissues, so that they are not forced to a drastic restriction of transpiration (and simultaneously of CO_2 uptake) so soon. This property reflects the adaptation of such plants to arid habitats. *Desert shrubs* in the dry season consume 20—40% of the amount of water given off in the wet season, and in *subshrubs and herbaceous species* this value can be as much as 60—80% (Table 5.7). Some ephemeroid species never restrict evaporation, continuing to transpire at a high rate until they dry up. By contrast, evergreen *sclerophyll trees and shrubs* in regions with periodic summer drought can very effectively limit water consumption, reducing it to 10—20% of that when water is readily available (Table 5.7). The most effective restriction of transpiration is found in *CAM plants*, the stomata of which are completely closed all day during dry periods; if they open at all, it is at night (cf. Fig. 3.48).

◁──

shifted downward, 1 scale division represents 30 mm); *q*, cold season, months with mean daily minimum below 0° C (*black bar*); *r*, period when daily minimum is below 0° C and daily maximum above 0° C, months with absolute minimum below 0° C (*cross-hatched bar*); *s*, number of days with mean temperatures above +10° C; *t*, number of days with mean temperatures above −10° C. These measures are representative of the climatic data important in plant ecology. After Walter and Lieth (1967)

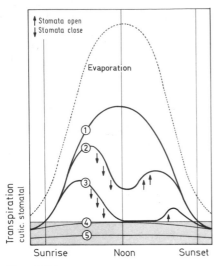

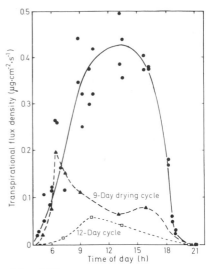

Fig. 5.27

Fig. 5.28

Fig. 5.27. Diagram of daily changes in transpiration as it becomes progressively more difficult (curves *1—5*) to maintain the water supply. The *arrows* indicate the stomatal movements elicited by changes in the water balance. The stippled area shows the range in which transpiration is exclusively cuticular. *1*, unrestricted transpiration; *2*, limitation of transpiration at noon as the stomata begin to close; *3*, full closure of the stomata at midday; *4*, complete cessation of stomatal transpiration by permanent closure of the stomata (only cuticular transpiration continues); *5*, considerably reduced cuticular transpiration as a result of membrane shrinkage. After Stocker (1956). Recent review on response and adaption of plants to water stress (Turner and Begg, 1981)

Fig. 5.28. Diurnal changes in transpiration of two-year-old seedlings of *Pinus radiata* during progressive drying of the soil. From Kaufmann (1977). For restriction of transpiration due to drying in eucalyptus trees see Gindel (1973); in avocado trees, Sterne et al. (1977)

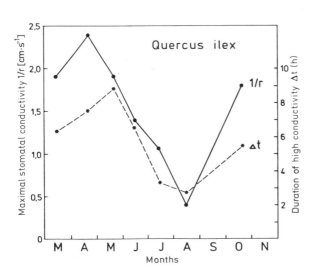

Fig. 5.29. Maximal diffusion conductivity (1/r, used as a measure of transpiration capacity) and daily duration of the period of maximal conductivity, for *Quercus ilex* leaves well supplied with water (February to June) and during the summer dry season (July to September) in southern France. As the soil becomes dryer, the stomata open less widely and for a shorter time (Lossaint and Rapp, 1978)

246

Table 5.7. Total daily transpiration of plants in periodically dry regions, comparing the rate under conditions of adequate water supply with that during prolonged drought, under the conditions of evaporation prevailing in the habitat in each case

Plant	Total daily transpiration ($g\ H_2O \cdot dm_2^{-2} \cdot d^{-1}$)		Level to which transpiration is reduced (%)	References
	Rainy season	Dry season		
Mediterranean maquis				
Olea europaea	10	2.8	28	a
Laurus nobilis	3.3	0.5	15	a
Phillyrea media	10.4	1.5	14	a
Quercus ilex	9.5	1.0	11	a
Australian sclerophyll woodland				
Acacia acuminata	12.4[a]	2.9[a]	23	b
Acacia craspedocarpa	18.6[a]	3.9[a]	21	b
Banksia menziesii	7.4	0.6	8	b
Australian bush				
Hakea preisii	19.8	13.4	68	b
Eremophila miniata	33.3	17.9	54	b
Halophyte desert in Utah				
Atriplex confertifolia (C_4)	14.2	2.7	19	g
Ceratoides lanata (C_3)	14.0	0.2	1	g
North African desert plants				
Nitraria retusa	19.1	15.0	78	c
Zilla spinosa	22	14	64	e
Zilla spinosa	38	15	39	d
Zygophyllum coccineum	15	7	47	e
Pennisetum dichotomum	15	6	40	e
Haloxylon persicum	26	10	38	d
Hammada scoparia	4[b]	1.5[b]	38	d
Anabasis articulata	3.1[b]	1.0[b]	32	d
Retama retam	25	7	28	d
Artemisia herba-alba	6[b]	1.6[b]	27	d
Noea mucronata	5.5[b]	1.0[b]	18	d
Cactus desert in California				
Ferocactus acanthodes	1.54	0.32	21	f

[a] $g\ H_2O \cdot g^{-1}$ dry weight $\cdot d^{-1}$.   [b] $g\ H_2O \cdot g^{-1}$ fresh weight $\cdot d^{-1}$.
References: (a) Rouschal, 1938; (b) Grieve and Hellmuth, 1968; Hellmuth, 1971; (c) Stocker, 1970; (d) Zohary, 1961; Zohary and Orshan, 1956; (e) Abd el Rahman and El Hadidi, 1958; (f) Nobel, 1977a; (g) Caldwell, pers. comm., 1978.

5.3.6.3 Winter Drought Effects

Precipitation in solid form cannot be taken up immediately by the plants. Winter, for plants projecting above the *snow cover*, is thus not only a cold but also a dry season. Wherever the snow is blown away in winter the young trees perish, and where the rooted soil strata are frozen too long, or permafrost rises too high, no trees can grow. The forest becomes sparse and gives way, in the north, to tundra, and in the mountains to scrub, dwarf shrubs, or alpine herbaceous communities.

Water deficiencies can thus occur *in winter* and can even cause injury if replenishment of the water supply is prevented by prolonged frost. In regions with long cold winters (mountains, the arctic and subarctic), the soil remains frozen near the surface for months, and the water ducts in shoots and roots become filled with ice (the water in the tracheary elements is frozen at temperatures below $-2°$ C). The twigs not protected by snow, however, continue to lose water. The water balance becomes negative, and the water content deteriorates progressively. The problem is greatest in late winter, when the soil has not yet thawed but the sun increasingly warms the branches and stimulates transpiration (Fig. 5.30).

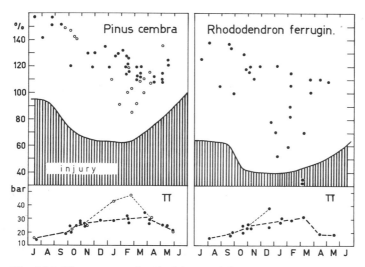

Fig. 5.30. Water content (as % of dry weight), osmotic pressure and drought resistance of pine needles and the leaves of *Rhododendron ferrugineum* at the alpine treeline. The *circles* in the pine diagram denote small trees (up to 2 m high), and the *dots* refer to mature trees. Pine needles and *Rhododendron* leaves acquire an enhanced protoplasmic resistance to desiccation at the beginning of their winter dormant period, and do not lose it until the beginning of the growing season in the spring. In late winter, when the ground is still frozen but the sun warms more strongly the branches not protected by the snow cover, the water content of the *Rhododendron* and of small pine trees falls, approaching and sometimes passing the limit for safety of the plant. Because of the water reserves in their trunks, larger trees are not as greatly endangered as are young trees and shrubs. After Larcher (1972). Winter-desiccation at the alpine timberline: Wardle (1974) and Tranquillini (1979)

5.3.7 Drought Resistance

Drought resistance is the *capacity of a plant to withstand periods of dryness*. This capacity is a complex characteristic. The prospects for survival of a plant under extreme stress due to drought are better, the longer a dangerous decrease in the water potential of the protoplasm can be delayed (the *avoidance* of desiccation) and the more the protoplasm can dry out without becoming damaged (the capacity to *tolerate* desiccation). In the terminology of Levitt (1958), drought resistance is the result of desiccation avoidance and desiccation tolerance. Among vascular plants desiccation tolerance is very slight, so that differences in the drought resistance of different species are chiefly due to differences in desiccation avoidance. In regions with regular dry periods, and in deserts, there are plants that evade drought by timing their growth and reproduction to occur in the brief period during which sufficient water is available. Figure 5.31 summarizes possibilities for survival of plants in dry regions (xerophytes).

5.3.7.1 Drought-Evading Xerophytes

Drought-evading plants as a rule are not drought-resistant. Survival of dry periods requires only the appropriately timed production of desiccation-resistant seeds or organs specially protected from drying out. *Pluviotherophytes* are short-lived vascular plants that germinate after fairly heavy rainfall and rapidly complete their cycles of development. They pass the dry season as seeds, unharmed by desiccation. *Geophytes* have underground organs full of water (rhizomes, tubers, bulbs) that can survive periods of drought, being protected from excessive water loss by the soil. When the rainy season begins they sprout immediately, utilizing stored carbohydrates, and soon flower and bear fruit.

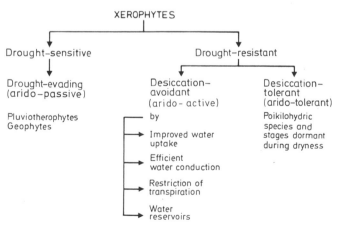

Fig. 5.31. Survival mechanisms of plants in dry regions. After the categorizations of Shantz (1927), Killian and Lemée (1956), Zohary (1952), Levitt (1958, 1972), Walter (1973b), Daubenmire (1974), Evenari et al. (1975a), and Sen (1982). For prevention and reduction of stress effects see Raper and Kramer (1983)

5.3.7.2 Desiccation Avoidance

Desiccation is delayed by all those mechanisms that enable the plant to maintain a favorable tissue water content as long as possible despite dryness of air and soil. This is achieved by improved *uptake* of water from the soil, by reduced *loss* of water (early and effective increase in diffusion resistance and reduction of the transpiring surface), by a high capacity for water *conduction* and by the *storage* of water. These behavioral measures for desiccation avoidance are also reflected in the morphology of the plant.

Water uptake by an extensive root system with a large active surface area is improved by rapid growth into deeper soil strata. There the roots reach horizons containing water on which the plants can draw for some time. The seedlings of woody plants in dry regions have taproots ten times as long as the shoot, while the grasses in such places elaborate a dense root system and also send their threadlike roots to depths measured in meters. The ratio between the masses of shoot and root is shifted further in favor of the roots, the more the exposure to drought (cf. Table 3.9). The situation becomes critical when there is not enough room for expansion of a root system. Plants with extensive roots (woody plants in particular) growing on shallow soils are especially endangered by drought. Analogous problems arise when one attempts to establish plants on and over man-made structures (flat roofs, tunnels and other underground construction). Even in humid regions only drought-resistant plants are suitable for such purposes, for they can manage with the small amounts of water stored in the limited volume of soil.

Increased diffusion resistance of the shoot surface is achieved by early closure of the stomata and effective cuticular protection against transpiration. As a *modulative* adaptation, this takes the form of a rise in the abscisic acid level in the plant, which enhances the readiness of the stomata to close (Fig. 5.32). A *modificative* change is the development, on leaves that grow when water is scarce, of smaller stomata more densely arrayed. Such leaves reduce transpiration more rapidly following the onset of stomatal regulation. Moreover, the epidermis of such leaves is made more impermeable by thicker cuticular layers and, in many cases, a dense covering of hairs. Restriction of transpiration by these natural means can be supplemented artificially by *antitranspirants*, which form surface films or trigger closing of the stomata. Substances that cause *stomatal closure* are abscisic acid, inhibitors of energy-supplying electron transfer (e.g., phenyl mercury acetate, acetyl salicylic acid, thiocarbamate, and other urea derivatives) and chemicals that affect permeability (partially oxidized fatty acids, sulfonates). Water-vapor release through the stomata, and to an even greater extent that through the cuticle, is diminished by spraying to form a *coating* of emulsions of polyethylene, acrylic, silicon and polyterpene compounds. Because of the undesirable side effects on CO_2 exchange and temperature regulation, the use of antitranspirants is practicable only as an emergency measure.

Reduction of the transpiring surface is brought about rapidly and reversibly by folding and *rolling up of the leaves*. This mechanism is widespread among grasses and sedges; in *Stipa tenacissima* it cuts transpiration to 40% of that in outspread leaves. A particularly effective reduction of the transpiring surface of a plant is the partial or com-

Fig. 5.32. Transpiration (*Tr*), water potential (*Ψ*) and abscisic-acid content (*ABA*) of the leaves of grapevines exposed to gradual soil drying for 6 days, after which watering was resumed. After Loveys and Kriedemann (1973). Reviews on phytohormone action on transpiration by Gale and Hagan (1966) and Coudret and Ferron (1977)

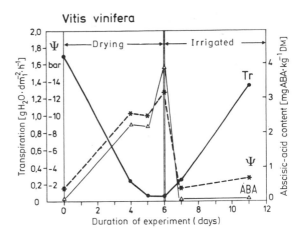

plete *abscission of the leaves*. A number of woody plants in dry regions shed their leaves regularly during the drought season. Through peridermal transpiration the trunks and branches of large trees lose only 1/300 to 1/3000 of the amount of water evaporated from the leaves when the water supply is good. Species of *Ephedra, Sarothamnus, Cytisus, Spartium, Retama*, and desert shrubs like *Zilla macroptera* and *Zollikoferia arborescens* shed their leaves as required, reducing the surface area to $\frac{1}{3}-\frac{1}{5}$ of its usual value. The relationship between transpiring surface and water stored in the tissues is expressed by the *specific leaf area* (*SLA*) of the leaves or shoots:

$$SLA = \frac{\text{Surface area (dm}^2)}{\text{Fresh weight of leaf (g)}}.$$

(5.18)

Alternatively, one may compare leaf area with the dry weight of the entire plant; this is the leaf area ratio (*LAR*).

Since the rate of water loss increases with the transpiring area, a limited leaf area conserves the water in the leaf. Leaves developed under conditions of poor water supply are, as a rule, correspondingly smaller and more deeply laciniated, and have a smaller specific leaf area (*xeromorphosis*).

The *capacity for water conduction* is increased by enlargement of the area of the water-conducting system (more vessels, dense leaf venation) and reducing the transport distance (shorter internodes). When the transpiring surface is reduced the *relative* area of the conducting system is increased even though the *absolute* area (conducting cross section) is unchanged.

With *water storage*, mechanisms for desiccation avoidance reach their peak of perfection — especially when they are coupled with surface reduction and high transpiration resistance of the epidermis. A measure of storage capacity is given by the degree of succulence:

$$\text{Degree of succulence} = \frac{\text{Water content at saturation (g)}}{\text{Surface area (dm}^2)}.$$

(5.19)

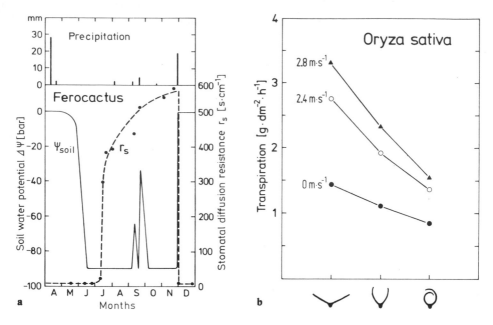

Fig. 5.33. a Soil drying (Ψ_{soil} at depth of 10 cm) and slowly rising stomatal diffusion resistance (r_s, daily minimum) of *Ferocactus acanthodes* in the Colorado desert during the months of drought between the relatively heavy rainfall in April and November. After Nobel (1977a). **b** Effect of leaf-rolling of rice on transpiration in still air and wind. After O'Toole et al. (1979). For relationship between water potential and degree of leaf-rolling see O'Toole (1982)

Under certain circumstances succulent plants can survive years of drought, and after the last rain sufficient water is stored that months can pass before the stomata are permanently closed (Fig. 5.33).

5.3.7.3 Desiccation Tolerance

Desiccation tolerance refers to the species-specific and adaptable *capacity of protoplasm to endure severe loss of water*. A water deficiency causes progressive loss of protoplasmic turgor and an increase in solute concentration. The initial result of these two effects is a disturbance of cell function (Fig. 5.34); then functional deficits appear, and finally protoplasmic structures (biomembranes in particular) are damaged. The resistance to desiccation is measured by equilibrating parts of plants or pieces of tissue (sections or cut-out disks), unprotected against transpiration, with air of known humidity. The lowest relative humidity (or the corresponding water-potential values) at which the cells are just capable of survival (the "critical limit") or are damaged by a certain amount (e.g., 50% injury is the *desiccation lethality*, DL_{50}) is used as the *measure of tolerance*. In the literature one also finds tolerance given in terms of water content — for example, critical and sublethal (5—10% injury) water content or water deficit. Such data, if carefully interpreted, are informative with respect to the detection of adaptive and seasonal changes in desiccation tolerance, but they should not be used for

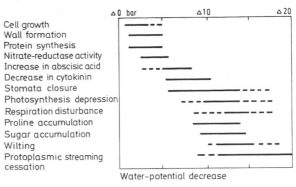

	Δ0 bar	Δ10	Δ20

Cell growth
Wall formation
Protein synthesis
Nitrate-reductase activity
Increase in abscisic acid
Decrease in cytokinin
Stomata closure
Photosynthesis depression
Respiration disturbance
Proline accumulation
Sugar accumulation
Wilting
Protoplasmic streaming cessation

Water-potential decrease

Fig. 5.34. Sensitivity of cell functions to water deficiency, and changes in the plant as it dries out. The *lines* indicate the range in which a clear effect begins to appear in most plant species. The measure of desiccation stress used here is the change in water potential as compared with that when there is a good supply of water. After Hsiao (1973), supplemented by data of Arvidsson (1951) and Kamiya (1959)

comparisons of different species since their absolute values are influenced by the anatomical peculiarities of the sample investigated. For example, the fraction of the measured weight associated with sclerenchymatous structures can alter the figure. The desiccation tolerance of plant protoplasm varies over a wide range (Table 5.8).

Desiccation Tolerant Species. All orders of *thallophytes* contain species capable of tolerating complete desiccation. Many representatives of the bacteria, cyanophytes and lichens can withstand a state of dryness (in equilibrium with the surrounding air) for months and even years, resuming metabolic activity as soon as they are moistened again. Some of them can survive weeks in absolutely dry air over concentrated sulfuric acid or phosphorus pentoxide. Other completely desiccation-tolerant plants include some fungus mycelia, various mosses, and some pteridophytes. Among the *flowering plants* there are relatively few desiccation-tolerant species — e.g., *Ramonda serbica* and some species of *Haberlea* (both Gesneriaceae) in the Balkan region, and species of Scrophulariaceae, Myrothamnaceae, Velloziaceae, Cyperaceae, and Poaceae in South Africa (lists of species and characteristics given by Gaff, 1977).

Desiccation-Sensitive Species. The protoplasm of most plants is extraordinarily sensitive to loss of water and dies after even a slight reduction in water content. The cells in the leaves of homoiohydrous vascular plants perish if they are exposed without protection for a few hours to a relative humidity between 92 and 96% (which in a closed system would be in equilibrium with a solution having $\pi = 55-110$ bar); roots are more sensitive, while buds are more resistant. Plants may become "hardened" to dry conditions, but this shifts the tolerance limit only slightly (3—4% *RH*). During growth periods the cells are especially sensitive to desiccation; during dormant periods they are somewhat more resistant (Fig. 5.35). Among the *thallophytes* the most sensitive are the planktonic algae and those seaweeds attached so far below the surface that they are normally always covered by water; mosses are also sensitive and are found only where the prevailing humidity is continuously high. Most mosses are adapted to a narrow

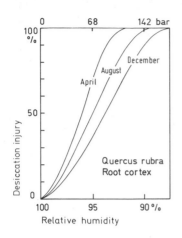

Fig. 5.35. Drought tolerance of the cortical parenchyma of oak roots at different seasons. Drought resistance was determined by placing sections in chambers with graded relative humidity until equilibrium was obtained. The measure of desiccation injury was the percentage of dead cells in each sample as compared with a control at 100% relative humidity. The relative humidity (or the corresponding absolute value of Ψ_{cell}, in bar) at which the desiccation injury amounts to 50% is the measure of drought tolerance. After Parker (1968)

Table 5.8. Desiccation tolerance of plant cells after 12—48 h in vapor chambers with different relative humidities

Plant		Tolerated without injury % RH	bar	Moderate injury % RH	bar	References
Marine algae						
Sublittoral algae		99—97	14—41			b
Algae of the ebb line		95—86	69—204			b
Intertidal algae		86—83	204—252			b
Liverworts						
Hygrophytes	usually	95—90	69—141	92—90	112—141	c
Mesophytes	usually	92—50	112—933	90—36	141—1400	c
Xerophytes	usually	(36)—0	(1400)—∞	0	∞	c
Mosses						
Water mosses and hygrophytes		95—90	69—141			d
Mesophytes	usually	90—50	141—933			d
	extreme	10	3000			d
Xerophytes	usually	5— 0	400—∞			d
Fern gametophytes						
Forest ferns		>90	>140	50—90	140—933	f
Rock ferns		40—60	690—1200	20—30	1575—2080	f
Tracheophytes (tissue sections)						
Leaf epidermis				96—92	55—112	a
Mesophyll		96	55	95—90	69—141	a, e
Root cortex				97—95	41—69	g

References: (a) Iljin, 1927, 1930; (b) Biebl, 1938; (c) Höfler, 1942, 1950; (d) Abel, 1956; (e) Sullivan and Levitt, 1959; (f) Kappen, 1965; (g) Parker, 1968.

range of water potential, so that the various species (there are differences even between species of a single genus) are of value as *indicators of the humidity of a habitat*. It should be said, of course, that "hardening" can considerably increase the resistance of mosses to desiccation.

5.3.7.4 Specific Survival Time

Drought can be so extreme that plants are no longer capable of extracting any water at all from the soil. Drought resistance after complete cessation of water replacement is termed *specific survival time (Überdauerungsvermögen)*. This is a measure of the characteristic degree to which a plant species can conserve the water stored in its shoots. According to A. Pisek, specific survival time is computed from the cuticular transpiration (E_c) and the available water (W_{av}) in the plant — that consumed between the time of stomatal closure and the appearance of the first signs of desiccation injury.

$$\text{Specific survival time} = \frac{W_{av}}{E_c} . \tag{5.20}$$

Survival time is measured in hours or days and indicates how long after stomatal closure the leaves of a plant species can remain undamaged without a supply of water, for a given evaporative power of the air. The amount of available water is greater if the stomata close as soon as there is a slight water deficit, if the tissues can store an abundance of water, and if desiccation injury is incurred only with large water deficits (Fig. 5.36). Survival time of the *leaves* thus depends both on the effectiveness of *desiccation avoidance* and of *desiccation tolerance*. From Table 5.9 it can be seen that of the factors prolonging life, the *cuticular surface protection* varies most among different species. Only the succulents have a remarkable specific survival time, attributable primarily to their *ability to store water*.

During a prolonged dry spell the water *stored in massive organs* (the trunks and larger branches of trees, and the underground storage organs of herbaceous plants) becomes very significant. Movement of water from the wood and cortex of 10–20-year-old winter-deciduous trees can prolong the survival of the *leaves* by factor of 5–10, depending on the amount of foliage, thickness of the cortex, and anatomical features of the wood; survival of young trees 0.5–1 m tall can be prolonged by a factor of 2–3. Once trees are fully grown, interspecific differences in ability to protect the leaves from drying by providing water from axial organs tend to disappear; at this stage the decisive factor is the *ratio of foliage volume to axial volume*. Redistribution of water, from axis and roots to the leaves, is also a useful mechanism in that it is fully effective even *after* the stomata have closed, when the leaves are consuming water most economically.

5.3.7.5 Relative Drought Index

Plants differ with respect to survival capacity; the extent to which individual plants, *in their habitats*, suffer from dry conditions depends not only upon their drought re-

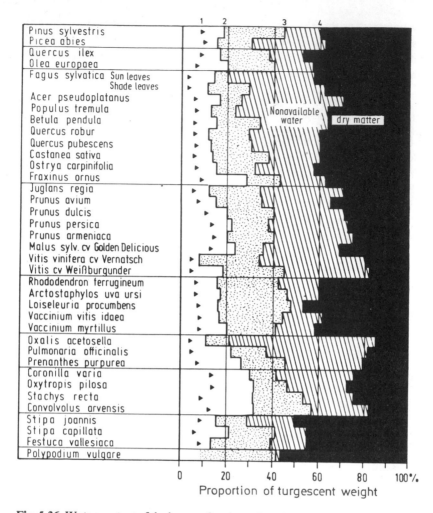

Fig. 5.36. Water content of the leaves of various plants in water-saturated state, and availability of the water. *1*, Water deficit (as compared with saturated weight) at onset of stomatal closing; *2*, saturation deficit after complete stomatal closure; *3*, saturation deficit at the appearance of slight desiccation damage (5%–10% of the leaf area dead). The water remaining at this stage is necessary for life and can be diminished only at the risk of severe injuries (*nonavailable water*). The "available water" (*dotted field*) can be utilized after stomatal closure until damage is incurred and thus represents an emergency reserve for times of drought. The species are grouped as follows, *top to bottom*: conifers, Mediterranean sclerophylls, winter-deciduous forest trees, fruit trees and grapevines, ericaceous dwarf shrubs, shade herbs, herbaceous dicotyledons of sunny habitats, steppe grasses, poikilohydric fern. After Pisek and Winkler (1953), supplemented by data of Y. Boyer (1968), Florineth (1975), Künz (1974), Larcher (1960), Mahlknecht (1976), Müller (1976), Rychnovská and Úlehlová (1975)

Table 5.9. Specific survival time of leaves cut from various plants; potential evaporation, as measured by Piche evaporimeter, was 0.4 ml $H_2O \cdot h^{-1}$. Computed from data given by Pisek and Berger (1938), Pisek and Winkler (1953), Larcher (1960), Weinberger et al. (1973), Müller (1976)

Plant group	Specific survival time (h)	Water consumption after stomatal closure (cuticular transpiration, mg H_2O $\cdot dm_2^{-2} \cdot h^{-1}$)	Water available from stomatal closure to appearance of damage; cf. Fig. 5.36 (mg H_2O $\cdot dm_2^{-2}$)	Specific leaf area (leaf area on both surfaces per saturation weight; $dm_2^{-2} \cdot g^{-1}$)
Evergreen conifers	20—50	10—15	350—650	0.4
Mediterranean sclerophylls	10—20	10—50	100—600	0.5
Evergreen broad-leaved woody plants	ca. 10	20—70	400—700	0.4—0.7
Deciduous trees	1.5—3.5	50—170	100—350	0.7—1.5
Heliophytic herbs	1—2	100—200	150—300	0.7—0.8
Sciophytes	0.5—1	ca. 250	100—150	1—2
Succulent leaves	ca. 20	ca. 150	ca. 3000	ca. 0.15
Succulent stems	>1000	2—5	>20,000	<0.025

sistance but also on the conditions prevailing in the habitat. Both of these factors are comprised in the *relative drought index* (*Trockenheitsbeanspruchung*; cf. Table 5.10). As defined by Höfler et al. (1941), this index expresses as a percentage the relative magnitudes of the actual water saturation deficit (WSD_{act}) measured at a given time in the habitat and the critical water saturation deficit (WSD_{crit}) for the species concerned:

$$\text{Relative drought index} = \frac{WSD_{act}}{WSD_{crit}} \cdot 100 . \tag{5.21}$$

Specific survival time and relative drought index are both useful in characterizing the *prospects of survival* of a species under conditions of water deficiency. *Specific survival time* is particularly suited to describe the probability that a plant will last through a season of dryness, for the effects of drought depend upon its *duration*. By reference to the *relative drought index* one can obtain additional information about *spatial differences* in severity of drought if several individuals of the same species, in different locations, are compared with one another. This measure reflects the degree to which the water stores in the soil and the local conditions for evaporation can vary; drought is thus appreciated as a habitat characteristic with a crucial influence upon the distribution of species.

Table 5.10. Average drought indices for different plant groups. From several authors cited by Larcher (1973b), including data from Bobrovskaya (1971) and Bannister (1976)

Plant group	Location	Average maximal water saturation deficit in the natural habitat, in % of WSD at 5—10% damage
Mediterranean	At aridity limit	80—85
maquis shrubs	At northern limit	40—70
Garrigue shrubs	Mediterranean region	90—105
Ericaceous dwarf shrubs	Atlantic heaths	15—40 (88)
Winter deciduous trees and shrubs	Northern Europe	10—40 (50)
Herbaceous forest undergrowth	Northern Europe	6—25 (50)
Dicotyledonous herbs of xerothermic habitats	Central Europe	40—85
Meadow grasses	Central Europe	20—40 (75)
Steppe grasses	Central Europe	50—90 (108)
Desert plants	Karakoum	50—55

Numbers in parentheses are extreme values.

5.4 Water Economy in Plant Communities

5.4.1 The Water Balance of Stands of Plants

5.4.1.1 The Water Balance Equation

The state of water balance in a stand of plants, and in the soil penetrated by its roots, can be expressed by the *water balance equation*, a formula similar in structure to those for the carbon balance (3.16) and mineral balance (4.2) of plant communities. All the quantities involved are given as precipitation equivalents in mm H_2O (i.e., liters per m² of ground):

$$Pr = \Delta W + L_E + L_O. \tag{5.22}$$

Under the simplifying assumption that the only input to the plant cover is precipitation, the situation is such that when averaged over years and decades, the water intake (the mean *total precipitation Pr*) is accounted for by evaporation from plants and soil (loss by *evapotranspiration*, L_E) and by *runoff and percolation* through the soil (L_O). Over shorter periods, however, the *water stores in the biomass* increase (positive ΔW) or decrease (negative ΔW), since occasionally more rain water falls than evaporates and

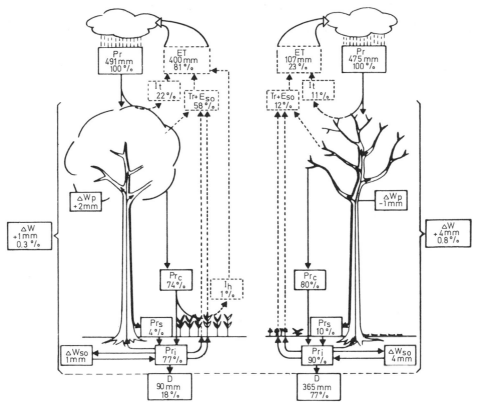

Fig. 5.37. Water balance of an oak forest in leaf and when bare in winter. *Pr*, Total incident precipitation; *Pr$_c$*, canopy throughfall; *Pr$_s$*, stemflow; *Pr$_i$*, infiltration (water soaking into the soil); *D*, drainage water; *ET*, evapotranspiration; *Tr*, transpiration of the stand; *E$_{so}$*, evaporation from the soil; *I$_t$*, interception by the tree stratum; *I$_h$*, interception by the herbaceous stratum; *ΔW*, total water content of the stand; *ΔW$_p$*, water content of the phytomass; *ΔW$_{so}$*, water content of the soil. On the average over the year 966 mm precipitation falls on this forest; 52,5% of this returns to the atmosphere by evaporation of intercepted water, transpiration and evaporation from the soil, 47% drains away, and 0,5% is retained in the increased biomass. After Schnock (1971), simplified. For the water balance of an evergreen conifer forest see Benecke as cited by Ellenberg (1978); of a tropical rain forest, Odum and Pigeon (1970)

drains off, or at times not enough falls to meet the requirements of the plants. In the hydrology literature, ΔW is considered to include only the water reserves in the soil — i.e., the amount of capillary water and available gravitational water. The amount of *water contained in the soil is greatest* in the temperate zone after the snow melts in the spring; during the summer the water content decreases steadily, despite occasional replenishment by precipitation, until a minimum is reached in late summer. In dry regions, the water reserves are filled up during the rainy season, a process requiring weeks, until even the deeper layers of the soil are thoroughly wet. From an ecological point of view, ΔW must also include the water stored in the phytomass and in the layer of litter. More than $^3/_4$ of the green plant mass, and half the mass of wood, is water; the water content of the layer of vegetation fluctuates in the course of the day and the year (being maximal when the plants bear leaves), and the total increases as the biomass grows.

5.4.1.2 Available Precipitation

The amount of precipitation available to the plants for maintenance of their water balance is that which reaches and penetrates the ground. In dense stands of plants not all of the *total precipitation* (*Pr*) actually reaches the ground; rather, the amount entering the ground is only that fraction that falls through holes in the plant canopy, or drops off the leaves or runs down the stems (Fig. 5.37). The resulting local unevenness in the distribution of precipitation is particularly pronounced in woodland, where the ground becomes wetter under holes in the crowns of the trees (throughfall), under the outer regions of the crown (as a result of dripping), and above all in the vicinity of the trunk. The amount of stem flow is greater, the steeper the angle of the branches and the smoother the bark; at the bases of beech trees more than 1.5 times as much water infiltrates the soil as in the open.

Some of the precipitated water adheres to the plant surfaces and evaporates. Trees in particular intercept large quantities of water in this way. In a forest, therefore, one can count as "intake" only the *net precipitation* (Pr_n) that falls through, drips from, or runs off the trees. Only a minuscule fraction of the water wetting the trees is taken in directly through the leaves and bark. By far the greater part of the intercepted water evaporates, so that for practical purposes all the water retained by vegetation is treated as a loss (the *loss by interception L_I*). Thus,

$$Pr_n = Pr - L_I. \tag{5.23}$$

Net precipitation = total precipitation − interception

Strictly speaking, then, the water balance equation for a stand of plants should read

$$Pr - L_I = \Delta W + L_E + L_O. \tag{5.24}$$

The *loss by interception* depends upon the composition and density of the plant cover and upon the meteorological conditions prevailing during the precipitation. Dense crowns of trees with small, easily wettable leaves or needles retain more precipitation than open crowns with large, smooth leaves; the degree to which leafing-out has progressed, of course, is also important. On the average, the loss by interception in coniferous forests amounts to 20—35% or as much as 50% in very dense stands, in mixed forests of the temperate zone to 15—30%, and in dense tropical forests to 35—70% of the total precipitation; the undergrowth intercepts an average of 10% (5—20%). Dwarf-shrub heaths intercept as much as 50%, whereas on cultivated fields and wasteland interception amounts to less than 10%. As weather conditions change, the fraction intercepted varies over a wide range according to amount and type of precipitation (rain, dew or snow), according to the prevailing temperature and according to the wind. In general more precipitation is intercepted, the finer the drops and the smaller the total amount of water that falls (Fig. 5.38). Clearly, a certain amount of

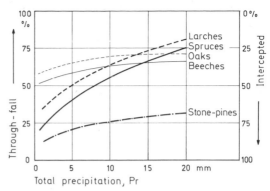

Fig. 5.38. Relative amount of water falling through (*left ordinate*) and intercepted (*right ordinate*) by the canopies of stands of trees under different amounts of precipitation; foliage trees are represented by *thin lines*, and conifers, by *thick lines*. The units on the ordinates are % of total precipitation. The crown of the stone-pine is very dense and even under heavy rainfall intercepts a great deal of the precipitated water. After Hoppe as cited by Walter (1960), Ovington (1954) and Turner as cited by Aulitzky (1968). For further data see Odum and Pigeon (1970), Rutter (1975), Likens et al. (1977), Ford and Deans (1978)

water is required to wet the plants thoroughly, and it is only thereafter that the water can begin to drip off the leaves and twigs. The water needed for wetting the plants sets a *threshold* for net precipitation; in evergreen coniferous forests it is about twice as high as in deciduous forests (with full foliage ca. 1 mm, when trees are bare ca. 0.5 mm); heath and pasture land retain 1—2 mm, and a cover of peat moss about 15 mm, before the water begins to reach the ground.

5.4.1.3 Evapotranspiration from a Stand

Water consumption by stands of plants is approximately *proportional to the green mass*, although the rate of transpiration of individual leaves decreases with increasing stand density because the microclimate within the stand tends to restrict evaporation (chiefly due to shielding from radiation and wind, and to the high humidity in the stand; cf. Fig. 6.1). Therefore in very dense stands the stand transpiration curve departs from proportionality (Fig. 5.39).

The most important procedures used to determine the evaporation from a stand of plants are the recording of water consumption in bricks of vegetation-bearing soil (by an automatic-weighing lysimeter) and the calculation of water turnover in homogeneous stands on the basis of their energy balance and water-vapor exchange (geophysical methods).

In computing average values for the growing season or the year, one must take into account the temporal variability in evaporation stress and water availability. Figure 5.40 gives an example of the *changes in stand transpiration over a year*, for a stand of reeds. This plant community is always adequately supplied with water, but the amounts transpired vary considerably from day to day, depending on the conditions for evaporation. Such fluctuations become much larger when stomatal regulation is

261

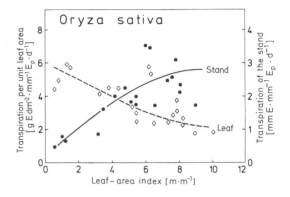

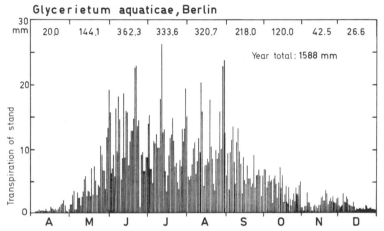

Fig. 5.39. Relative transpiration (E/E_p) of rice plantations in relation to the foliage density (LAI), and the effect of foliage density on the relative transpiration of individual leaves. After Sugimoto (1973)

Fig. 5.40. Daily and monthly totals for transpiration of a stand of reeds during one growing season. The daily water consumption fluctuates according to the weather conditions. After Kiendl (1953)

involved due to *water deficiency* — a complication to be expected in land plants. In this case it is necessary to determine not only the average daily transpiration but also minimal water turnover when the water supply is restricted. These limiting values are particularly important data when the annual water consumption of the vegetation in regions with dry seasons is to be determined. In stands bearing *foliage periodically* (for example, deciduous forests, grassland, agricultural plantations) the fluctuations in transpiration over the year are also affected by the development of the foliage mass, the degree of differentiation and age of the leaves, and growth processes in shoot and root systems. Water consumption by broad-leaved trees increases rapidly in spring when the leaves unfold, and reaches a maximum toward the end of the main growth period. The situation is similar in stands of evergreen plants in the temperate zone, for the new shoots transpire more strongly than the leaves and needles from the previous year. In autumn and winter transpiration falls to a minimum. This lower rate is due primarily to the lower potential evaporation of the atmosphere, but an additional factor is that evergreen woody plants do not open their stomata during winter dormancy and especially in freezing weather. Table 5.11 summarizes the *water consumption of stands* in various climatic regions. Under similar climatic conditions, forests (probably because of their greater mass) transpire appreciably more than grasslands, and these in turn transpire more than heath. The greatest water turnover is always found in stands of plants growing in marshy habitats (or in any sort of wet land) or in places where the plants have ready access to groundwater. These plants transpire much more water than is brought in by precipitation, and sometimes even more than would evaporate from an open water surface. The most limited water turnover is exhibited by stands in dry regions. As water becomes less plentiful, the stands are more open; but the separation of individual plants is increased only above the ground — in the soil, the root systems extend over wide areas. Thus the distance between the trees is determined not by meeting of the crowns but rather by the radii of the root systems of the individual trees. For regions with dry seasons, there is little point in giving annual data; more useful is an average value for the rainy period and the minimal value for the dry period.

Quantitative data for the average and minimal water consumption of stands of plants are valuable as a basis for decision-making in forestry and landscape management, as well as for irrigation projects, if they are interpreted in connection with the expected precipitation. For example, it has been possible to calculate that open stands of trees can exist only where they receive at least 110 mm precipitation in a year (10—12 mm per month). The water-balance equation (5.22) indicates just where forestation becomes uneconomical. On the other hand, the enormous transpiration of trees that tap the groundwater (e.g., poplars and eucalyptus) can be employed to reduce the water table and to increase air humidity.

5.4.1.4 Runoff and Percolation

Not all the water reaching the ground is available for evapotranspiration. Some of it runs off the *surface* of the ground, and another fraction *percolates* into the deeper layers of the ground and joins the groundwater, which in humid regions is a subterranean form of drainage, appearing on the surface here and there as springs, maintaining connections with the rivers, and eventually falling to sea level. The surface runoff is relatively easy to measure, particularly when circumscribed catchment areas of a river are studied. Subterranean drainage, however, must be estimated indirectly.

The amount of *water drained away* depends primarily on the slope of the terrain and the type and density of vegetation. In Table 5.12 precipitation, evapotranspiration and drainage are compared for sections of terrain each covered by uniform vegetation. In dry regions the water is soaked up by the soil and little drains off. In regions of heavy precipitation, it is important for water retention that the falling water penetrate rapidly the layers covering the ground and the soil itself. In loose forest soils with a thick layer of litter, the water infiltrates the soil most rapidly. There is no superficial runoff from woods on flat land; all the water percolates and flows away as groundwater. Precipitated water penetrates much more slowly into the soils of meadows with a dense mat of roots and into pasture soils, compacted by the weight of grazing animals. Frozen soil, too, slows the percolation of melted snow, which accumulates in depressions or, if the land is sloped, runs off.

Table 5.11. Transpiration by stands of plants. From the data of numerous authors, taken from summaries in Pisek and Cartellieri (1941), Stocker (1956, 1972), Rutter (1968), Zelniker (1968), Mitscherlich (1971), Rychnovská et al. (1972); as well as measurements by Ringoet (1952), Ringoet and Mittenaere (1961), Karschon and Heth (1967), Sugimoto (1971), Florineth (1975), Nakhuzrishvili (1971), Körner (1978), Berger et al. (1978), Rychnovská (1978)

Stand	Region	Transpiration mm per year	mm per day	Precipitation mm per year	Transpiration in % of total precipitation
Forests and stands of trees					
Eucalyptus plantation	Israel	466		640	73
Eucalyptus plantation	S. Africa	1200		760	160
Tree plantations	Java	2300—3000		4200	55—72
Tree plantations	Brazil	600		1400	43
Oil palm plantation	Zaire	350		1870	19
Evergreen rainforest	Kenya	1570		1950	80
Bamboo forest	Kenya	1150		2160	53
Mixed forest	Europe, Japan, USA	500—860		1000—1600	50—54
Coniferous forest	Central Europe	580		1250	46
Coniferous forest	Northern taiga	290		525	55
Forested steppe	Russia	(110) 200—400		400—500	(25) 50—80
Maquis	Israel	500		650	77
Chaparral	California	400—500		500—600	80—83
Heath and tundra					
Ericaceous heath under pine wood	Russia	115—130		500	24—26
Alpine dwarf-shrub heath	Central Alps	100—200	2—5	870	11—23
Lichen tundra with moss cover	Siberia	80—100		500	16—20

264

Grassland				
Sedge and reed	Central Europe	1300–1600	6–12 (20)	800	160–190
Wet meadow	Austria	1160	15.5	860	135
Wet meadow	S. Moravia	–	8.3	560	–
Grain fields	Germany	ca. 400		800	50
Upland rice, irrigated	Zaire	530		1870	28
	Malaysia		7.4–9.6		
Upland rice, not irrigated	Zaire	430		1870	23
Pasture (clover and turf)	Central Europe	ca. 400	3–5	800	50
Meadow	Austria	320	4.3	860	37
Dry grassland	Austria	200	2.6	860	30
Dry grassland	S. Moravia	–	2.0	560	–
Steppe	Bechuanaland	200		430	46
Steppe	S. Moravia	–	0.5–1	560	
Alpine grassland	Austria	50	3–5	1100	5
Alpine grassland	Caucasus	–	6	790	–
Open vegetation					
Desert	Mauretania	–	0.03–0.2		–
Desert	Israel	–	0.01–0.4		–
Alpine cushion and rosette plants on calcareous ground	Northern Alps	18	0.4	1100	1.7
Upper alpine cushion and rosette plants on silicaceous ground	Central Alps	11	0.3>	1100	<1.0
Halophyte community	Southern France	–	2–5	–	–

Table 5.12. Water balance in extensive stands of plants. From summaries by Duvigneaud (1967), Stanhill (1970), Mitscherlich (1971), Grin (1972), Likens et al. (1977), Pleiss (1977), Körner et al. (1978); supplemented by data of Rydén (1977) and Geyger (1978)

Stand	Region	Precipitation, mm $\cdot$ yr^{-1}	Evapotranspiration L_E in % of precipitation	Drainage L_O (surface and ground water), in % of precipitation
Forested regions				
Tropical rain forest	Congo basin	1900	73	27
Savanna	Congo basin	1250	82	18
Deciduous forests	Central Europe	600	67	33
(flatland)	NE Asia	700	72	28
Coniferous forests	Central Europe	730	60	40
(flatland)	NE Europe	800	65	35
Mountain forests	Southern Andes	2000	25	75
	Alps	1640	52	48
	Central Europe	1000	43	57
	N. America	1300	38	62
Grassland				
Savannas	Semi-arid tropics	700	85	15
Reeds	Central Europe	800	>150	
Pastureland	Central Europe	700	62	38
Mountain meadows	Alps			
	year:	1000—1700	10—20	80—90
	growing season:	500— 600	25—40	60—75
Steppes	Eastern Europe	500	95	5
Wasteland				
Semideserts	Subtropics	200	95	5
Arid deserts	Subtropics	50	>100	0
Tundra	N. America	180	55	45
Puna	S. America	370	70—80	20—30

Where there is a *steep slope*, more than half the precipitation flows off over the surface, and where the precipitation is heavy and woodland is limited (e.g., mountain meadows) as much as $^2/_3$–$^4/_5$ is lost in this way. Because of the marked effects of gravity in mountainous regions, water balance here is of considerable importance to man; white-water conditions, rockslides, erosion, and avalanches are intimately related to the drainage and water-storage capacity of the soil and its plant cover.

5.4.1.5 Additional Water Supplies to the Plant Cover

Because precipitation is not the exclusive source of water for plants, the simplified water-balance equation must be extended by terms that take into account the additional sources of water in the surroundings of a stand. Chief among these are the *stream of groundwater*, which can be utilized by deep roots, the provision of *surface water* from watercourses, and artificial *irrigation*. In dry regions a permanent cover of vegetation can become established only by drawing upon the groundwater, and even in humid climates trees soak up a great deal of groundwater. By exploiting deep-lying stores of water, plants accelerate the circulation of water in the biosphere, for they pump directly back into the atmosphere water which otherwise would have to flow to the sea and return via a much longer pathway.

6 Synopsis

The goal of ecology is a comprehensive view of the diverse and numerous interactions of organisms and their environments, in their full complexity and ranges of variability (Fig. 6.1). The ecologist must try to understand the dynamic interplay of environmental factors, the reactions and adaptations of the organisms, and the regulatory mechanisms underlying them. All environmental influences, of course, act continuously and conjointly upon the plant, which must adjust to them if it is to maintain itself in the face of competition and survival. The study of these responses is the task of ecophysiology. An illustrative example of the complex interrelationships of environmental influences and physiological processes is given in Fig. 6.2.

6.1 Analysis of Ecological Factors

In any analysis of plant ecology, it is necessary to bear in mind a number of fundamental considerations, which may be summarized according to Evans (1972) as follows:

1. *The response of a plant to a changing environmental factor is usually not linear.* Example: Shifts in the optimum of temperature-dependence curves and in the saturation level of various physiological processes.

2. *When two independent environmental factors change simultaneously, the overall effect may not be a simple addition of the individual responses.* Example: Either SO_2 or NaCl presented alone in low concentration under certain conditions increases transpiration in some plants, whereas when the two are presented together transpiration is reduced.

3. *The organs of a plant may respond differently to a given environmental stimulus, so that the appearance of an effect on the plant as a whole may be delayed.* Example: Water deficiency in the soil at first stimulates root growth while simultaneously inhibiting growth of the shoot.

4. *The action of a particular environmental factor causes different forms of response at different phases of the life cycle.* Example: The differing temperature requirements and sensitivity of seedlings, young plants in the extension phase, and reproductive processes.

5. *Plants that have grown under certain environmental conditions respond in a modified manner to the later action of environmental factors.* Example: Metabolic differences between high-radiation- and shade-adapted plants.

Moreover, it must be taken into account that

6. *in every plant population there are individual genetic differences*, and often — particularly along climatic and edaphic gradients — *ecotype differentiation* occurs.

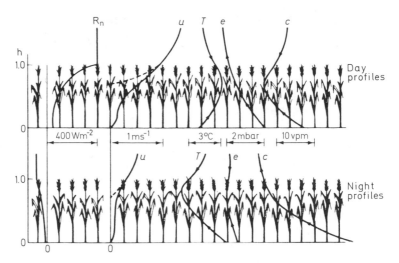

Fig. 6.1. Example of a comprehensive study of microclimatic habitat factors in a field of grain. R_n, net radiation; u, wind speed; T, air temperature; e, vapor pressure; c, CO_2 concentration; h, relative height (Monteith, 1973). For variations in space and time in a coniferous forest see Jarvis et al. (1976)

Fig. 6.2. Daily course of CO_2 exchange, transpiration and diffusion resistance in leaves of the grapevine as compared with the environmental conditions in an extremely dry habitat — an example of an ecological experiment including simultaneous measurement of several critical variables. I, illuminance; F_n, net photosynthesis; R, respiration; E, transpiration; Δ_e, difference in water-vapor concentration between leaf and surrounding air (mg $H_2O \cdot l^{-1}$); r, diffusion resistance for water vapor (s $\cdot$ cm^{-1}); T_L, leaf temperature; T_A, air temperature; RH, relative humidity in % (E. D. Schulze et al., 1972). For a classical example of simultaneous measurements see Stocker et al. (1938)

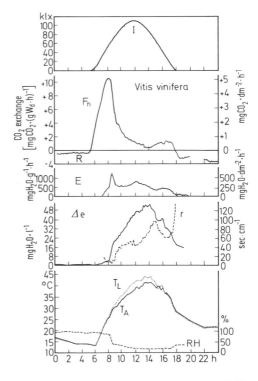

6.2 Special Features of Ecological Methodology

Because of the complexity of environmental influences, and the large number of external and internal variables on which a plant's responses depend, experimental ecophysiological data must be as comprehensive and as thoroughly analyzed as possible. The sequence of steps in an ecological research program diagrammed in Fig. 6.3 is designed to make clear the scope of the subject and the variety of methods employed in such a study.

As a first step, one must select the *methods of measurement* most suited to the object of the investigation and to the peculiarities of the site of the study. Not uncommonly it turns out that some measurement devices must be newly developed for the purpose. The *schedule of measurements* should include the spatial distribution and time course of habitat factors and the physiological behavior of the plants; data should be taken over a long period, at least for an entire growing season, so that automatic recording systems are essential. Ecophysiological measurements in the field can be supplemented by related laboratory experiments. *Laboratory experiments* yield reproducible data concerning the physiological behavior of the plants under a variety of environmental conditions, with the added advantage that the conditions of interest can be varied at will, singly, or in groups. The information so obtained facilitates the interpretation of field studies, and is particularly useful in determining limiting factors.

Ecological research can be done at various *levels of integration*, depending on the nature of the problem under study and the structure and extent of the system investigated.

First Level of Integration: Environmental Influences on Single Plants. In open plant communities, which as a rule are found in extreme habitats, the individual plants are directly exposed to abiotic environmental conditions. Bioclimatological, edaphic and ecophysiological measurements provide the data required for a causal analysis of the way of life and the possibilities for survival of these plants in their habitat. The ecological — that is, integrative — approach consists in relating the behavior of the plant to the combinations of environmental parameters most commonly encountered in the habitat, and to the probability of occurrence of potentially lethal situations.

Second Level of Integration: the Stand and the Environment. Closed stands of plants exhibit a special bioclimate and affect soil formation. The action of abiotic environmental factors on the individual plants in the stand is diminished and modified. Effects specific to the stand appear — for example, competition for space and nutrients, and morphogenetic influences — which alter the behavior of the individual. The ecological study of stands of plants, then, must take into account that the situation of a plant growing singly is different from that of a plant in a stand.

Third Level of Integration: Biocenosis and Environment. A complete program of ecological research includes all the organisms comprising the biocenosis and all the interactions, in their full complexity, among the individual groups of organisms and the abiotic environment. In the foreground of such an investigation are those functions and relationships that determine the nature of the whole: energy flows and nutrient transfer,

270

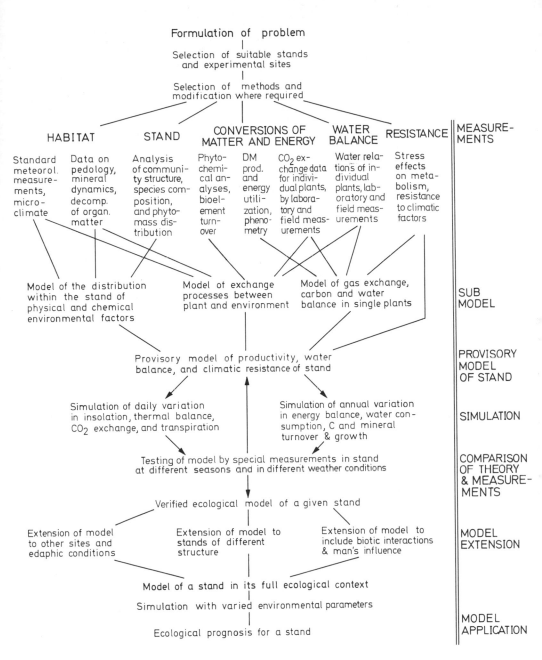

Fig. 6.3. Flow diagram of a program of research in experimental ecology, from the formulation of the problem to the ecological prognosis. Based in part on Larcher et al. (1973), Cernusca (1973)

the cycling of materials, information and interference processes, and regulatory mechanisms. As a member of an interdisciplinary group, the phyto-ecophysiologist has the task of quantitatively analyzing the functional relationships between the plants and their environment and of establishing the principles underlying the behavior of the plants.

6.3 Data Synthesis, Ecological Models, and Computer Simulation

When experimental findings are *evaluated*, the individual items of information must be put together to form a whole. With the large amounts of data usually obtained in field work, this is facilitated by the use of computers. The final results of such processing are sufficiently complicated that they cannot be adequately expressed in the traditional verbal form, but are better presented as graphic models such as flow diagrams (e.g., Fig. 2.9) or correlation diagrams (e.g., Fig. 2.51).

An *ecological model* is the quantitative description of the relationships between biological processes and the environmental factors that affect them, or among different biological processes. The purpose of modeling is to assist understanding of the causes of these processes and to help the experimenter to predict the course of events in particular situations. The first approach (*data synthesis*) generates models describing the behavior of single plants and certain aspects of the observed situation in the habitat (for example, the radiation budget or exchange processes). This subsystem model is extended by computer experiments (*simulations*; cf. Figs. 3.27 and 3.28) and tested by appropriate new measurements. The verified model can be *expanded* spatially by carefully planned short-term measurements under typical weather conditions, so that the ecological amplitude of the entire population or of a complex plant community can be determined with minimal effort. The overall ecological model that eventually results facilitates the computation of causal and cybernetic relationships in the system vegetation/environment, and can be used in making *ecological predictions*, which offer an objective basis for critical planning decisions.

As is the case with all *generalizations and extrapolations*, the value of the inferences drawn from simulation models depends chiefly on conscientious selection of the primary data. Not only must the measurements be sufficiently precise, they must also be appropriate to the problem in hand. Concrete data always apply to a certain plant in a certain state of development and to a state of adaptation that depends on prior conditions. Often, though not always, generalization will be possible — within limits. Only if these limits are observed, and if intermediate results of the model are tested repeatedly by well-designed experiments, can mathematical deductions and logical implications give reliable information. To offer assistance in this regard will be an important task of plant ecophysiology.

Literature

Methods of Ecophysiological Research

Allen, S. E., Grimshaw, H. M., Parkinson, J. A., Quarmby, Ch.: Chemical analysis of ecological materials. Oxford: Blackwell 1974 (E)

Beideman, I. N.: Metodika izučenija fenologii rastenii i rastitelnych soobščestv. Novosibirsk: Nauka 1974

Böhm, W.: Methods of studying root systems. Ecological Studies, Vol 33. Berlin, Heidelberg, New York: Springer 1979

Chapman, S. B. (ed.): Methods in plant ecology. Oxford: Blackwell 1976

Downs, R. J., Hellmers, H.: Environment and the experimental control of plant growth. London, New York: Academic Press 1975

Evans, G. C.: The quantitative analysis of plant growth. Oxford: Blackwell 1972

Fritschen, L. J., Gay, L. W.: Environmental instrumentation. Berlin, Heidelberg, New York: Springer 1979

Grodzinski, W., Klekowski, R. Z., Duncan, A.: Methods for ecological bioenergetics. IBP Handbook No 24, Oxford: Blackwell 1975

Hunt, R.: Plant growth analysis. Studies in Biology, Vol. 96. London: Arnold 1978

Kreeb, K.: Methoden der Pflanzenökologie. Stuttgart, New York: Fischer 1977

Lieth: Phenology and seasonality modeling. Ecological Studies, Vol. 8. Berlin, Heidelberg, New York: Springer 1974

Milner, C., Hughes, R. E.: Methods for the measurement of the primary production of grassland. IBP Handbook No 6. Oxford: Blackwell 1968

Montheith, J. L.: Survey of instruments for micrometeorology. IBP Handbook No 22. Oxford: Blackwell 1972

Newbould, P. J.: Methods for estimating the primary production of forests. IBP Handbook No 2. Oxford: Blackwell 1967

Philipson, J.: Methods of study in quantitative soil ecology. Population, production and energy flow. IBP Handbook No 18. Oxford: Blackwell 1971

Schnelle, F.: Pflanzenphänologie. Leipzig: Akad. Verl. Ges. 1955

Schuurman, J. J., Goedewaagen, M. A. J.: Methods for the examination of root systems and roots. Wageningen: Centre Agric. Publ. Doc. 1965

Šesták, Z., Čatský, J., Jarvis, P. G.: Plant photosynthetic production. Manual and methods. Den Haag: Junk 1971

Slavík, B.: Methods in studying plant water relations. Ecological Studies, Vol. 9. Berlin, Heidelberg, New York: Springer 1974

Steubing, L.: Pflanzenökologisches Praktikum. Berlin, Hamburg: Parey 1965

Vollenweider, R. A.: A Manual on methods for measuring primary production in aquatic environments. Oxford: Blackwell 1969

Vosnesensky, V. L., Zalensky, O. V., Semichatova, O. A.: Methody issledovnija fotosinteza i dychanija rastenii. Moscow-Leningrad: Nauka 1965

References

Space limitations have prohibited citation of the titles of these papers, which were referred to in tables and figures. The language of each reference is denoted by the letter after the year, as follows: CS = Czechoslovakian; D = Dutch; E = English; F = French; G = German; I = Italian; J = Japanese; R = Russian; SP = Spanish; SV = Swedish. Despite the difficulties, it is worth taking the trouble to read original works in that language in which the author can express himself most precisely. Most of the publications cited at least contain an English summary.

Abd El-Rahman, A. A., El-Hadidi, M. N.: Egypt. J. Bot. **1,** 19–38 (1958) (E)
Abd El-Rahman, A. A., El-Monayeri, M.: Flora B **157,** 229–238 (1967) (E)
Abd El-Rahman, A. A., Ezzat, N. H., Hussein, A. H.: Flora **164,** 73–84 (1975) (E)
Abel, W. O.: Österr. Akad. Wiss. Math.-Naturwiss. Kl. Sitzungsber. Abt. I. **165,** 619–707 (1956) (G)
Addicott, F. T.: Plant Physiol. **43,** 1471–1479 (1968) (E)
Albert, R., Popp, M.: Oecologia **27,** 157–170 (1977) (E)
Albert, R., Popp, M.: Oecol. Plant. **13,** 27–42 (1978) (G)
Alekseev, V. A.: Svetovoi rezhim lesa. Leningrad: Nauka 1975 (R)
Alexandrov, V. Ya.: Q. Rev. Biol. **39,** 35–77 (1964) (E)
Alexandrov, V. Ya.: Cells, molecules and temperature. Ecological Studies Vol. 21. Berlin, Heidelberg, New York: Springer 1977 (E)
Allen, L. H., Lemon, E. R.: In: Vegetation and the atmosphere. Monteith, J. L. (ed.), Vol. II, pp. 265–308. London, New York: Academic Press 1976 (E)
Allen, L. H., Yocum, C. S., Lemon, E. R.: Agronomy **56,** 253–259 (1964) (E)
Allen, L. H., Lemon, E. Jr., Müller, L.: Ecology **53,** 102–111 (1972) (E)
Altman, L., Dittmer, S.: Biology Data Book, Vol. I and II. Fed. Am. Soc. Exp. Biol. Bethesda 1972, 1973 (E)
Alvim, P. de T.: In: The formation of wood in forest trees. Zimmermann, M. H. (ed.), pp. 479–495. London, New York: Academic Press 1964 (E)
Alvim, P. de T.: In: Ecophysiology of tropical crops de T. Alvim, P., Kozlowski, T. T. (eds.), 279–313. London, New York: Academic Press 1977 (E)
Alvim, P. de T., Kozlowski, T. T.: Ecophysiology of tropical crops. London, New York: Academic Press 1977 (E)
Alvim, P. de T., Machado, A. D., Vello, F.: Publ. Centro Pesqu. Cacau Itabuna 1973 (E)
André, M., Massimo, D., Daguenet, A.: Physiol. Plant **43,** 397–403 (1978) (E)
Antonovics, J., Bradshaw, A. D., Turner, R. G.: Adv. Ecol. Res. **7,** 1–85 (1971) (E)
Arvidsson, I.: Oikos Kopenhagen Suppl. **1,** 5–181 (1951) (G)
Ashton, D. H.: Aust. J. Bot. **6,** 154–176 (1958) (E)
Aulitzky, H.: Arch. Meteorol. Geophys. Bioklimatol. Ser. B **10,** 445–532 (1961) (G)
Aulitzky, H.: Arch. Meteorol. Geophys. Bioklimatol. Ser. B. **11,** 301–362 (1962) (G)
Aulitzky, H.: Zentralbl. Gesamte Forstwes. **85,** 2–32 (1968) (G)
Babb, T. A., Whitfield, D. W. A.: In: Truelove Lowland, Devon Island, Canada: A high arctic ecosystem Bliss, L. C. (ed.), pp. 587–620. Edmonton: Univ. Alberta Press 1977 (E)
Baker, N. R., Hardwick, K.: Photosynthetica **10,** 361–366 (1976) (E)
Baldy, Ch.: Ann. Agron. **24,** 507–532 (1973) (F)
Bannister, P.: Introduction to physiological plant ecology. Oxford: Blackwell 1976 (E)

Baross, J. A., Morita, R. Y.: In: Microbial life in extreme Environments. Kushner, D. J. (ed), pp. 9—71. London: Academic Press 1978 (E)

Barrett, E. C., Curtis, L. F.: Introduction to environmental remote sensing. London: Chapman & Hall 1976 (E)

Barrs, H. D.: In: Water deficits and plant growth. Kozlowski, T. T. (ed.), Vol. I, pp. 235—368. London, New York: Academic Press 1968 (E)

Barrs, H. D.: C. S. I. R. O. Div. Irrig. Res. Annu. Rep. 1968—1969, 7. Canberra (1969) (E)

Bartholomew, D. P., Kadzimin, S. B.: In: Ecophysiology of tropical crops. de Taede Alvim, P., Kozlowski, T. T. (eds.), pp. 113—157. London, New York: Academic Press 1977 (E)

Bartkov, B. I., Zvereva, G.: Fiziol. Biochim. Kult. Rast. **6**, 502—505 (1974) (R)

Bassham, J. A.: Science **197**, 630—638 (1977) (E)

Bauer, H.: Diss. Innsbruck 1970 (G)

Bauer, H., Senser, M.: Z. Pflanzenphysiol. **91**, 359—369 (1979) (E)

Bauer, H., Larcher, W., Walker, R. B.: In: Photosynthesis and productivity in different environments. Cooper, J. P. (ed.), pp. 557—586. Cambridge: Cambridge Univ. Press 1975 (E)

Baumeister, W., Ernst, W.: Mineralstoffe und Pflanzenwachstum, 2. Aufl. Stuttgart: Fischer 1978 (G)

Baxter, P., West, D.: Ann. Appl. Biol. **87**, 95—101 (1977) (E)

Bazilevich, N. I., Rodin, L. Y.: Geographical regularities in productivity and the circulation of chemical elements in the earth's main vegetation types. In: Soviet geography (Rev. & Translation). New York: Am. Geogr. Soc. 1971 (E)

Bazzaz, F. A., Carlson, R. W., Rolfe, G. L.: Environ. Pollut. **7**, 241—246 (1974) (E)

Beadle, C. L., Jarvis, P. G.: Physiol. Plant. **41**, 7—13 (1977) (E)

Beevers, L.: Nitrogen metabolism in plants. London: Arnold 1976 (E)

Begg, J. E., Jarvis, P. G.: Agric. Meteorol. **5**, 91—109 (1968) (E)

Begg, J. E., Turner, N. C.: Plant Physiol. **46**, 343—346 (1970) (E)

Beideman, I. N.: Metodika izucheniya fenologii rastenii i rastitelnykh soobshchestv. Novosibirsk: Nauka 1974 (R)

Bell, C. J., Rose, D. A.: Plant, Cell and Environm. **4**, 89—96 (1981) (E)

Berge, H.: Immissionsschäden. Handbuch der Pflanzenkrankheiten, Bd. 11, Liefg. 4. Berlin, Hamburg: Parey 1970 (G)

Berger, W.: Beih. Bot. Zentralbl. **48**, I, 364—390 (1931) (G)

Berger, A., Corré, J. J., Heim, G.: Terre Vie **32**, 241—278 (1978) (F)

Berger-Landefeldt, U.: Der Wasserhaushalt der Alpenpflanzen. Bibl. Bot. 115. Stuttgart: Schweizerbart 1936 (G)

Berry, J., Björkman, O.: Ann. Rev. Plant Physiol. **31**, 491—543 (1980) (E)

Berry, J. A., Raison, J. K.: In: Encyclopedia of plant physiology, Vol. 12 A, pp. 277—338. Berlin, Heidelberg, New York: Springer 1981 (E)

Biebl, R.: Jahrb. Wiss. Bot. **86**, 350—386 (1938) (G)

Biebl, R.: Sitzungsber. Österr. Akad. Wiss. **155**, 145—157 (1947) (G)

Biebl, R.: Protoplasmatische Ökologie der Pflanzen. Wasser und Temperatur. Protoplasmatologia XII/1. Wien: Springer 1962 (G)

Biebl, R., Maier, R.: Österr. Bot. Z. **117**, 176—194 (1969) (G)

Bierhuizen, J. F.: In: Plant response to climatic factors. Slatyer, R. O. (ed.), pp. 89—98 Paris: UNESCO 1973 (E)

Bierhuizen, J. F., Wagenvoort, W. A.: Sci. Hortic. **2**, 213—219 (1974) (E)

Billings, W. D., Godfrey, P. J., Chabot, B. F., Bourquet, D. P.: Arct. Alp. Res. **3**, 277—290 (1971) (E)

Biscoe, P. V., Unsworth, M. H., Pinckney, H. R.: New Phytol. **72,** 1299—1306 (1973) (E)

Björkman, O.: In: Photosynthesis and photorespiration. Hatch, M. D., Osmond, C. B., Slatyer, R. O. (eds.), pp. 18—32. New York: Wiley 1971 (E)

Björkman, O.: In: Encyclopedia of plant physiology, Vol. 12 A, pp. 57–107. Berlin, Heidelberg, New York: Springer 1981 (E)

Björkman, O., Holmgren, P.: Physiol. Plant. **16,** 889—914 (1963) (E)

Björkman, O., Ludlow, M. M., Morrow, P. A.: Carnegie Inst. Yearb. **71,** 94—102 (1972) (E)

Björkman, O., Boynton, J., Berry, J.: Carnegie Inst. Yearb. **75,** 400—407 (1976) (E)

Black, C. C.: Advances in ecological research. Vol. 7, pp. 87—114. London, New York: Academic Press 1971 (E)

Black, C. C.: Annu. Rev. Plant Physiol. **24,** 253—286 (1973) (E)

Blackmann, G. E., Black, J. N.: Ann. Bot. **23,** 131—145 (1959) (E)

Blinks, L. R.: Manual of phycology. Mass: Waltham 1951 (E)

Bliss, L. C.: Annu. Rev. Ecol. Syst. **2,** 405—438 (1971) (E)

Bliss, L. C.: Devon Island, Canada. In: Structure and function of tundra ecosystems. Rosswall, T., Heal, O. W. Ecol. Bull. **20,** 17—60, NFR, Stockholm 1975 (E)

Bliss, L. C.: Truelove Lowland, Devon Island, Canada: A high arctic ecosystem. Edmonton: Univ. Alberta Press 1977 (E)

Blüthgen, J.: Allgemeine Klimageographie. Berlin: W. de Gruyter 1964 (G)

Boardman, N. K.: Annu. Rev. Plant Physiol. **28,** 355—377 (1977) (E)

Bobrovskaya, N. I.: Bot. Zhurn. **56,** 361—368 (1971) (R)

Böhning, R. H., Burnside, C.: Am. J. Bot. **43,** 557—561 (1956) (E)

Bonhomme, R.: C. R. Assoc. Intercaraibe Plantes Alimentaires, 7. Congr. 279—293 (1969) (F)

Bonner, J., Varner, J. E. (eds.): Plant biochemistry, 3. ed. London, New York: Academic Press 1976 (E)

Bowling, D. J. F.: Uptake of ions by plant roots. London: Chapman & Hall 1976 (E)

Boyer, J. S.: Plant Physiol. **46,** 233—235 (1970a) (E)

Boyer, J. S.: Plant Physiol. **46,** 236—237 (1970b) (E)

Boyer, J. S.: Plant Physiol. **48,** 532—536 (1971) (E)

Boyer, J. S.: Planta **117,** 187—207 (1974) (E)

Boyer, J. S.: In: Water deficits and plant growth. Kozlowski, T. T. (ed.), Vol. IV, pp. 154—190. London, New York: Academic Press 1976 (E)

Boyer, J. S., Bowen, B. L.: Plant Physiol. **45,** 612—615 (1970) (E)

Boyer, Y.: Vie Milieu Ser. C **19,** 331—344 (1968) (F)

Boysen-Jensen, P.: Die Stoffproduktion der Pflanzen. Jena: Fischer 1932) (G)

Boysen-Jensen, P., Müller, D.: Jahrb. Wiss. Bot. **70,** 493—502 (1929) (G)

Bradshaw, A. D., McNeilly, T.: Evolution and Pollution: Studies in Biology 130. London: Arnold (1981) (E)

Braun, H. J.: Z. Pflanzenphysiol. **74,** 91—94 (1974) (G)

Braun, H. J.: Z. Pflanzenphysiol. **84,** 459—462 (1977) (G)

Breckle, S. W.: Diss. Bot. Bd. 35. Vaduz: Cramer 1976 (G)

Bremner, I.: Q. Rev. Biophys. **7,** 75—124 (1974) (E)

Briggs, L. J., Shantz, H. L.: Bot. Gaz. **53,** 20—37 (1913) (E)

Brix, H.: Physiol. Plant **15,** 10—20 (1962) (E)

Brock, Th. D.: Science **158,** 1012—1019 (1967) (E)

Brock, Th. D.: Thermophilic microorganisms and life at high temperatures. Berlin, Heidelberg, New York: Springer 1978 (E)

Brouwer, R., Kuiper, P. J. C.: Leerboek der Plantenfysiologie, Deel 3. Oecofysiologische relaties. Oosthoek, Utrecht 1972 (D)

Buesa, R. J.: Aquat. Bot. **3**, 203–216 (1977) (E)

Bunnell, F. L., Maclean, S. F., Brown, Jr., Brown, J.: In: Structure and function of tundra ecosystems. Rosswall, T., Heal, O. W., (eds.), Ecol. Bull. Vol. 20, pp. 73–124. Stockholm: NFR 1975 (E)

Bünning, E.: Entwicklungs- und Bewegungsphysiologie der Pflanze, 3. Aufl. Berlin, Göttingen, Heidelberg: Springer 1953 (G)

Burian, K.: In: Ökosystemforschung. Ellenberg, H. (ed.), pp. 61–78. Berlin, Heidelberg, New York: Springer 1973 (G)

Burke, M. J., Gusta, L. V., Quamme, K. A., Weiser, C. J., Li, P. H.: Annu. Rev. Plant Physiol. **27**, 507–528 (1976) (E)

Burrows, F. J., Milthorpe, F. L.: In: Water deficits and plant growth. Kozlowski, T. T. (ed.), Vol. IV, pp. 103–152. New York: Academic Press 1976 (E)

Burnside, C. A., Böhning, R. H.: Plant Physiol. **32**, 61–63 (1957) (E)

Caldwell, M. M.: In: Research in photobiology. Castellani, A. (ed.), pp. 597–602. New York: Plenum Publ. Corp. 1977 (E)

Caldwell, M. M., White, R. S., Moore, R. T., Camp, L. B.: Oecologia **29**, 275–300 (1977) (E)

Cannon, H. L.: Science **132**, 591–598 (1960) (E)

Cappelletti, C.: Acc. Naz. Lincei Cl. Fis. Mat. Nat. **30**, 331–342 (1961) (I)

Caprio, J. M., Hopp, R. J., Williams, J. S.: In: Phenology and seasonality modeling. Lieth, H. (ed.), pp. 77–82. Berlin, Heidelberg, New York: Springer 1974 (E)

Cernusca, A.: In: Ökosystemforschung. Ellenberg, H. (ed.), pp. 195–201. Berlin, Heidelberg, New York: Springer 1973 (G)

Cernusca, A.: Oecol. Plant. **11**, 71–102 (1976) (G)

Cernusca, A.: Bonn: Veröff. Fonds Umweltstudien 1977 (G)

Ceulemans, R., Impens, I.: Biol. Plant. **21**, 302–306 (1979)

Chabot, B. F., Billings, W. D.: Ecol. Mongr. **42**, 163–199 (1972) (E)

Chapin, F. S., Cleve, K., Tieszen, L. L.: Arct. Alp. Res. **7**, 209–226 (1975) (E)

Chapman, S. B., Hibble, J., Rafarel, C. R.: J. Ecol. **63**, 233–258 (1975) (E)

Chartier, Ph., Bethenod, O.: In: Les processus de la production végétale primaire. Moyse, A. (ed.) pp. 77–112 Paris: Gauthier-Villars 1977 (F)

Chollet, R., Ogren, R. W. L.: Bot. Rev. **41**, 137–179 (1975) (E)

Christophersen, J.: In: Temperatur und Leben. Precht, H., Christophersen, J., Hensel, H. (eds.), Berlin, Göttingen, Heidelberg: Springer 1955 (G)

Cintron, G.: In: A tropical rain forest Odum, H. T., Pigeon, F. F. (eds.), Vol. 3, H-133. US. Atomic Energy Comm., Oak Ridge 1970 (E)

Clark, J.: State Univ. Coll For. Syracuse N. Y. (1961) (E)

Clebsch, E. E., Billings, W. D.: Arc. Alp. Res. **8**, 255–262 (1976) (E)

Collatz, J., Ferrar, P. J., Slatyer, R. O.: Oecologia **23**, 95–105 (1976) (E)

Cooper, J. P.: Ann. Appl. Biol. **87**, 237–242 (1977) (E)

Coudret, A., Ferron, F.: Ann. Amelior, Plant. **27**, 613–638 (1977) (F)

Cowan, I. R.: J. Appl. Ecol. **2**, 221–239 (1965) (E)

Cowan, I. R.: Planta **106**, 185–219 (1972) (E)

Crawford, R. M. M.: New Phytol. **79**, 511–517 (1977) (E)

Dässler, H. G.: Einfluß von Luftverunreinigungen auf die Vegetation. VEB Jena: Gustav Fischer 1976 (G)

Dässler, H. G., Ranft, H.: Flora B **158**, 454–461 (1969) (G)

Dalton, F. N., Gardner, W. R.: Agron. J. **70**, 404–406 (1978) (E)

Daubenmire, R. F.: Plants and environment. 3rd ed. New York: Wiley 1974 (E)

Davies, W. J., Kozlowski, T. T.: Plant Soil **46**, 435—444 (1977) (E)

Dawes, C. J., Moon, R. E., Davis, M. A.: Estuarine and Coastal Mar. Sci. **6**, 175—185 (1978) (E)

Decker, J. P.: J. Sol. Energy **1**, 30—33 (1957) (E)

Deevey, E. S.: Sci. Am. **223**, 148—158 (1970) (E)

De Wit, C. T.: Versl. Landbouwkd. Onderz. **64**, 65—88 (1958) (E)

De Wit, C. T., Brouwer, R., Penning de Vries, F. W. T.: Proc. IBP/PP Techn. Meeting Treboň, 47—70. Centre Agric. Publ. Doc. Wageningen 1970 (E)

Dinger, B. E., Patten, D. T.: Photosynthetica **6**, 345—353 (1972) (E)

Döring, B.: Bot. Z. **28**, 305—383 (1935) (G)

Dormling, I., Gustafsson, A., Wettstein, D.: Silvae Genet. **17**, 44—64 (1968) (E)

Downs, R. J., Hellmers, H.: Environmental control of plant growth. London, New York: Academic Press 1975 (E)

Downton, W. J. S.: Photosynthetica **9**, 96—105 (1975) (E)

Downton, W. J. S.: Aust. J. Plant Physiol. **4**, 183—192 (1977) (E)

Driessche, R. Van den, Tunstall, B. R., Connor, D. J.: Photosynthetica **5**, 210—217 (1971) (E)

Duvigneaud, P.: Écosystèmes et biosphère. Bruxelles: Minist. Educ. Nat. Cult. 1967 (F)

Duvigneaud, P., Denaeyer-De Smet, S.: Bull. Soc. R. Bot. Belge **96**, 93—224 (1963) (F)

Duvigneaud, P., Denaeyer-De Smet, S.: In: Analysis of temperate forest ecosystems. Reichle, D. E. (ed.). Ecological Studies, Vol. 1, pp. 199—225. Berlin, Heidelberg, New York: Springer 1970 (E)

Duvigneaud, P., Denaeyer-De Smet, S.: Oecol. Plant **5**, 1—32 (1970) (F)

Duvigneaud, P., Denaeyer-De Smet, S.: Oecol. Plant. **8**, 219—246 (1973) (F)

Duvigneaud, P., Denaeyer-De Smet, S., Ambroes, P., Timperman, J., Marbaise, J. L.: Bull Soc. R. Bot. Belge **102**, 317—327, 339—354 (1969) (F)

Dykyjová, D., Hradečká, D.: Folia Geobot. Phytotaxon. **11**, 23—61 (1976) (E)

Dykyjová, D., Květ, J.: Pond littoral ecosystems. Ecological Studies, Vol. 28. Berlin, Heidelberg, New York: Springer 1978 (E)

Eagles, C. F., Wilson, D.: In: Handbook of Agricultural productivity, Vol. I. Rechcigl, M. (ed.), pp. 213—247. Boca Raton: CRC Press 1982 (E)

Eberhardt, E.: Planta **45**, 57—68 (1955) (G)

Eckhardt, F. E., Heim, G., Methy, M., Saugier, B., Sauvezon, R.: Oecol. Plant. **6**, 51—100 (1971) (F)

Eckhardt, F. E., Berger, A., Methy, M., Heim, G., Sauvezon, R.: In: Les processus de la production végétale primaire. Moyse, A. (ed.), pp. 1—75. Paris: Gauthier-Villars 1977 (F)

Egunjobi, J. K.: Oecol. Plant. **9**, 1—10 (1974) (E)

Ehleringer, J. R.: Oecologia **31**, 255—267 (1978) (E)

Elfving, D. C., Kaufmann, M. R., Hall, A. E.: Physiol. Plant. **27**, 161—168 (1972) (E)

Ellenberg, H.: Mitt. Floristisch-soziol. Arbeitsgem. Niedersachsen **5**, 3—135 (1939) (G)

Ellenberg, H.: In: Handbuch der Pflanzenphysiologie, Vol. IV, pp. 638—708. Berlin, Heidelberg, New York: Springer 1958 (G)

Ellenberg, H.: Vegetation Mitteleuropas mit den Alpen. Stuttgart: Ulmer 1963, 1978 (G)

El-Sharkawy, M. A., Hesketh, J. D.: Crop. Sci. **4**, 514—518 (1964) (E)

Epstein, E.: Mineral nutrition of plants. London: Wiley 1971 (E)

Ernst, W.: Kirkia **8/2**, 125—145 (1972a) (E)

Ernst, W.: Schwermetallresistenz und Mineralstoffhaushalt. Opladen: Westdeutscher Verlag 1972b (G)

Ernst, W.: In: Effects of air pollutants on plants. Mansfield, I. A. (ed.), pp. 115–133. Cambridge: Cambridge Univ. Press 1976 (E)

Ernst, W.H.O.: Int. Conf. Heavy Metals, Toronto 1975, 121–136 (1975) (E)

Esterbauer, H., Grill, D., Zotter, M.: Biochem. Physiol. Pflanz. **172**, 155–159 (1978) (G)

Etherington, J. R.: Environment and plant ecology. London: Wiley 1975 (E)

Evans, G. C.: The quantitative analysis of plant growth. Oxford: Blackwell 1972 (E)

Evenari, M., Schulze, E. D., Kappen, L., Buschbom, U., Lange, O. L.: In: Physiological adaptation to the environment. Vernberg, F. J. (ed.), pp 111–130. New York: Intext. Educ. Publ. 1975a (E)

Evenari, M., Bamberg, S., Schulze, E.-D., Kappen, L., Lange, O. L., Buschbom, U.: In: Photosynthesis and productivity in different environments. Cooper, J. P. (ed.), pp. 121–127. Cambridge: Cambridge Univ. Press 1975b (E)

Farrar, J. F., Relton, J., Rutter, A. J.: J. appl. Ecol. **14**, 861–875 (E)

Fereres, E., Acevedo, E., Henderson, D. W., Hsiao, T. C.: Plant Physiol. **44**, 261–267, 1978 (E)

Ferri, M. G., Labouriau, L. G.: Rev. Bras. Biol. **12**, 301–312 (1952) (E)

Ferry, B. W., Baddeley, M. S., Hawksworth, D. L. (eds.): Air pollution and lichens. London: Athlone Press 1973 (E)

Finck, A.: Z. Pflanzenernähr. Bodenkd. **119**, 197–208 (1968) (G)

Finck, A.: Pflanzenernährung in Stichworten. Kiel: F. Hirt 1969 (G)

Fischer, A.: Jahrb. Wiss. Bot. **22**, 73–160 (1891) (G)

Fischer, R. A., Turner, N. C.: Annu. Rev. Plant. Physiol. **29**, 277–317 (1978) (E)

Flint, H. L.: Ecology **53**, 1163–1170 (1972) (E)

Florineth, F.: Oecol. Plant. **9**, 295–315 (1975) (G)

Flowers, T., Troke, P. F., Yeo, A. R.: Annu. Rev. Plant Physiol. **28**, 89–121 (1977) (E)

Fogg, G. E.: Algal cultures and phytoplankton ecology, 2. ed. Madison: Univ. Wisconsin Press 1975 (E)

Fonteno, W. C., McWilliams, E. L.: J. Am. Hortic., Sci. **102**, 52–56 (1978) (E)

Ford, E. D., Deans, J. D.: J. appl. Ecol. **15**, 905–917 (1978) (E)

Fortescue, J. A. C., Marten, G. G.: In: Analysis of temperate forest ecosystems. Reichle, D. E. (ed.). Ecological Studies, Vol. 1, pp. 173–198. Berlin, Heidelberg, New York: Springer 1970 (E)

Foy, C. D., Chaney, R. L., White, M. C.: Annu. Rev. Plant Physiol. **29**, 511–566 (1978) (E)

Franco, C. M.: IBEC Res. Inst. Bull. **16**, 1–21 (1958) (ref. Maestri and Barros, 1977)

Francois, L. E., Clark, R. A.: J. Am. Soc. Hortic. Sci. **103**, 280–283 (1978) (E)

French, C. S., Young, V. M. K.: Radiat. Biol. **3**, 343–391 (1956) (E)

Frey-Wyssling, A.: Stoffwechsel der Pflanzen, 2. ed. Zürich: Gutenberg 1949 (G)

Friend, D. J. C.: Photosynthetica **9**, 157–164 (1975) (E)

Friend, D. J. C., Helson, V. A., Fisher, J. E.: Can J. Bot. **40**, 1299–1311 (1962) (E)

Fritschen, L. J., Hsia, J., Doraiswamy, P.: Water Resour. Res. **13**, 145–148 (1977) (E)

Fritts, H. C.: Tree ring and climate. London, New York: Academic Press 1976 (E)

Froment, A., Tanghe, M., Duvigneaud, P., Galoux, A., Denaeyer-De Smet, S., Schnok, G., Grulois, J., Mommaerts-Billiet, F., Vanseveren, J. P.: In: Productivity of forest Ecosystems. Duvigneaud, P. (ed.), pp. 635–665. Paris: UNESCO 1971 (F)

Gaastra, P.: Med. Landbouwhogesch. Wageningen **59**, 1–68 (1959) (E)

Gäumann, E.: Ber. Schweiz. Bot. Ges. **44**, 157–334 (1935) (G)

Gäumann, E., Jaag, O.: Ber. Schweiz. Bot. Ges. **49**, 178–238, 555–626 (1939) (G)

Gaff, D. F.: Oecologia **31**, 95–109 (1977) (E)

Galazii, G. I., Beideman, I. N.: Vodnyi obmen b osonovnyh tipakh rastitelnost SSSR. Novosibirsk: Nauka 1975 (R)

Gale, J., Hagan, R. M.: Annu. Rev. Plant Physiol. **17**, 269–282 (1966) (E)

Galoux, A.: In: Productivity of forest ecosystems. Duvigneaud, P. (ed.), pp. 21–40. Paris: UNESCO 1971 (F)

Gamper, A.: Thesis Padua-Innsbruck 1975 (G/I)

Garber, K.: Luftverunreinigung und ihre Wirkungen. Berlin: Gebr. Borntraeger 1967 (G)

Garber, K.: Eidg. Anst. Forst. Vers. Birmensdorf, Ber.-Nr. **102**, (1973) (G)

Gardner, W. R.: In: Water deficits and plant growth. Kozlowski, T. T. (ed.), Vol. I, pp. 107–135. New York: Academic Press 1968 (E)

Gates, D. M.: Proc. Semicent. Celebr. Univ. Michigan, 31–52 (1959) (E)

Gates, D. M.: Ecology **46**, 1–14 (1965) (E)

Gaudet, J. J.: J. Ecol. **63**, 483–491 (1975) (E)

Gauhl, E.: Oecologia **39**, 61–70 (1979) (E)

Gaussman, H. W., Allen, W. A.: Plant Physiol. **52**, 57–62 (1973) (E)

Geiger, R.: Das Klima der bodennahen Luftschicht. 4th ed. Braunschweig: Vieweg 1961 (G)

Geiger, R.: Die Atmosphäre der Erde. Darmstadt: Perthes 1965 (G)

George, M. F., Burke, M. J.: Plant Physiol. **59**, 319–325 (1977) (E)

George, M. F., Burke, M. J., Weiser, C. J.: Plant Physiol. **54**, 29–35 (1974) (E)

Gerwick, B. C., Williams III, G. J., Uribe, E. G.: Plant Physiol. **60**, 430–432 (1977) (E)

Gerwick, B. C., Williams III, G. J.: Oecologia **35**, 149–159 (1978) (E)

Gessner, F.: Hydrobotanik. Vol. I: Energiehaushalt. Berlin: VEB Deutscher Verlag d. Wiss. 1955 (G)

Gessner, F.: Hydrobotanik, Bd. II: Stoffhaushalt. Berlin: VEB Deutscher Verlag d. Wiss. 1959 (G)

Geyger, E.: Habilitationsschrift, Göttingen 1978 (G)

Gifford, R. M.: Aust. J. Plant. Physiol. **1**, 107–117 1974 (E)

Gill, C. J.: For. Abstr. **31**, 671–688 (1970) (E)

Gimingham, C. H.: Ecology of heathlands. London: Chapman & Hall 1972 (E)

Gindel, I.: A new ecophysiological approach to forest-water relationships in arid climates. The Hague: Junk 1973 (E)

Gloser, J.: Acta Sci. Nat. Brno **10**, 1–39 (1976) (E)

Gloser, J.: Photosynthetica **11**, 139–147 (1977) (E)

Golley, F. B., McGinnis, J. T., Clements, R. G., Child, G. I., Duever, M. J.: Mineral cycling in a tropical moist forest ecosystem. Athens: Univ. Georgie Press 1975 (E)

Goryshina, T. K.: Rannevesennie efemeroidy lesostepnykh dubrav. Izd. Leningradskogo Univ. 1969 (R)

Grace, J.: Plant response to wind. London, New York: Academic Press 1977 (E)

Grace, J., Woolhouse, H. W.: J. Appl. Ecol. **10**, 77–91 (1973) (E)

Granhill, U., Lid-Torsvik, V.: In: Fennoscandian tundra ecosystems. Wigolaski, F. E., Kallio, P., Rosswall, T. (eds.). Ecological Studies, Vol. 16 (1), pp. 305–315. Berlin, Heidelberg, New York: 1975 (E)

Gregory, F. G.: Ann. Bot. **40**, 1–26 (1926) (E)

Grieve, B. J.: J. R. Soc. West Aust. **40**, 15–30 (1956) (E)

Grieve, B. J., Hellmuth, E. O.: Proc. Ecol. Soc. Aust. **3**, 46–54 (1968) (E)

Griffin, D. M.: Ecology of soil fungi. London: Chapman & Hall 1972 (E)

Grill, D., Liegl, E., Windisch, E.: Phytopath. Ztschr. **94,** 335–342 (1979) (G)

Grime, J. P., Hodgson, J. G.: In: Ecological aspects of the mineral nutrition of plants. Rorison, I. H. (ed.), pp. 67–99. Oxford: Blackwell 1969 (E)

Grin, A. M.: Umschau **72** (17), 551–554 (1972) (G)

Grouzis, M., Heim, G., Berger, A.: Oecol. Plant **12,** 307–322 (1977) (F)

Guderian, R.: Air pollution. Ecological Studies, Vol. 22. Berlin, Heidelberg, New York: Springer 1977 (E)

Hadfield, W.: In: Light as an ecological factor II. Evans, G. C., Bainbridge, R., Rackham, O. (eds.), pp. 477–496. Oxford: Blackwell Sci. Publ. 1975 (E)

Härtel, O.: Umschau **76,** 347–350 (1976) (G)

Halbwachs, G.: Ber. Dtsch. Bot. Ges. **84,** 507–514 (1971) (G)

Hall, A. E., Kaufmann, M. R.: In: Perspectives of biophysical ecology. Gates, D. M., Schmerl, R. B. (eds.), pp. 187–202. Berlin, Heidelberg, New York: Springer 1975 (E)

Hällgren, J.-E.: In: Sulfur in the environment, Part II. Nriagu, J. O. (ed.), pp. 163–209. Chichester: Wiley 1978 (E)

Hanscom, Z., Ting, I. P.: Plant Physiol. **61,** 327–330 (1978) (E)

Harder, R.: Z. Bot. **15,** 305–355 (1923) (G)

Harder, R.: Jahrb. Wiss. Bot. **64,** 169–200 (1925) (G)

Harder, R., Filzer, P., Lorenz, A.: Jahrb. Wiss. Bot. **75,** 45–194 (1931) (G)

Harrasser, J.: Diss. Innsbruck 1969 (G)

Hartsema, A. M., Luyten, I., Blaauw, A. H.: Verh. K. Akad. Wet. Amsterdam, 2. Sect., XXVII, 1, Med. **30,** 37–45 (1930) (D)

Hatch, M. D., Osmond, C. B.: In: Encyclopedia of plant physiology, N. S. Vol. 3/III. Stocking, C. R., Heber, U. (eds.), pp. 144–184. Berlin, Heidelberg, New York: Springer 1976 (E)

Hattersley, P. W.: Oecologia **57,** 113–128 (1983) (E)

Hawksworth, D. L., Rose, F.: Nature (London) **227,** 145–148 (1970) (E)

Hawksworth, D. L., Rose, F., Coppins, B. J.: In: Air pollution and lichens. Ferry, B. W., et al. (eds.), pp. 330–367. London: Athlone Press 1973 (E)

Heber, U., Santarius, K. A.: In: Temperature and life. Precht, H., Christophersen, J., Hensel, H., Larcher, W. (eds.), pp. 232–292. Berlin, Heidelberg, New York: Springer 1973 (E)

Heintzeler, I.: Arch. Mikrobiol. **10,** 92–132 (1939) (G)

Hellmers, H.: For. Sci. **12,** 275–283 (1966) (E)

Hellmuth, E. O.: Flora B **157,** 265–286 (1967) (E)

Hellmuth, E. O.: J. Ecol. **59,** 225–259 (1971) (E)

Hellriegel, F.: Beiträge zu den naturwissenschaftlichen Grundlagen des Ackerbaues. Braunschweig: Vieweg und Sohn 1883 (G)

Helms, J. A.: Ecology **46,** 698–708 (1965) (E)

Helms, J. A.: Photosynthetica **4,** 243–253 (1970) (E)

Henkel, P. A., Shakhov, A. A.: Bot. Zh. **30,** 154–166 (1945) (R)

Hesketh, J., Baker, D.: Crop. Sci. **7,** 285–293 (1967) (E)

Hess, D.: Plant physiology. Berlin, Heidelberg, New York: Springer 1975 (E)

Hew, Ch.-S., Krotkov, G., Canvin, D. T.: Plant Physiol. **44,** 662–670 (1969) (E)

Hinckley, T. M., Schroeder, M. O., Roberts, J. E., Bruckerhoff, D. N.: For. Sci. **22,** 201–211 (1975) (E)

Hinckley, T. M., Lassoie, J. P., Running, S. W.: Forest Sci. Mon. **20,** 1–72 (1978) (E)

Hiroi, T., Monsi, M.: J. Fac. Sci. Tokyo **9,** 241–285 (1966) (E)

Hoarsman, D. C., Wellburn, A. R.: In: Effects of air pollutants on plants. Mansfield, T. A. (ed.), pp. 185–199. Cambridge: Cambridge Univ. Press 1976 (E)

Höfler, K.: Ber. Dtsch. Bot. Ges. **60** (94)–(10) (1942) (G)

Höfler, K.: Ber. Dtsch. Bot. Ges. **63,** 3—10 (1950) (G)
Höfler, K., Migsch, H., Rottenburg, W.: Forschungsdienst **12,** 50—61 (1941) (G)
Höhne, H.: Arch. Forstwes. **12,** 791—805 (1963) (G)
Hoffmann, G.: Flora **161,** 303—319 (1972) (G)
Holmgren, P., Jarvis, P. G., Jarvis, M. S.: Physiol. Plant. **18,** 557—573 (1965) (E)
Horak, O., Kinzel, H.: Österr. Bot. Z. **119,** 475—495 (1971) (G)
Hsiao, T. C.: Annu. Rev. Plant Physiol. **24,** 519—570 (1973) (E)
Huang, Ch.-Y., Bazzaz, F. A., Vanderhoff, L. N.: Plant Physiol. **54,** 122—124 (1974) (E)
Huber, B.: In: Handbuch der Pflanzenphysiologie, Vol. 3, pp. 509—513. Berlin: Springer 1956 (G)
Iljin, W. S.: Jahrb. Wiss. Bot. **66,** 947—964 (1927) (G)
Iljin, W. S.: Protoplasma **10,** 379—414 (1930) (G)
Incoll, L. D., Long, S. P., Ashmore, M. R.: Current Adv. Plant Sci. **28,** 331—343 (1977) (E)
Innis, G. S.: Grassland simulation model. Ecological Studies, Vol. 26. Berlin, Heidelberg, New York: Springer 1978 (E)
Iwaki, H., Midorikawa, B.: In: Methods of productivity studies in root system and rhizosphere organisms. Ghilarov, M. S., Kovda, V. A., Novichkova-Ivanova, L. N., Rodin, L. E., Sveshnikova, V. M. (eds.), pp. 72—78. Leningrad: Nauka 1968 (E)
Jacobson, J. S., Hill, A. C.: Recognition of air pollution injury to vegetation: a pictorial atlas. Pittsburgh: Air Pollut. Contr. Assoc. 1970 (E)
Jaffré, T., Brooks, R. R., Reeves, R. D.: Science **193,** 579—580 (1976) (E)
Jäger, H. J., Klein, H.: Phytopathol. Ztschr. **89,** 128—134 (1977) (G)
Janzen, D. H.: Ecology of plants in the tropics. London: Arnold 1975 (E)
Jarvis, P. G., Jarvis, M. S.: Physiol. Plant **17,** 654—666 (1964) (E)
Jarvis, P. G., James, G. B., Landsberg, J. J.: In: Vegetation and the atmosphere. Monteith, J. L. (ed.), Vol. 2, pp. 171—240. London: Academic Press 1976 (E)
Johnson, D. W., Henderson, G. S., Huff, D. D., Lindberg, S. E., Richter, D. D., Shriner, D. S., Todd. D. E., Turner, J.: Oecologia **54,** 141—148 (1982) (E)
Johnson, Ph. L., Atwood, D. M.: In: A tropical rain forest. Odum, H. T., Pigeon, R. F. (eds.), Book 1, B-5. USAEC Techn. Inf. Ctr. Oak Ridge 1970 (E)
Jones, M. B., Mansfield, T. A.: Planta **103,** 134—146 (1972) (E)
Jones, M. B., Leafe, E. L., Stiles, W., Collett, B.: Ann. Bot. **42,** 693—703 (1978) (E)
Jordan, C. F., Kline, J. R.: J. appl. Ecol. **14,** 853—860 (1977) (E)
Joshi, M. C., Boyer, J. S., Kramer, P. J.: Bot. Gaz. **126,** 174—179 (1965) (E)
Kacperska-Palacz, A.: In: Plant cold hardiness and freezing stresses. Li, P. H., Sakai, A. (ed.), pp. 139—152. New York: Academic Press 1978 (E)
Kainmüller, Ch.: Diss. Innsbruck 1974 (G)
Kairiukštis, L. A.: In: Svetovoi rezhim fotosintez i produktivnost lesa. Tselniker, Ju. L. (ed.), pp. 151—166. Moscow: Nauka 1967 (R)
Kalle, K.: In: Handbuch der Pflanzenphysiologie, Vol. 4, pp. 170—178. Berlin: Springer 1958 (G)
Kallio, P.: In: Structure and function of tundra ecosystems. Rosswall, T., Heal, O. W., (eds.), Ecol. Bull. Vol. 20, pp. 193—223 Stockholm: NFR 1975 (E)
Kallio, P., Heinonen, S.: Rep. Kevo Subarct. Res. Sta. **8,** 63—72 (1971) (E)
Kallio, P., Heinonen, S.: Rep. Kevo Subarct. Res. Sta. **10,** 43—54 (1973) (E)
Kamiya, N.: Protoplasmic streaming. Protoplasmatologia VIII 3a. Wien: Springer 1959 (E)
Kappen, L.: Flora **156,** 101—116 (1965) (G)

Kappen, L.: In: Encyclopedia of plant physiology, Vol. 12 A, pp. 439–474. Berlin, Heidelberg, New York: Springer 1981 (E)

Kappen, L., Lange, O. L., Schulze, E. D., Evenari, M., Buschbom, U.: Oecologia **23**, 323–334 (1976) (E)

Karschon, R., Heth, D.: Contr. Eucalyptus Isr. **III,** 7–34 (1967) (E)

Kaufmann, M. R.: Plant Physiol. **56**, 841–844 (1975) (E)

Kaufmann, M. R.: Can. J. Bot. **55**, 2413–2418 (1977) (E)

Kausch, W.: Planta **45,** 217–265 (1955) (G)

Keck, R. W., Boyer, J. S.: Plant Physiol. **53**, 474–479 (1974) (E)

Keller, Th.: Allg. Forst. Jagdztg. **142** (4) (1971) (G)

Keller, Th.: Eur. J. For. Pathol. **4,** 11–19 (1974) (E)

Keller, Th.: Mitt. Eidg. Anst. Forst. Vers. **51**, 301–331. Zürich: Birmensdorf 1975 (G)

Kelly, P. C., Brooks, R. R., Dilli, S., Jaffré, T.: Proc. R. Soc. London B**189**, 6980 (1975) (E)

Ketner, P.: Primary production of salt-marsh communities on the island of Terschelling in the Netherlands. Verh. Rijksinst. Natuurbeheer No. **5**, (1972) (E)

Khairi, M. A., Hall, A. E.: J. Am. Soc. Hortic. Sci. **101**, 337–341 (1976) (E)

Kiendl, J.: Ber. Dtsch. Bot. Ges. **66**, 246–263 (1953) (G)

Killian, Ch.: Bull. Soc. Bot. Fr. **78**, 460–501 (1931) (F)

Killian, Ch.: Bull. Soc. Bot. Fr. **79**, 185–220 (1932) (F)

Killian, Ch., Lemée, G.: In: Handbuch der Pflanzenphysiologie, Vol. 3, pp. 787–824. Berlin: Springer 1956 (F)

Kimura, M.: Bot. Mag. Tokyo **82**, 6–19 (1969) (E)

Kinzel, H.: Ber. Dtsch. Bot. Ges. **82**, 143–158 (1969) (G)

Kinzel, H.: Pflanzenökologie und Mineralstoffwechsel. Stuttgart: Ulmer (1980) (G)

Kira, T., Shidei, T.: Jpn. J. Ecol. **17**, 70–87 (1967) (E)

Kira, T., Shinozaki, K., Hozumi, K.: Plant Cell Physiol. **10**, 129–142 (1969) (E)

Kislyuk, I. M.: Issledovanie povezhdayushchego lejstviya ochlazhdeniya na kletki listev rastenii chuvstvitelnych k kholodu. Moscow-Leningrad: Nauka 1964 (R)

Kislyuk, I. M., Alexandrov, V. Ya., Denko, E. I., Feldman, N. L., Kamentseva, I. E., Lutova, M. I., Shukhtina, H. G., Vaskovsky, M. D.: Phytotronic Newslett. **15**, 59–64 (1977) (E)

Klein, H., Jägcr, H. J., Domes, W., Wong, C. H.: Oecologia **33**, 203–208 (1978) (E)

Klinge, H.: Biogeographica **7**, 59–77 (1976) (G)

Kluge, M.: Ber. Dtsch. Bot. Ges. **84**, 417–424 (1971) (G)

Kluge, M.: In: Water and plant life. Lange, O. L., Kappen, L., Schulze, E. D. (eds.). Ecological Studies, Vol. 19, pp. 313–322. Berlin, Heidelberg, New York: Springer 1976

Kluge, M.: Oecologia **29,** 77–83 (1977) (E)

Kluge, M., Ting, J. P.: Crassulacean acid metabolism: Analysis of an ecological adaptation. Ecological Studies, Vol. 30. Berlin, Heidelberg, New York: Springer 1978 (E)

Kluge, M., Lange, O. L., Eichmann, M., Schmid, R.: Planta **112**, 357–372 (1973) (G)

Koch, W.: Flora B **158**, 402–428 (1969) (G)

Körner, Ch., Hoflacher, H., Wieser, G.: In: Ökologische Analysen von Almflächen im Gasteiner Tal. Cernusca, A. (ed.), pp. 67–79. Innsbruck: Wagner 1978 (G)

Körner, Ch., Scheel, J. A., Bauer, H.: Photosynthetica **13,** in press (1979) (E)

Koh, S., Kumura, A., Murata, Y.: Jpn. J. Crop. Sci. **47**, 63–68 (1978a) (E)

Koh, S., Kumura, A., Murata, Y.: Jpn. J. Crop. Sci. **47**, 69–74 (1978b) (E)

Kol, E.: Kryobiologie, I. Kryovegetation. Stuttgart: Schweizerbart 1968 (G)

Kozlowski, T. T.: Growth and development of trees, Vol. 1: Seed germination, ontogeny, and shoot growth. London, New York: Academic Press 1971a (E)

Kozlowski, T. T.: Growth and development of trees, Vol. 2: Cambial growth, root growth, reproductive growth. London, New York: Academic Press 1971b (E)

Kozlowski, T. T.: Water deficits and plant growth, Vol. IV. New York: Academic Press 1976 (E)

Kramer, P. J.: Plant Physiol. **10,** 87—112 (1940) (E)

Kramer, P. J.: Am. J. Bot. **39,** 828—832 (1942) (E)

Kramer, P. J.: Plant and soil water relationship. New York, Toronto, London: McGraw-Hill 1949 (E)

Kramer, P. J.: In: The physiology of forest trees. Thimann, V. (ed.). New York: Ronald Press Co. 1958 (E)

Kramer, P. J., Kozlowski, T. T.: Physiology of trees, 2. ed. New York: McGraw-Hill 1979 (E)

Krasavtsev, O. A.: Kalorimetriya rastenii pri temperaturakh nizhe nulja. Moscow: Nauka 1972 (R)

Kreeb, K.: Angew. Bot. **39,** 1—15 (1965) (G)

Kreeb, K.: Naturwissenschaften **61,** 337—343 (1974a) (G)

Kreeb, K.: Ökophysiologie der Pflanzen. Jena: Fischer 1974b (G)

Kucera, C. L., Dahlman, R. C., Koelling, M. R.: Ecology **48,** 536—541 (1967) (E)

Künstle, E., Mitscherlich, G.: Allg. Forst. Jagdztg. **148,** 227—239 (1977) (G)

Künz, M.: Dissertation Padua-Innsbruck 1974 (G/I)

Kuroiwa, S.: In: JIBP Synthesis, Vol. 19. Monsi, M., Saeki, T. (ed.), pp. 111—123. Tokyo: Univ. of Tokyo Press 1978 (E)

Kushner, D. J. (ed.): Microbial life in extreme environments. London: Academic Press 1978 (E)

Kusumoto, T.: Bull. Educ. Res. Inst. Kagoshima Univ. **9,** 21—25 (1957) (J)

Kwakwa, R. S.: Thesis Univ. Ghana. Ref.: Longman and Jenik, 1964 (E)

Kwolek, A. V., Woolhouse, H. W.: Ann. Bot. **47,** 435–442 (1981) (E)

Kyriakopoulos, E., Larcher, W.: Z. Pflanzenphysiol. **77,** 268—271 (1976) (G)

Laatsch, W.: Dynamik der mitteleuropäischen Mineralböden. Dresden: Steinkopf 1954 (G)

Ladefoged, K.: Physiol. Plant **16,** 378—414 (1963) (E)

Laetsch, W. M.: Annu. Rev. Plant Physiol. **25,** 27—52 (1974) (E)

Läuchli, A.: In: Transport und transfer processes in plants. Wardlaw, I. F., Passioura, J. B. (eds.), pp. 101—112. London, New York: Academic Press 1976 (E)

Lane, S. D., Martin, E. S., Garrod, J. F.: Planta **144,** 79—84 (1978) (E)

Lang, A. R. G., Klepper, B., Cumming, M. J.: Plant Physiol. **44,** 826—830, (1969) (E)

Lange, O. L.: Flora **147,** 595—651 (1959) (G)

Lange, O. L.: Ber. Dtsch. Bot. Ges. **75,** 351—352 (1962) (G)

Lange, O. L.: Planta **64,** 1—19 (1965) (G)

Lange, O. L.: In: The cell and environmental temperature. A. S. Troshin (ed.), pp. 131–141. Oxford: Pergamon Press 1967 (E)

Lange, O. L.: Flora, B **158,** 324—359 (1969) (G)

Lange, O. L., Kappen, L.: In: Antarctic terrestrial biology. Llano, G. A. (ed.), pp. 83—95. Washington: Am. Geophys. 1972 (E)

Lange, O. L., Lange, R.: Flora **153,** 387—425 (1963) (G)

Lange, O. L., Zuber, M.: Oecologia **31,** 67—72 (1977) (E)

Lange, O. L., Schulze, E. D., Koch, W.: Flora **159,** 38—62 (1970) (G)

Lange, O. L., Schulze, E. D., Kappen, L., Buschbom, U., Evenari, M.: In: Perspectives of biophysical ecology. Gates, D. M., Schmerl, R. B. (eds.). Ecological Studies, Vol. 12, pp. 121—143. Berlin, Heidelberg, New York: Springer 1975 (E)

Larcher, W.: Bull. Res. Counc. Isr. Sect D **8,** 213—224 (1960) (E)

Larcher, W.: Planta **56,** 575—606 (1961) (G)

Larcher, W.: Planta **60,** 1—18 (1963) (G)

Larcher, W.: In: Water stress in plants. Slavík, B. (ed.), pp. 184—194. Prague: Academia 1965 (E)

Larcher, W.: Photosynthetica **3,** 167—198 (1969a) (E)

Larcher, W.: Ber. Dtsch. Bot. Ges. **82,** 71—80 (1969b) (G)

Larcher, W.: Oecol. Plant. **5,** 267—286 (1970) (G)

Larcher, W.: Oecol. Plant. **6,** 1—14 (1971) (G)

Larcher, W.: Ber. Dtsch. Bot. Ges. **86,** 315—327 (1972) (G)

Larcher, W.: In: Temperature and life, 2. ed. Precht, H., Christophersen, J., Hensel, H., Larcher, W. (eds.). Berlin, Heidelberg, New York: Springer 1973a (E)

Larcher, W.: Ökologie der Pflanzen. UTB 232. Stuttgart: Ulmer 1973b (G)

Larcher, W.: Anz. Österr. Akad. Wiss. Math. Naturwiss. Kl. 194—213 (1975) (G)

Larcher, W.: Sitzungsber. Österr. Akad. Wiss. Math. Naturwiss. Kl., Abt. I, **186,** 301—371 (1977) (G)

Larcher, W.: Plant Syst. Evol. **137,** 145—180 (1981a) (G)

Larcher, W.: In: Physiological processes limiting plant productivity. Johnson, C. B. (ed.), pp. 253—269. London: Butterworths 1981b (E)

Larcher, W., Bauer, H.: In: Encyclopedia of plant physiology, Vol. 12 A, pp. 403—437. Berlin, Heidelberg, New York: Springer 1981 (E)

Larcher, W., Eggarter, H.: Protoplasma **51,** 595—619 (1960) (G)

Larcher, W., Mair, B.: Oecol. Plant. **4,** 347—376 (1969) (G)

Larcher, W., Cernusca, A., Schmidt, L.: In: Ökosystemforschung. Ellenberg, H. (ed.), pp. 175—194. Berlin, Heidelberg, New York: Springer 1973 (G)

Larcher, W., De Moraes, J. A. P. V., Bauer, H.: In: Components of productivity of mediterranean-climate regions — Basic and applied aspects. Margaris, N. S., Mooney, H. A. (eds.), pp. 77—84. The Hague: Junk 1981 (E)

Laurie, A., Kiplinger, D. C.: Commercial flower forcing, 4. ed. Philadelphia: Saunders 1944 (E)

Ledig, F. T., Drew, A. P., Clark, J. G.: Ann. Bot. **40,** 289—300 (1976) (E)

Lemée, G.: Rev. Gén. Bot. **63,** 41—95 (1956) (F)

Lenz, F.: Proc. Ist. Int. Citrus Symp., Vol. 1, pp. 333—338 1969 (E)

Leopold, A. C., Kriedemann, P. E.: Plant growth and development, 2. ed. New York: McGraw-Hill 1975 (E)

Lerch, G.: Kulturpflanze **24,** 53—63 (1976) (G)

Levitt, J.: Frost, Drought, and heat resistance. Protoplasmatologia VIII, 6. Wien: Springer 1958 (E)

Levitt, J.: Responses of plants to environmental stresses. London, New York: Academic Press 1972 (E)

Lewis, D. A., Nobel, P. S.: Plant Physiol. **60,** 609—616 (1977) (E)

Liese, W., Schneider, M., Eckstein, D.: Eur. J. For. Path. **5,** 152—161 (1975) (G)

Lieth, H.: Die Stoffproduktion der Pflanzendecke. Stuttgart: Fischer 1962 (G)

Lieth, H.: In: Analysis of temperate forest ecosystems. Reichle, D. E. (ed.). Ecological Studies, Vol. 1, pp. 29—46. Berlin, Heidelberg, New York: Springer 1970 (E)

Lieth, H.: Angew. Bot. **46,** 1—37 (1972) (G)

Lieth, H.: Phenology and seasonality modeling. Ecological Studies, Vol. 8. Berlin, Heidelberg, New York: Springer 1974 (E)

Lieth, H., Whittaker, R. H.: Primary productivity of the biosphere. Ecological Studies, Vol. 14. Berlin, Heidelberg, New York: Springer 1975 (E)

Likens, G. E., Bormann, F. H., Pierce, R. S., Eaton, J. S., Johnson, N. M.: Biogeochemistry of a forested ecosystem. Berlin, Heidelberg, New York: Springer 1977 (E)

Linzon, S. N.: In: Sulphur in the environment, Vol. II, pp. 109—162. London: Wiley 1978 (E)

Lloyd, E. J.: Z. Pflanzenphysiol. **78,** 1—12 (1976) (E)

Lloyd, N. D. N., Canvin, D. T., Bristow, J. M.: Can. J. Bot. **55,** 3001—3005 (1977) (E)

Lockwood, J. G.: World climatology. London: Arnold 1974 (E)

Lösch, R.: Oecologia **29,** 85—97 (1977) (E)

Logan, K. T.: Dep. For. Can. For. Serv. Publ. 1256, Ottawa (1969) (E)

Logan, K. T.: Dept. Environ. Can. For. Serv. Publ. 1323, Ottawa (1973) (E)

Longman, K. A., Jenik, J.: Tropical forest and its environment. London: Longman 1974 (E)

Loomis, R. S., Gerakis, P. A.: In: Photosyntesis and productivity in different environments. Cooper, J. P. (ed.) Cambridge: Cambridge Univ. Press 1975 (E)

Lossaint, P., Rapp, M.: Productivity of forest ecosystems. Duvigneaud, P. (ed.), pp. 597—617. Paris: UNESCO 1971 (F)

Lossaint, P., Rapp, M.: In: Problemes d'Écologie. Écosystémes terrestres. Lamotte, M., Bourliére, C. (eds.), pp. 129—185. Paris: Masson 1978 (F)

Louwerse, W., Zweerde, W. v. d.: Photosynthetica **11,** 11—21 (1977) (E)

Loveys, B. R., Kriedemann, P. E.: Physiol. Plant **28,** 476—479 (1973) (E)

Ludlow, M. M.: In: Environmental and biological control of photosynthesis. Marcelle, R. (ed.), pp. 123—134. The Hague: Junk 1975 (E)

Ludlow, M. M.: In: Water and plant life. Lange, O. L., Kappen, L., Schulze, E. D. (eds.). Ecological Studies, Vol. 19, pp. 364—386. Berlin, Heidelberg, New York: Springer 1976 (E)

Ludlow, M. M., Wilson, G. L.: Aust. J. Biol. Sci. **24,** 449—470 (1971a) (E)

Ludlow, M. M., Wilson, G. L.: Aust. J. Biol. Sci. **24,** 1065—1075 (1971b) (E)

Ludlow, M. M., Wilson, G. L.: Aust. J. Biol. Sci. **24,** 1077—1087 (1971c) (E)

Lüttge, U.: Stofftransport der Pflanzen. Berlin, Heidelberg, New York: Springer 1973 (G)

Lundegårdh, H.: Flora **121,** 273—300 (1927) (G)

Luyten, I., Versluys, M. C., Blaauw, A. H.: Verh. K. Akad. Wet. Amsterdam, 2. Sect. XXIX, 5, Med. **36,** 57—64 (1932) (D)

Lyons, J. M., Raison, J. K.: Plant Physiol. **45,** 386—389 (1970) (E)

Lyr, H., Polster, H., Fiedler, H. J.: Gehölzphysiologie. Jena: Fischer 1967 (G)

Mächler, F., Nösberger, J.: Oecologia **31,** 73—78 (1977) (E)

Mächler, F., Nösberger, J.: Oecologia **35,** 267—276 (1978) (E)

Maestri, M., Barros, R. S.: In: Ecophysiology of tropical crops. de T. Alvim, P., Kozlowski, T. T. (eds.), pp. 249—278. London, New York: Academic Press 1977 (E)

Magalhaes, A. C.: Whats new in plant physiol. **7,** 1—5, (1975) (E)

Magalhaes, A. C., Neyra, C. A., Hageman, R. H.: Plant Physiol. **53,** 411—415 (1974) (E)

Mahlknecht, A.: Dissertation Padua-Innsbruck 1976 (G/I)

Mair, B.: Planta **82,** 164—169 (1968) (G)

Majernik, O., Mansfield, T. A.: Nature (London) **227,** 377—378 (1970) (E)

Malaisse, F. P.: In: Phenology and seasonality modeling. Lieth, H. (ed.). Ecological Studies, Vol. 8, pp. 269—286. Berlin, Heidelberg, New York: 1974 (E)

Malkina, I. S.: Fiziol. Rast. **25,** 792—797 (1978) (R)

Malkina, I. S., Tselniker, Yu. L., Jakshina, A. M.: Fotosintez i dykhanie podrosta. Moscow: Nauka 1970 (R)

Malone, C., Koeppe, D. E., Miller, R. J.: Plant Physiol. **53,** 388—394 (1974) (E)

Manohar, M. S.: Z. Pflanzenphysiol. **84,** 227—235 (1977) (E)

Mansfield, T. A.: Effects of air pollutants on plants. Cambridge: Cambridge Univ. Press 1976 (E)

Mar-Möller, C., Müller, D., Nielsen, J.: Forstl. Forsøgsvals. Dan. **21**, 327–335 (1954) (E)

Margaris, N. S.: J. Biogeography **3**, 249–259 (1976) (E)

Mathys, W.: Physiol. Plant **33**, 161–165 (1975) (E)

Maximov, N. A.: Jahrb. Wiss. Bot. **53**, 327–420 (1914) (G)

Maximov, N. A.: Jahrb. Wiss. Bot. **62**, 128–144 (1923) (G)

Maximov, N. A.: Protoplasma **7**, 259–291 (1929) (E)

Mayer, A. M., Poljakoff-Mayber, A.: The germination of seeds, 2. ed. Oxford: Pergamon Press 1975 (E)

McCune, C. C.: Establishment of air quality criteria, with reference to the effects of atmospheric fluorine on vegetation. New York: Aluminum Assoc. 1968 (E)

McGregor, W. H. D., Kramer, P. J.: Am. J. Bot. **50**, 760–765 (1963) (E)

McManmon, M., Crawford, R. M. M.: Phytologia **70**, 299–306 (1971) (E)

Medina, E., Delgado, M., Troughton, J. H., Medina, J. D.: Flora **166**, 137–152 (1977) (E)

Meidner, H., Mansfield, T. A.: Physiology of stomata. London: McGraw-Hill 1968 (E)

Meidner, H., Sheriff, D. W.: Water and plants. Glasgow: Blackie 1976 (E)

Mengel, K.: Ernährung und Stoffwechsel der Pflanze, 3. Aufl. Jena: Fischer 1968 (G)

Merino, J., García Novo, F., Sánchez Díaz, M.: Oecol. Plant **11**, 1–11 (1976) (E)

Meyer, F. M.: Bäume in der Stadt. Stuttgart: Ulmer 1978 (G)

Meyer, N.: Flora A **159**, 215–232 (1968) (G)

Miller, P. C., Hom, J., Poole, D. K.: Oecol. Plant. **10**, 355–367 (1975) (E)

Milner, H. W., Hiesey, W. M.: Plant Physiol. **39**, 208–213 (1964) (E)

Milthorpe, F. L., Moorby, J.: An introduction to crop physiology London: Cambridge Univ. Press 1974 (E)

Mitscherlich, G.: Wald, Wachstum und Umwelt. Vol. 2. Frankfurt/Main: Sauerländer 1971 (G)

Mokronosov, A. T., Nekrasova, G. F.: Fiziol. Rast. **24**, 458–465 (1977) (R)

Molisch, H.: Der Einfluß einer Pflanze auf die andere — Allelopathie. Jena: Fischer 1937 (G)

Monsi, M., Saeki, T.: Jpn. J. Bot. **14**, 22–52 (1953) (G)

Monsi, M., Uchijima, Z., Oikawa, T.: Annu. Rev. Ecol. Syst. **4**, 301–327 (1973) (E)

Monteith, J. L.: Ann. Bot. **29**, 17–37 (1965) (E)

Monteith, J. L.: Principles of environmental physics. London: Edward Arnold 1973 (E)

Monteith, J. L.: Vegetation and the atmosphere, Vol. 1: Principles. London: Academic Press 1975 (E)

Monteith, J. L.: Vegetation and the atmosphere, Vol. 2: Case studies. London: Academic Press 1976 (E)

Monteith, J. L.: Exp. Agric. **14**, 1–5 (1978) (E)

Monzigo, H. N., Comanor, P. L.: Cact. Suculentas J. Suppl. Vol. 1975, 22–28 (1975) (E)

Mooney, H. A.: Ann. Rev. Ecol. Syst. **3**, 315–346 (1972) (E)

Mooney, H. A., Hays, R. I.: Flora **162**, 295–304 (1973) (E)

Mooney, H. A., Ehleringer, J., Berry, J. A.: Science **194**, 322–324 (1976) (E)

Mooney, H. A., Weisser, P. J., Gulmon, S. L.: Flora **166**, 117–124 (1977) (E)

Mooney, H. A., Björkman, O., Collatz, G. J.: Plant Physiol. **61**, 406–410 (1978) (E)

Moraes, V. H. F.: In: Ecophysiology of tropical crops. de T. Alvim, P., Kozlowski, T. T. (eds.), pp. 315–331. New York: Academic Press 1977 (E)

Morain, S. A.: In: Phenology and seasonality modeling. Lieth, H. (ed.). Ecological Studies, Vol. 8, pp. 55—75. Berlin, Heidelberg, New York: Springer 1974

Moser, W., Brzoska, W., Zachhuber, K., Larcher, W.: Sitzungsber. Österr. Akad. Wiss. Math. Naturwiss. Kl., Abt. I, **186**, 387—419 (1977) (G)

Moss, D.: Conn. Agric. Exp. Stn. Bull. **664**, 86—101 (1963) (E)

Mousseau, M.: In: Les processus de la production végétale primaire Moyse, A. (ed.), pp. 157—181. Paris: Gauthier-Villars 1977 (F)

Mudd, J. B., Kozlowski, T. T.: Responses of plants to air pollution. New York: Academic Press 1975 (E)

Müller, D.: Handbuch der Pflanzenphysiologie, Part 2, Vol. 12, pp. 934—948. Berlin: Springer 1960 (G)

Müller, D., Nielsen, J.: Forstl. Forsoegsvaes. Dan. **29**, 69—160 (1965) (F)

Müller, E., Löffler, W.: Mykologie, 2. ed. Stuttgart: Thieme 1971 (G)

Müller, H.: Thesis Padua-Innsbruck 1976 (G/I)

Münch, E.: Die Stoffbewegungen in der Pflanze. Jena: Fischer 1930 (G)

Murata, Y.: In: JIBP Synthesis, Vol. 20. Tamiya, H. (ed.), pp. 57—62. Tokyo: Univ. of Tokyo Press 1978 (E)

Myers, J.: J. Gen. Physiol. **29**, 429—440 (1946) (E)

Naegele, J. A.: Air pollution damage to vegetation. Adv. Chem. Ser. **122**, Washington D.C. 1973 (E)

Nakhuzrishvili, G. Sh.: Ekologiya vysokogornykh travjanistykh rastenii i fitozenozov zentralnogo Kavkaza. Vodnyi rezhim. Tbilisi: Mezniereba 1971 (R)

Nakhuzrishvili, G. Sh.: Ekologiya vysokogornykh rastenii i fitozenozov zentralnogo Kavkaza. Ritmika razvitiya, fotosintez, ekobiomorfy. Tbilisi: Mezniereba 1974 (R)

Napp-Zinn, K.: Handbuch der Pflanzenanatomie VIII/2A. Berlin: Borntraeger 1973 (G)

National Research Council (USA): Biologic effects of atmospheric pollutatns. Natl. Acad. Sci. USA (1971) (E)

Nátr, L.: Photosynthetica **6**, 80—99 (1972) (E)

Nátr, L.: In: Photosynthesis and productivity in different environments. Cooper, J. P. (ed.), pp. 537—555. Cambridge: Cambridge Univ. Press 1975 (E)

Neales, T. P.: In: Environmental and biological control of photosynthesis. Marcelle, R., (ed.), pp. 299—310. The Hague: Junk 1975 (E)

Negisi, K.: Bull. Tokyo Univ. For. **62**, 1—115 (1966) (E)

Neilson, R. E., Ludlow, M. M., Jarvis, P. G.: J. Appl. Ecol. **9**, 721—745 (1972) (E)

Nelson, N. D., Dickmann, D. I., Gottschalk, K. W.: Photosynthetica **16**, 321–333 (1982) (E)

Neuwirth, G.: Biol. Zentralbl. **78**, 560—584 (1959) (G)

Neuwirth, G.: Arch. Forstwes. **17**, 613—620 (1968) (G)

Neyra, C. A., Hageman, R. H.: Plant Physiol. **62**, 618—621 (1978) (E)

Nobel, P. S.: Plant Physiol. **58**, 576—582 (1976) (E)

Nobel, P. S.: Oecologia **27**, 117—133 (1977a) (E)

Nobel, P. S.: Physiol. Plant. **40**, 137—144 (1977b) (E)

Noland, T. L., Kozlowski, T. T.: Can. J. Forest Res. **9**, 57—62 (1979) (E)

Nordhausen, M.: Ber. Dtsch. Bot. Ges. **21**, 30—44 (1903) (G)

O'Connor, M., Woodford, F. P.: Writing scientific papers in English. Amsterdam: Elsevier 1976 (E)

Odening, W. R., Strain, B. R., Oechel, W. C.: Ecology **55**, 1086—1095 (1974) (E)

Odum, E. P.: Fundamentals of ecology. Philadelphia, London: Saunders Comp. 1953, 1959, 1971 (E)

Odum, H. T., Pigeon, R. F.: A tropical rain forest. Oak Ridge: US Atomic Energy Commun. 1970 (E)

Oechel, W. C., Lawrence, W. T.: Oecologia **39**, 321–335 (1979) (E)

Olien, C. R.: Annu. Rev. Plant Physiol. **18**, 387–408 (1967) (E)

Olien, C. R.: Plant Physiol. **53**, 764–767 (1974) (E)

Olien, C. R.: In: Li, P. H., Sakai, A. (eds.): Plant cold hardiness and freezing stresses, pp. 37–48. New York: Academic Press 1978 (E)

Osmond, C. B.: Annu. Rev. Plast. Physiol. **29**, 379–414 (1978) (E)

Osmond, C. B., Winter, K., Ziegler, H.: In: Encyclopedia of plant physiology, Vol. 12 B, pp. 479–547. Berlin, Heidelberg, New York: Springer 1982 (E)

O'Toole, J. C.: In: Drought resistance in crops with emphasis on rice, pp. 195–213. Los Baños: IRRI 1982 (E)

O'Toole, J. C., Cruz, R. T., Singh, T. N.: Plant Sci. Letters **16**, 111–114 (1979) (E)

Ovington, J. D.: Forestry **27**, 41–53 (1954) (E)

Ovington, J. D.: Ann. Bot. **23**, 75–88 (1959a) (E)

Ovington, J. D.: Ann. Bot. **23**, 229–239 (1959b) (E)

Parcevaux, S.: Oecol. Plant. **7**, 371–401 (1972) (F)

Parcevaux, S.: Oecol. Plant. **8**, 41–62 (1973) (F)

Parker, J.: Ecology **42**, 372–380 (1961) (E)

Parker, J.: NE. For. Exp. Stu. Upper Darby, Pa., US-For. Serv. Res. Pap. NE-94 (1968) (E)

Pate, J. S.: In: Transport and transfer processes in plants. Wardlaw, I. F., Passioura, J. B. (eds.), pp. 447–462. New York: Academic Press 1976 (E)

Pate, J. S., Sharkey, P. J., Atkins, C. A.: Plant Physiol. **59**, 506–510 (1977) (E)

Pearcy, R. W., Björkamn, O., Harrison, A. T., Mooney, H. A.: Carnegie Inst. Yearb. **70**, 540–550 (1971) (E)

Penning De Vries, F. W. T.: In: Rees, Cockshull, Hand, Hurd (eds.): Crop processes in controlled environments, pp. 327–347. London: Academic Press 1972 (E)

Pereira, J. S., Kozlowski, T. T.: Physiol. Plant. **41**, 184–192 (1977) (E)

Perkins, D. F.: In: Production ecology of british moors and montane grasslands. Heal, O. W., Perkins, D. F. (eds.). Ecological Studies, Vol. 27, pp. 375–416. Berlin, Heidelberg, New York: Springer 1978 (E)

Perry, Th.: Science **171**, 29–36 (1971) (E)

Pichler, J.: Thesis Padua-Innsbruck 1975 (G/I)

Pisek, A.: Gartenbauwissenschaft **23**, 54–74 (1958) (G)

Pisek, A., Berger, E.: Planta **28**, 124–155 (1938) (G)

Pisek, A., Cartellieri, E.: Jahrb. Wiss. Bot. **90**, 256–291 (1941) (G)

Pisek, A., Kemnitzer, R.: Flora B, **157**, 314–326 (1968) (G)

Pisek, A., Schiessl, R.: Ber. Naturwiss. Med. Ver. Innsbruck **47**, 33–52 (1947) (G)

Pisek, A., Tranquillini, W.: Physiol. Plant. **4**, 1–27 (1951) (G)

Pisek, A., Winkler, E.: Planta **42**, 253–278 (1953) (G)

Pisek, A., Winkler, E.: Planta **51**, 518–543 (1958) (G)

Pisek, A., Winkler, E.: Planta **53**, 532–550 (1959) (G)

Pisek, A., Larcher, W., Moser, W., Pack, I.: Flora B **158**, 608–630 (1969) (G)

Pisek, A., Knapp, H., Ditterstorfer, J.: Flora **159**, 459–479 (1970) (G)

Pleiss, H.: Der Kreislauf des Wassers in der Natur. Jena: Fischer 1977 (G)

Polster, H.: Die physiologischen Grundlagen der Stofferzeugung im Walde. München: Bayer. Landwirtschaftsverlag 1950 (G)

Polster, H.: In: Gehölzphysiologie. Lyr, H., Polster, H., Fiedler, H. J. (eds.). Jena: Fischer 1967 (G)

Polster, H., Fuchs, S.: Arch. Forstwes. **12**, 1011–1024 (1963) (G)

Polster, H., Neuwirth, G.: Arch. Forstwes. **7**, 749–785 (1958) (G)

Ponomareva, M. M.: Exp. Bot. (Moscow-Leningrad) **14,** 53–72 (1960) (R)

Poole, D. K., Miller, P. C.: Decologia **13,** 289–299 (1978) (E)

Porter, E. K., Peterson, P. J.: Sci Total Environ. **4,** 365–371 (1975) (E)

Pospišilová, J.: Biol. Plant. **17,** 392–399 (1975) (E)

Pospišilová, J., Solárová, J.: Biol. Plant. **20,** 435–457 (1978) (E)

Pospišilová, J., Tichá, I., Čatský, J., Solárová, J.: Biol. Plant. **20,** 368–372 (1978) (E)

Queiroz, O.: In: Photosynthesis II. Gibbs, M., Latzko, E. (ed.). Encyclopedia of Plant Physiology, N.S., Vol. 6, pp. 126–139. Berlin, Heidelberg, New York: Springer 1979

Rabe, R., Kreeb, K. H.: Environm. Poll. **19,** 119–137 (1979) (E)

Rabotnov, T. A.: Trudy Bot. Inst. Akad. Nauk SSSR Ser. **3,** (6) (1950) (R)

Rabotnov, T. A.: Fitocenologija. Izd. Moskovsk. Univ. 1978a (R)

Rabotnov, T. A.: Verh. K. Ned. Akad. Wet. II **70,** 1–26 (1978b) (E)

Raeuber, A., Meinl, G., Engel, K. H.: Wiss. Z. Univ. Leipzig **17,** 295–301 (1968) (G)

Raghavendra, A. S., Das, V. S. R.: Z. Pflanzenphys. **87,** 379–393 (1978a) (E)

Raghavendra, A. S., Das, V. S. R.: Photosynthetica **12,** 200–208 (1978b) (E)

Raper, C. D., Kramer, P. J. (eds.): Crop reactions to water and temperature stresses in humid, temperature climates. Boulder: Westview Press 1983 (E)

Rapp, M.: Oecol. Plant. **4,** 377–410 (1969) (F)

Rapp, M.: Cycle de la matière organique et des éléments minèraux dans quelques écosystèmes méditerrannées. C. N. R. S. **40,** 19–184 (1971) (F)

Raschke, K.: Planta **48,** 200–239 (1956) (G)

Raschke, E., Von der Haar, T. H., Bandeen, W. R., Pasternak, M.: J. Atmos. Sci. **30,** 341–364 (1973) (E)

Raunkiaer, C.: The life forms of plants and statistical plant geography. Oxford: Clarendon Press 1934 (E)

Raven, J. A.: Advances in botanical research. Vol. 5, pp. 154–240. London: Academic Press 1977 (E)

Raven, J. A., Smith, F. A.: Photosynthetica **11,** 48–55 (1977) (E)

Rawson, H. M., Turner, N. C., Begg, J. E.: Aust. J. Plant Physiol. **5,** 195–209 (1978) (E)

Retter, W.: Diss. Innsbruck 1965 (G)

Richards, P. W.: The tropical rain forest, 6. ed. Cambridge: Cambridge Univ. Press 1976 (E)

Richter, H.: J. Exp. Bot. **24,** 983–994 (1973) (E)

Richter, H.: In: Water and plant life. Lange, O. L., Kappen, L., Schulze, E. D. (eds.). Ecological Studies, Vol. 19, p. 19. Berlin, Heidelberg, New York: Springer 1976 (E)

Richter, H., Halbwachs, G., Holzner, W.: Flora **161,** 401–420 (1972) (G)

Ringoet, A.: Recherches sur la transpiration et la bilan d'eau de quelques plantes tropicales. Publ. Ist. Natl. Etude Agron. Congo Belge Sér. Sci. **56,** 1952 (F)

Ringoet, A., Mittenaere, C. O.: Évapotranspiration et croissance du riz de montagne (Oryza sativa L.) dans les sols à humidité variable. Publ. Inst. Natl. Étude Agron. Congo Ser. Sci. **92,** 95–145 (1961) (F)

Ritchie, G. A., Hinckley, Th. M.: Advances in ecological research, Vol. 9, pp. 165–254. London: Academic Press 1975 (E)

Robberecht, R., Caldwell, M. M.: Oecologia **32,** 277–287 (1978) (E)

Rodin, L. Y., Bazilevich, N. I.: Dinamika organicheskogo veshchestva i biologicheskii krugovorot zolnykh elementov i azota v osnovykh tipakh rastitelnost zemnogo shara. Moscow: Nauka 1965 (R)

Rodin, L. Y., Bazilevich, N. I.: Production and mineral cycling in terrestial vegetation. Edinburgh, London: Oliver & Boyd 1967 (E)

Rook, D. A.: N. Z. J. Bot. **7,** 43–55 (1969) (E)

Rorison, I. H.: Ecological aspects of the mineral nutrition of plants. pp. 155—175. Oxford-Edinburgh: Blackwell 1969 (E)

Rose, D. A.: J. Soil Sci. **30**, 1—15 (1979) (E)

Rosenberg, N. J.: Microclimate: the biological environment. New York: Wiley 1974 (E)

Ross, J.: In: Vegetation and the atmosphere. Monteith, J. L. (ed.), Vol. 1, pp. 13—55. London: Academic Press 1975 (E)

Rosswall, T., Heal, O. W.: Structure and function of tundra ecosystems. Ecol. Bull. 20, N. F. R., Stockholm (1975) (E)

Rouschal, E.: Jahrb. Wiss. Bot. **87**, 436—523 (1938) (G)

Roussel, L.: Photologie forestrière. Paris: Masson & Cie 1972 (F)

Rowe, R. N., Beardsell, D. V.: Hort Abstr. **43**, 533—548 (1973) (E)

Rustamov, I. G.: In: Eco-physiological foundation of ecosystems productivity in arid zone. Rodin, L. E. (ed.), pp. 129—132. Leningrad: Nauka 1972 (E)

Rutter, A. J.: In: Water deficit and plant growth Kozlowski, T. T. (ed.), Vol. II, pp. 23—84. London, New York: Academic Press 1968 (E)

Rutter, A. J.: In: Vegetation and the atmosphere. Monteith, J. L. (ed.), Vol. I, pp. 111—154. London, New York: Academic Press 1975 (E)

Rychnovská, M.: Preslia **37**, 42—52 (1965) (E)

Rychnovská, M.: In: Pond littoral ecosystems. Dykyjová, D., Kvet, J. (eds.). Ecological Studies, Vol. 28, pp. 246—256. Berlin, Heidelberg, New York: Springer 1978 (E)

Rychnovská, M., Úlehlová, B.: Autökologische Studie der tschechoslowakischen Stipa-Arten. Vegetace CSSR, A 8. Prague: Academia 1975 (G)

Rychnovská, M., Květ, J., Gloser, J., Jakrlová, J.: Acta Sci. Natl. Acad. Sci. Bohemoslovacae Brno, VI (5) (1972) (E7

Rydén, B. E.: In: Truelove Lowland, Devon Island, Canada: A high arctic ecosystem. Bliss, L. C. (ed.), pp. 107—136. Edmonton: Univ. Alberta Press 1977 (E)

Sakai, A.: Contr. I. L. T. Sci. B **11**, 1—40 (1962) (E)

Sakai, A.: Ecology **51**, 485—491 (1970) (E)

Sakai, A., Wardle, P.: New Zeal. J. Ecol. **1**, 51—61 (1978) (E)

Sakai, A., Weiser, C. J.: Ecology **54**, 118—126 (1973) (E)

Sale, P. J. M.: Aust. J. Plant Physiol. **1**, 283—296 (1974) (E)

Sale, P. J. M.: Aust. J. Plant Physiol. **4**, 555—569 (1977) (E)

Salisbury, E. I.: J. Ecol. **4**, 83—117 (1916) (E)

Salisbury, F. B., Spomer, G. G.: Planta **60**, 497—505 (1964) (E)

Samsuddin, Z., Impens, I.: Experimental Agric. **14**, 173—177 (1978) (E)

Saunders, P. J. W., Wood, C. M.: In: Air pollution and lichens. Ferry, B. W. et al. (eds.), pp. 6—37. London: Athlone Press 1973 (E)

Savage, M. J.: HortSci. **14**, 492—495 (1979) (E)

Sawada, S., Iwaki, H.: In: Ecophysiology of photosynthetic productivity, JIBP. Vol. 19, 11—18. 1978 (E)

Sawada, S., Miyachi, S.: Plant & Cell Physiol. **15**, 111—120 (1974) (E)

Schennikow, A. P.: In: Handbuch der biologischen Arbeitsmethoden. Abderhalden, E., (ed.), Vol. 11, pp. 251—266. Berlin: Springer 1932 (G)

Schmidt, L.: Oecol. Plant. **12**, 195—213 (1977) (G)

Schnarrenberger, C., Fock, H.: Encyclopedia of plant physiology (N. S.). Transport in plants. Vol. III, pp. 185—234. Berlin, Heidelberg, New York: 1976 (E)

Schnelle, F.: Pflanzenphänologie. Leipzig: Akad. Verlagsges. 1955 (G)

Schnock, G.: Productivity of forest ecosystems. Duvigneaud, P. (ed.), pp. 41—42. Paris: UNESCO 1971 (F)

Schölm, H. E.: Protoplasma **65**, 97—118 (1968) (G)

Scholander, P. F., Hammel, H. T., Bradstreet, E. D., Hemmingsen, E. A.: Science **148**, 339—346 (1965) (E)

Scholefield, P. B., Walcott, J. J., Ramadasan, A., Kriedemann, P. E.: C. S. I. R. O. Div. Hortic. Res. Rep. 1975—77, 60—61, Adelaide 1977 (E)

Schulz, J. P.: Ecological studies on rain forest in Northern Suriname. Amsterdam: N.-Holl. Uitg. Maatsch. 1960 (E)

Schulze, E. D.: Flora **159**, 177—232 (1970) (G)

Schulze, E.-D.: In: Encyclopedia of plant physiology, Vol. 12 B, pp. 615—676. Berlin, Heidelberg, New York: Springer 1982 (E)

Schulze, E.-D., Hall, A. E.: In: Encyclopedia of plant physiology, Vol 12 B, , pp. 181—230. Berlin, Heidelberg, New York: Springer 1982 (E)

Schulze, E. D., Lange, O. L., Koch, W.: Oecologia **9**, 317—340 (1972) (G)

Schulze, E. D., Lange, O. L., Evenari, M., Kappen, L., Buschbom, U.: Oecologia **22**, 355—372 (1976) (E)

Schulze, E. D., Fuchs, M. I., Fuchs, M.: Oecologia **29**, 43—61 (1977) (E)

Schulze, R.: Strahlenklima der Erde. Darmstadt: Steinkopff 1970 (G)

Semikhatova, O. A.: Energetika dykhaniya rastenii pri povyshennoi temperature. Leningrad: Nauka 1974 (R)

Sen, D. N.: Environment and plant life in Indian desert. Jodhpur, Geobios 1982 (E)

Senft, W. H.: Limnol. Oeanogr. **23**, 709—718 (1978) (E)

Serre, F.: Oecol. Plant. **11**, 143—171 (1976a) (F)

Serre, F.: Oecol. Plant. **11**, 201—224 (1976b) (F)

Šesták, Z.: Photosynthetica **11**, 367—474 (1977) (E)

Šesták, Z.: Photosynthetica **12**, 89—109 (1978) (E)

Šesták, Z., Čatský, J., Jarvis, P. G.: Plant photosynthetic production. Manual and methods. The Hague: Junk 1971 (E)

Shanmugan, K. T., O'Gara, F., Andersen, K., Valentine, R. C.: Annu. Rev. Plant Physiol. **29**, 263—276 (1978) (E)

Shantz, H. L.: Ecology **8**, 145—157 (1927) (E)

Shantz, H. L., Piemeisel, N.: J. Agric. Res. **34**, 1093—1190 (1927) (E)

Sheehy, J. E., Green, R. M., Robson, M. J.: Ann. Bot. **39**, 387—401 (1977) (E)

Sheridan, R. P., Ulik, T.: J. Phycol. **12**, 255—261 (1976) (E)

Siegelman, H. W., Butler, W. L.: Annu. Rev. Plant Physiol. **16**, 383—392 (1965) (E)

Sionit, N., Kramer, P. J.: Plant Physiol. **58**, 537—540 (1976) (E)

Sirén, G., Sivertsson, E.: Dept. Reforest. Stockholm Res. Note No. 83 (1976) (SV)

Slatyer, R. O.: Plant-water relationships. London, New York: Academic Press 1967 (E)

Slatyer, R. O.: Planta **93**, 175—189 (1970) (E)

Slatyer, R. O.: Aust. J. Bot. **26**, 111—121 (1978) (E)

Slatyer, R. O., Taylor, S. A.: Nature (London) **187**, 922—924 (1960) (E)

Smeets, L.: Meded. Inst. Vered. Tuinbouwgew. No. 283 Wageningen 1968 (D)

Smeets, N.: Diss. Innsbruck 1977 (G)

Smith, A. P.: Biotropica **6**, 263—265 (1974) (E)

Smith, P. F.: Annu. Rev. Plant Physiol. **13**, 81—108 (1962) (E)

Stålfelt, M. G.: Planta **23**, 715 (1935) (G)

Stålfelt, M. G.: Planta **27**, 30—60 (1937) (G)

Stanhill, G.: In: Analysis of temperate forest ecosystems. Reichle, D. E. (ed.). Ecological Studies, Vol. 1, pp. 247—256. Berlin, Heidelberg, New York: Springer 1970 (E)

Stanhill, G., Fuchs, M.: J. Appl. Ecol. **14**, 317—322 (1977) (E)

Steiner, M.: Jahrb. Wiss. Bot. **81**, 94—202 (1934) (G)

Sterne, R., Kaufmann, M. R., Zentmyer, G.: Physiol. Plant **41**, 1—6 (1977) (E)

Steshenko, A. P.: Problemy botaniki. Vol. XI, pp. 284. Leningrad: Nauka 1969 (R)

Steubing, L.: Landsch. u. Stadt **8,** 97—144 (1976) (G)

Steubing, L., Dapper, H.: Ber. Dtsch. Bot. Ges. **77,** 71—74 (1964) (G)

Stewart, W. S., Bannister, P.: Flora **162,** 134—155 (1973) (E)

Stocker, O.: Planta **7,** 382—387 (1929) (G)

Stocker, O.: Tabulae biologicae, Vol. V, pp. 510—686. Berlin: Junk 1929 (G)

Stocker, O.: Planta **24,** 402—445 (1935a) (G)

Stocker, O.: Jahrb. Wiss. Bot. **81,** 464—496 (1935b) (G)

Stocker, O.: Handbuch der Pflanzenphysiologie, Vol. 3, pp. 436—488. Berlin: Springer 1956 (G)

Stocker, O.: Flora **159,** 539—572 (1970) (G)

Stocker, O.: Flora **160,** 445—494 (1971) (G)

Stocker, O.: Flora **161,** 46—110 (1972) (G)

Stocker, O.: Flora **163,** 480—529 (1974) (G)

Stocker, O., Rehm, S., Paetzold, I.: Jahrb. Wiss. Bot. **86,** 556—580 (1938) (G)

Stoy, V.: Physiol. Plant. **4,** 1—125 (1965) (E)

Stoy, V.: Ber. Arbeitstag. Gumpenstein 1966, pp. 29—49 1966 (G)

Strain, B. R.: Ecology **50,** 511—513 (1969) (E)

Strain, B. R.: In: Perspectives in biophysical ecology. Gates, D. M., Schmerl, R. B. (eds.). Ecological Studies, Vol. 12, pp. 145—158. Berlin, Heidelberg, New York: Springer 1975 (E)

Strain, B. R.: Report on the workshop on anticipated plant responses to global carbon dioxide enrichment. Duke Environ. Center, Durham 1978 (E)

Strain, B. R., Higginbotham, K. O., Mulroy, J. C.: Photosynthetica **10,** 47—53 (1976) (E)

Sucoff, E.: Tech. Bull. 303, For. Ser. 20, Minnesota Agric. Exp. St. pp. 3—49, 1975 (E)

Sugimoto, K.: Tech. Bull. Trop. Agric. Res. Ctr. T. A. R. C. Tokyo Vol. 1, pp. 1—80, 1971 (E)

Sugimoto, K.: Jpn. J. Trop. Agric. **16,** 260—264 (1973) (J)

Sullivan, C. J., Levitt, J.: Physiol. Plant **12,** 299—305 (1959) (E)

Svoboda, J.: Truelove Lowland, Devon Island, Canada: A high arctic ecosystem. Bliss, L. C. (ed.), pp. 185—216. Edmonton: Univ. Alberta Press 1977 (E)

Sweeney, F. C., Hopkinson, J. M.: Trop. Grassland **9,** 209—217 (1975) (E)

Szarek, S. R.: J. Arid Env. **2,** 187—209 (1979) (E)

Szarek, S. R., Ting, I. P.: Plant Physiol. **54,** 76—81 (1974) (E)

Szarek, S. R., Ting, I. P.: In: Environmental and biological control of photosynthesis. Marcelle, R. (ed.), pp. 289—297. The Hague: Junk 1975 (E)

Szarek, S. R., Ting, I. P.: Photosynthetica **11,** 330—342 (1977) (E)

Szarek, S. R., Woodhouse, R. M.: Oecologia **26,** 225—234 (1974) (E)

Szarek, S. R., Woodhouse, R. M.: Oecologia **35,** 285—294 (1978) (E)

Szarek, S. R., Jonson, H. B., Ting, I. P.: Plant Physiol. **53,** 539—541 (1973) (E)

Szeicz, G.: J. Appl. Ecol. **11,** 617—636 (1974) (E)

Talling, J. F.: Prediction and measurement of photosynthetic productivity. Proc. IBP/PP. Tech. Meet. Trebon 1969. pp. 431—445. Wageningen: Centre Agric. Publ. 1970 (E)

Tanaka, A., Kawano, K.: Plant Soil **24,** 128—144 (1966) (E)

Taylor, A. O., Rowley, J. A.: Plant Physiol. **47,** 713—718 (1971) (E)

Taylor, G. E.: New Phytol. **80,** 523—534 (1978) (E)

Taylor, O. C., Thompson, C. R., Tingey, D. T., Reinert, R. A.: In: Responses of plants to air pollution. Mudd, J. B., Kozlowski, T. T. (eds.), pp. 122—140. London, New York: Academic Press 1975 (E)

Tenhunen, J. D., Westrin, S. S.: Oecologia **41**, 145–162 (1979) (E)

Thienemann, A. F.: Leben und Umwelt. Hamburg: Rowohlt 1956 (G)

Thomas, W.: Plant Physiol. **2**, 109–137 (1927) (E)

Thomas, W.: Biological indicators of environmental quality. Ann. Arbor Sci. Publ. Mich. 1973 (E)

Thomaser, W.: Thesis Padua-Innsbruck 1975 (G/I)

Thompson, K., Grime, J. P., Mason, G.: Nature (London) **267**, 147–149 (1977) (E)

Thompson, P. A.: Nature (London) **217**, 1156–1157 (1968) (E)

Thompson, P. A.: Ann. Bot. **34**, 427–449 (1970) (E)

Tieszen, L. L. (ed.): Vegetation and production ecology of an Alaskan arctic tundra. Ecological Studies, Vol. 29. Berlin, Heidelberg, New York: Springer 1978 (E)

Tieszen, L. L., Wieland, N. K.: Physiological adaptation to the environment. Vernberg, F. J. (ed.), pp. 157–200. New York: Intext. Educ. Publ. 1975 (E)

Tiku, B. L.: Physiol. Plant **37**, 23–28 (1976) (E)

Tinus, R. W.: Contr. No. 176. Congr. A. A. P. P. 1973 (E)

Tolbert, N. E.: Photosynthesis and photorespiration. Hatch, M. D., Osmond, C. B., Slatyer, R. O. (eds.), pp. 458–471. New York: Wiley 1971 (E)

Tolbert, N. E.: In: Photosynthesis II. Gibbs, M., Latzko, E. (ed.). Encyclopedia of Plant Physiology, N.S., Vol. 6, pp. 338–352. Berlin, Heidelberg, New York: Springer 1979

Totsuka, T.: Theoretical analysis of the relationship between water supply and dry matter production of plant communities. Fac. Sci. Univ. Tokyo **8**, 341–375 (1963) (E)

Tranquillini, W.: Planta **40**, 612–661 (1957) (G)

Tranquillini, W.: Planta **54**, 130–151 (1959) (G)

Tranquillini, W.: Physiological ecology of the alpine timberline. Ecological Studies, Vol. 31. Berlin, Heidelberg, New York: Springer 1979

Tranquillini, W., Schütz, W.: Zentralbl. Gesamte Forstwes. **87**, 42–60 (1970) (G)

Treharne, K. J., Eagles, C. F.: Photosynthetica **4**, 107–117 (1970) (E)

Treharne, K. J., Nelson, C. J.: In: Environmental and biological control of photosynthesis. Marcelle, R. C. (ed.), pp. 61–69. The Hague: Junk 1975 (E)

Treshow, M.: Environment and plant response. New York: McGraw-Hill 1970 (E)

Troll, C.: Stud. Gen. **8**, 713–733 (1955) (G)

Truog, E.: Soil Sci. Soc. Am. (Proc.) **II**, 305–308 (1947) (E)

Tselniker, Yu. L.: Fiziologitcheske osnovy tenevynoslivosti drevesnykh rastenii. Moscow: Nauka 1978 (R)

Tselniker, Yu. L.: Photosynthetica **13**, 124–129 (1979) (E)

Tsuno, Y., Hirayama, T.: In: JIBP Synthesis, Vol. 19. Monsi, M., Saeki, T. (ed.), pp. 100–111. Tokyo: Univ. Press of Tokyo 1978 (E)

Türk, R., Wirth, V., Lange, O. L.: Oecologia **15**, 33–64 (1974) (G)

Tumanov, I. I.: In: The cell and environmental temperature. Troshin, A. S. (ed.), pp. 6–14. Oxford: Pergamon Press 1967 (E)

Tunstall, B. R., Connor, D. J.: Aust. J. Plant Physiol. **2**, 489–499 (1975) (E)

Turner, N. C.: Plant Physiol. **55**, 932–936 (1975) (E)

Turner, N. C., Begg, J. E.: In: Plant relations in pastures. Wilson, J. R. (ed.), pp. 50–66. Melbourne: CSIRO 1978 (E)

Turner, N. C., Begg, J. E.: Plant and Soil **58**, 97–131 (1981) (E)

Tyree, M. T., Jarvis, P. G.: In: Encyclopedia of plant physiology, Vol. 12 B, pp. 35–77. Berlin, Heidelberg, New York: Springer 1982 (E)

Tyurina, M. M., Gogoleva, G. A., Jegurasdova, A. S., Bulatova, T. G.: Acta Hort. **81**, 51–60 (1978) (E)

Udagawa, T., Ito, A., Uchijima, Z.: Proc. Crop Sci. Soc. Jpn. **43**, No. 2 (1974) (J)

Ullrich, H., Mäde, A.: Planta **31,** 251–263 (1940) (G)

Ulrich, B.: Allg. Forsztg. **47,** (1968) (G)

Ulrich, B., Meyer, R.: In: Ökosystemforschung. Ellenberg, H. (ed.), pp. 165–174. Berlin, Heidelberg, New York: Springer 1973 (G)

Unsworth, M. H.: In: Physiological processes limiting plant production. Johnson, C. B. (ed.), pp. 293–306. London: Butterworths 1981 (E)

Url, W.: Protoplasma **46,** 768–793 (1956) (G)

Ursprung, A., Blum, G.: Ber. Dtsch. Bot. Ges. **36,** 599–618 (1918) (G)

Valovitch, E. M., Grif, V. G.: Fiziol. Rast. **21,** 1258–1264 (1974) (R)

Vareschi, V.: Planta **40,** 1–35 (1951) (G)

Vareschi, V.: Bol. Soc. Venezol. C. Nat. **14,** 121–173 (1953) (SP)

Vareschi, V.: Bol. Soc. Venezol. C. Nat. **96,** 128–134 (1960) (SP)

Vegis, A.: In: Temperature and life. Precht, H., Christophersen, J., Hensel, H., Larcher, W. (eds.), pp. 145–170. Berlin, Heidelberg, New York: Springer 1973 (E)

Villiers, T. A.: Dormancy and survival of plants. London: Edward Arnold 1975 (E)

Vincent, W. A.: In: Microbes and biological productivity. Hughes, D. E., Rose, A. H. (eds.), pp. 47–76. Cambridge: Cambridge Univ. Press 1971 (E)

Vinš, B.: Lesnictvi **8,** 263–280 (1962) (CS)

Vong, N. Q., Murata, Y.: Jpn. J. Crop. Sci. **46,** 45–52 (1977) (E)

Vong, N. Q., Murata, Y.: Jpn. J. Crop. Sci. **47,** 90–99 (1978) (E)

Voznesenskii, V. L.: Fotosintez pustynnych rastenii. Leningrad: Nauka 1977 (R)

Waisel, Y.: Biology of halophytes. London, New York: Academic Press 1972 (E)

Wallace, T.: Mineral deficiencies in plants. London: Stationery Office 1951 (E)

Wallace, W., Pate, J. S.: Ann. Bot. **31,** 213–228 (1967) (E)

Walter, H.: Die Hydratur der Pflanze und ihre physiologisch-ökologische Bedeutung. Jena: Fischer 1931 (G)

Walter, H.: Einführung in die Phytologie, III/2nd. Standortslehre. Stuttgart: Ulmer 1960 (G)

Walter, H.: Z. Pflanzenphysiol. **56,** 170–185 (1967) (G)

Walter, H.: Die Vegetation der Erde in ökophysiologischer Betrachtung, Vol. 2. Die gemäßigten und arktischen Zonen. Jena: Fischer 1968 (G)

Walter, H.: Vegetation of the earth in relation to climate and the ecophysiological conditions. Berlin, Heidelberg, New York: Springer 1973a (E)

Walter, H.: Allgemeine Geobotanik. Stuttgart: Ulmer 1973b (G)

Walter, H., Lieth, H.: Klimadiagramm-Weltatlas. Jena: Fischer 1967 (G)

Wardlaw, I. F.: In: Mechanisms of regulation of plant growth. Bieleski, R. L., Ferguson, A. R., Cresswell, M. M. (eds.), pp. 533–538. New Zealand: Wellington 1974 (E)

Wardlaw, I. F.: In: Transport and transfer processes in plants. Warlaw, I. F., Passioura, I. B. (eds.), pp. 381–391. London, New York: 1976 (E)

Wardle, P.: In: Arctic and alpine environments. Ives, I. D., Barry, R. G. (ed.), pp. 371–402. London: Methuen 1974 (E)

Wareing, P. F., Phillips, I. D. J.: The control of growth and differentiation in plants, 2. ed. Oxford: Pergamon Press 1978 (E)

Wareing, P. F., Saunders, P. F.: Annu. Rev. Plant Physiol. **22,** 261–288 (1971) (E)

Wassink, E. C.: Meded. Landbouwhogesch. Wageningen **72/31,** 1–22, 1972 (E)

Watson, D. J.: Ann. Bot. **11,** 41–76 (1947) (E)

Waughman, G. J.: J. Exp. Bot. **28,** 949–960 (1977) (E)

Weatherley, P. E.: In: Ecological aspects of the mineral nutrition of plants. Rorison, I. H. (ed.), pp. 323–340. Oxford-Edinburgh: Blackwell Sci. Publ. 1969 (E)

Weatherley, P. E.: In: Encyclopedia of plant physiology, Vol. 12 B, pp. 79–109. Berlin, Heidelberg, New York: Springer 1982 (E)

Weidner, M., Ziemens, C.: Plant Physiol. **56,** 590–694 (1975) (E)

Weinberger, P., Romero, M., Oliva, M.: Vegetatio **28,** 75–98 (1973) (G)

Weiser, C. J.: Science **169,** 1269–1278 (1970) (E)

Went, F. W.: Ecology **30,** 1–13 (1949) (E)

Went, F. W.: The experimental control of plant growth. Mass.: Waltham 1957 (E)

West, N. E., Skujinš, J.: Oecol. Plant. **12,** 45–54 (1977) (E)

Whittaker, R. H., Likens, G. E.: In: Primary Productivity of the biosphere. Lieth, H., Whittaker, R. H. (eds.). Ecological Studies, Vol. 14, pp. 305–328. Berlin, Heidelberg, New York: Springer 1975 (E)

Whittaker, R. H., Woodwell, G. M.: J. Ecol. **56,** 155–174 (1968) (E)

Whittaker, R. H., Woodwell, G. M.: In: Productivity of forest ecosystems. Duvigneaud, P. (ed.), pp. 159–175. Paris: UNESCO 1971 (E)

Wiebe, H. H., Brown, R. W., Daniel, T. W., Campell, E.: Bio Sci. 225–226 (1970) (E)

Wielgolaski, F. E., Kjelvik, S., Kallio, P.: In: Fennoscandian tundra ecosystems, Part 1. Wielgolaski, F. E. (ed.). Ecological Studies, Vol. 16, pp. 316–332. Berlin, Heidelberg, New York: Springer 1975 (E)

Wiesner, J.: Der Lichtgenuß der Pflanzen. Leipzig: Engelmann 1907 (G)

Wild, H.: Kirkia **9,** 233–241 (1974) (E)

Williams, G. J. III., McMillan, C.: Am. J. Bot. **58,** 24–31 (1971) (E)

Williams, G. J. III., Lazor, R., Yourgrau, P.: Photosynthetica **9,** 35–39 (1975) (E)

Williams, W. A., Loomis, R. S., Lepley, C. R.: Crop Sci. **5,** 211–219 (1965) (E)

Wilson, Ch. Ch.: Plant Physiol. **23,** 5–35 (1948) (E)

Winkler, E.: Flora **151,** 621–662 (1961) (G)

Winkler, E., Gamper, L., Schwienbacher-Mascotti, M.: Veröff. Mus. Ferdinandeum Innsbruck **55,** 253–292 (1975) (G)

Winkler, S.: In: Beiträge zur Biologie der niederen Pflanzen. Frey, W., Hurka, H., Oberwinkler, F. (eds.), pp. 155–176. Stuttgart: Fischer 1977 (G)

Winter, K., Troughton, J. H.: Flora **167,** 1–34 (1978) (E)

Woledge, J., Leafe, E. L.: Ann. Bot. **40,** 773–783 (1976) (E)

Wolverton, B. C., McDonald, R. C.: Economic Bot. **33,** 1–10 (1979) (E)

Woodman, J. N.: Photosynthetica **5,** 50–54 (1971) (E)

Woodward, R. G.: Photosynthetica **10,** 274–279 (1976) (E)

Woodwell, G. M., Whittaker, R. H., Houghton, R. A.: Ecology **56,** 318–332 (1975) (E)

Woolley, J. T.: Plant Physiol. **47,** 656–662 (1971) (E)

Wuenscher, J. E., Kozlowski, T. T.: Physiol. Plant. **24,** 254–259 (1971) (E)

Yocum, C. S., Allen, L. H., Lemon, E. R.: Agron. J. **56,** 249–253 (1964) (E)

Yoda, K.: Nature Life Southeast Asia **5,** 83–148 (1967) (E)

Zachhuber, K., Larcher, W.: Photosynthetica **12,** 436–439 1978 (E)

Zavodnik, N.: Bot. Mar. **18,** 245–250 (1975) (E)

Zelniker, J. L.: Dtsch. Akad. Landwirtschaftswiss. Berlin, Tagungsber. 100, 131–140 (1968) (G)

Ziegler, I.: Planta **103,** 155–163 (1972) (E)

Zima, J., Šesták, Z.: Photosynthetica **13,** 83–106 (1979) (E)

Zimmermann, G., Butin, H.: Flora **162,** 393–419 (1973) (G)

Zimmermann, M. H., Brown, C. L.: Trees, structure and function. Berlin, Heidelberg, New York: Springer 1971 (E)

Zohary, M.: Symp. Biol. Prod. Hot Cold Deserts. pp. 56–67. London 1952, (E)

Zohary, M.: Proc. UNESCO Symp. Arid Zone Res. Madrid Vol. 16, pp. 199–212, 1961 (E)

Zohary, M., Orshan, G.: Vegetatio **7,** 15–37 (1956) (E)

Subject Index

metabolic activity 56, 59
metallophytes 193, 194
micronutrients 158, 168
minerals
- accumulation 165, 168, 170–172, 179, 190, 197
- deficiency 67, 121, 154, 165, 168, 175
- growth 158, 173
- mobilization 161, 162, 165–169
- photosynthesis 126, 127, 166, 168, 176
- toxicity 87, 183, 191
- turnover 205
- uptake 171, 202
mineral cycling 204
mitochondria 82, 83, 206
montmorillonite 159
mycorrhiza 146

NaCl see salt
necrosis 14, 168, 169, 187, 196
net assimilation rate 134, 135, 145
net photosynthesis 84
net precipitation 260
night frost 117
nitrate 83, 163, 175, 176, 179
- reductase 166, 175, 178, 192, 253
nitrite reductase 176
nitrogen
- assimilation 175, 176, 187
- atmosphere 1, 2
- deficiency 175
- fixation 169, 178, 180–182, 192
- translocation 170, 177–179
- turnover 179
nitrogenase 180–182, 192
nitrophytes 175, 179
nucleic acids 35, 45, 165, 168, 175
nutrient requirements 158, 173

osmoregulation 90, 92, 166
osmotic
- potential 185, 186, 209, 210, 215, 235, 237, 240
- stress 159
oxalate 183, 184, 186
oxygen 1, 2, 86, 87
ozone 9, 14, 195

PAN 195
PEP carboxylase 76, 78–80, 89, 102, 118
percolation 214, 258, 263
peridermal transpiration 223, 229, 251

periodicity 49–51, 53, 56, 94, 98, 131
permafrost 248
permanent wilting percentage 214, 216
permeability 53, 56, 167, 196, 250
peroxisomes 82
pH 85, 160, 161, 175, 183
PhAR 12, 13
phenology 61, 63–65, 68
phenometry 70, 71
phloem transport 170
photoconversion 13
photocybernetic processes 15, 91
photodestruction 13, 14
photoenergetic processes 11
photoinduction 16
photolysis 74, 126
photomodulation 16
photomorphogenesis 15, 16
photon flux 75
photooxidation 13
photoperiod 59, 61, 67, 72
photoperiodism 16, 51–53, 68
photoreceptors 13
photorespiration 14, 80, 81, 96, 98, 121, 176
photostimulation 15
photosynthate budget 139
photosynthesis
- adaptation 104, 105, 118–120, 123
- energetics 75, 102, 157
- stress effects 34, 116–118, 123–125, 187, 192, 196, 197, 253
photosynthetic
- capacity 17, 53, 80, 94, 99–102
- efficiency 82, 98, 110
photosynthetically active radiation see PhAR
photosystem I 73, 118
- II 17, 74, 118
phototaxis 14
phototropism 16
phycobilins 13, 16, 74
phytochrome 13, 16, 59
phytohormones
- growth and development 54, 59, 60
- metabolism 86, 101, 136, 137, 168
- stomata 91, 94
- water relations 251
phytomass 4, 149, 154, 202, 203, 208, 259
phytotron 61
plankton 28, 95, 104, 109, 110, 137, 146, 147, 150, 156, 172, 253
pollution injury 195, 196, 197, 200
pore area 90, 91

H. Remmert

Ecology

A Textbook
Translated from the German by
M. A. Biederman-Thorson
1980. 189 figures, 12 tables. VIII, 289 pages
ISBN 3-540-10059-8

Contents: Ecology: the Basic Concept. – Autecology. – Population Ecology. – Ecosystems. – Outlook. – References. – General Books on Subjects Related to Ecology. – Subject Index.

The second edition of this outstanding textbook is now available in translation to English-speaking readers. Revised and expanded from the first edition, it brings into even greater focus the relationship between ecology and sensory physiology.

From the reviews:
"The literature is not exactly poor in attempts to describe ecological relationships in textbook form. What makes Remmert's book stand out from the rest is the author's dynamic perspective, his simple, flowing style, and, in many instances, the interpretations themselves... Appropriately enough, the coevolution of the various of an ecosystem is given special emphasis. This point usually receives inadequate attention in comparable textbooks. The author tackles generally accepted or postulated trends, discusses them, reformulates them, and, whereever possible, authenticates and explains them with individual analyses and case studies..."
translated from:
Helgoländer wissenschaftliche Meeresuntersuchungen

"We would expect an ecology textbook from one of Germany's best known ecologists to be highly original, and we certainly weren't disappointed. Professor Remmert has charged this overview of modern ecology with wit and expertise. Designed especially for students, it will prove equally valuable for all those truly interested in the subject... The descriptions and choice of examples are extraordinarily original. This book will be a joy for the expert, a treasure-trove for the student, and food for the thought for environmental administrators."
translated from:
Anzeiger Ornithologische Gesellschaft

L. van der Pijl

Principles of Dispersal in Higher Plants

3rd revised and expanded edition. 1982. 30 figures. X, 215 pages
ISBN 3-540-11280-4

Contents: Introduction. – General Terminology. – The Units of Dispersal. – The Relation Between Flowers, Seeds and Fruits. – Ecological Dispersal Classes, Established on the Basis of the Dispersing Agents. – Dispersal Strategy and the Biocoenosis. – Establishment. – The Evolution of Dispersal Organs in General. – Ecological Developments in Leguminous Fruits. – Dispersal and the Evolution of Grasses. – Man and His Plants in Relation to Dispersal. – References. – Subject Index. – Index of Scientific Plant Names. – Index of Scientific Animal Names.

The third edition of this classic work has been completely revised and expanded to include research results obtained over the last 15 years. As in previous editions, the emphasis remains on principles and ecology (especially synecology), evolution and establishment after transport.

"... the present work is an invaluable addition to the literature on reproductive biology of plants... Few botanists tody are better qualified than van der Pijl to write on dispersal (and pollution) biology... an excellent up-todate treatment of a long neglected subject... this splendid volume is unlikely to be surpassed for quite some time..." *Science*

"... are so fascinating and excellent in scope and treatment that they should appeal to a broad spectrum of amateur, teaching and professional botanists and biologists of all kinds everywhere. ... The illustrations are especially valuable since they cover so many more examples than the limited few that get repeated from one botany text to another... Actually, the book is really a gem..." *Phytologia*

Springer-Verlag Berlin Heidelberg New York Tokyo

Encyclopedia of Plant Physiology

New Series

Editors: A. Pirson, M. H. Zimmermann

Volume 12, Part A

Physiological Plant Ecology I

Responses to the Physical Environment
Editors: O. L. Lange, P. S. Nobel, C. B. Osmond, H. Ziegler
With contributions by numerous experts
1981. 110 figures. XV, 625 pages
ISBN 3-540-10763-0

Contents: Introduction: Perspectives in Ecological Plant Physiology. - Fundamentals of Radiation and Temperature Relations. - Photosynthetically Active Radiation. - Responses to Different Quantum Flux Densities. - Nonphotosynthetic Responses to Light Quality. - Responses to Photoperiod. - Plant Response to Solar Ultraviolet Radiation. - Responses to Ionizing Radiation. - The Aquatic Environment. - Responses to Light in Aquatic Plants. - Responses of Macrophytes to Temperature. - Responses of Microorganisms to Temperature. - Responses to Extreme Temperatures. - Cellular and Sub-Cellular Bases. - Ecological Significance of Resistance to Low Temperature. - Ecological Significance of Resistance to High Temperature. - Wind as an Ecological Factor. - Fire as an Ecological Factor. - The Soil Environment. - Author Index. - Taxonomic Index. - Subject Index.

Volume 12, Part B

Physiological Plant Ecology II

Water Relations and Carbon Assimilation
Editors: O. L. Lange, P. S. Nobel, C. B. Osmond, H. Ziegler
With contributions by numerous experts
1982. 153 figures. XI, 747 pages
ISBN 3-540-10906-4

Contents: Introduction. - Water in the Soil-Plant-Atmosphere Continuum. - Water in Tissues and Cells. - Water Uptake and Flow Roots. - Water Uptake by Organs Other Than Roots. - Transport and Storage of Water. - Resistance of Plant Surfaces to Water Loss: Transport Properties of Cutin, Suberin and Associated Lipids. - Stomatal Responses, Water Loss and CO_2 Assimilation Rates of Plants in Contrasting Environments. - Mathematical Models of Plant Water Loss and Plant Water Relations. - Physiological Responses to Moderate Water Stress. - Desiccation-Tolerance. - Frost-Drought and Its Ecological Significance. - Water Relations in the Germination of Seeds. - Environmental Aspects of the Germination of Spores. - Physiological Responses to Flooding. - Functional Significance of Different Pathways of CO_2 Fixation in Photosynthesis. - Modelling of Photosynthetic Response to Environmental Conditions. - Regulation of Water Use in Relation to Carbon Gain in Higher Plants. - Plant Life Forms and Their Carbon, Water and Nutrient Relations. - Author Index. - Taxonomic Index. - Subject Index.

Volume 12, Part C

Physiological Plant Ecology III

Responses to the Chemical and Biological Environment
Editors: O. L. Lange, P. S. Nobel, C. B. Osmond, H. Ziegler
With contributions by numerous experts
1983. 104 figures. XI, 799 pages
ISBN 3-540-10907-2

Contents: Introduction. - The Ionic Environment and Plant Ionic Relations. - Osmoregulation. - Halotolerant Eukaryotes. - Halophilic Prokaryotes. - Physiology and Ecology of Nitrogen Nutrition. - Influence of Limestone, Silicates and Soil pH on Vegetation. - Toxicity and Tolerance in the Responses of Plants to Metals. - Ecophysiology of Nitrogen-Fixing Systems. - Ecophysiology of Mycorrhizal Symbioses. - Ecophysiology of Lichen Symbioses. - Interactions Between Plants and Animals in Marine Systems. - Ecophysiology of Carnivorous Plants. - Host-Parasite Interactions in Higher Plants. - Virus Ecology "Struggle" of the Genes. - Ecophysiology of Zoophilic Pollination. - Physiological Ecology of Fruits and Their Seeds. - Physiological and Ecological Implication of Herbivory. - Interactions Between Plants. - Author Index. - Taxonomic Index. - Subject Index.

Volume 12, Part D

Physiological Plant Ecology IV

Ecosystem Processes:
Mineral Cycling, Productivity and Man's Influence
Editors: O. L. Lange, P. S. Nobel, C. B. Osmond, H. Ziegler
With contributions by numerous experts
1983. Approx. 60 figures, approx. 67 tables.
Approx. 690 pages
ISBN 3-540-10908-0

Contents: Introduction. - Nutrirent Allocation in Plant Communities: Mineral Cycling in Terrestrial Ecosystems. - Nutrient Cycling in Freshwater Ecosystems. - Nutrient Cycling in Marine Ecosystems. - Modelling of Growth and Production. - Productivity of Agricultural Systems. - Productivity of Grassland and Tundra. - Productivity of Desert and Mediterranean-Climate Plants. - Productivity of Temperate Deciduous and Evergreen Forests. - Productivity of Tropical Forests and Tropical Woodlands. - Phytoplankton Productivity in Aquatic Ecosystems. - Effects of Biocides and Growth Regulators: Physiological Basis. - Effects of Biocides and Growth Regulators: Ecological Implications. - Eutrophication Processes and Pollution of Freshwater Ecosystems Including Waste Heat. - Ecophysiological Effects of Atmospheric Pollutants. - Ecophysiological Effects of Changing Atmospheric CO_2 Concentration. - Man's Influence on Ecosystem Structure, Operation, and Ecophysiological Processes. - Author Index. - Taxonomic Index. - Subject Index.

Springer-Verlag
Berlin Heidelberg New York Tokyo